T0419533

INTRODUCTION TO PHYSICS OF ELEMENTARY PARTICLES

INTRODUCTION TO PHYSICS OF ELEMENTARY PARTICLES

O.M. BOYARKIN

Nova Science Publishers, Inc.
New York

For permission to use material from this book please contact us:
Telephone 631-231-7269; Fax 631-231-8175
Web Site: http://www.novapublishers.com

NOTICE TO THE READER

Library of Congress Cataloging-in-Publication Data
Available upon request

ISBN 1-60021-200-X

Published by Nova Science Publishers, Inc. ✣ New York

In the textbook all the known types of fundamental interactions are considered. The main directions of their unification are viewed. The basic theoretical ideas and the basic experiments, which allow to establish a quark-lepton level of a matter structure, are discussed. The general scheme of building up the theory of interacting fields with the help of the local gauge invariance principle is given. This scheme is used under presentation of the basic aspects of the quantum chromodynamics and the electroweak theory by Weinberg-Salam-Glashow. Principles of operation and designs of accelerators, neutrino telescopes, and elementary particle detectors are considered. The modern theory of the Universe evolution is described.

The textbook is primarily meant for Physics Department students. The book also will be useful to teachers, researches, post-graduate students and to all who are interested in problems of a modern physics.

Contents

Preface

The history of Physics is as abundant in shocks as a political life of a typical banana republic, where babies learn the word "revolution" immediately after the word "mama". At quiet times, however, physicists often have the illusion of complete understanding of the world. It was also the case in the beginning of the XXth century, when in the clear skies of Classical Physics there were only two small clouds, two unsolved problems, namely ether hypothesis and radiation spectrum of blackbody. The results of experiments by Mickelson and Morlet demonstrated that ether possesses no observable properties. They did not only ruin the ether theory altogether, but they also made the foundation of the special theory of relativity (STR), created by Einstein in 1905. In our opinion, there are two following aspects of the STR which form nowadays an education standard of any physicist and that is the highest price for any physical theory. The former aspect deals with understanding of properties belonging to the four dimensional space-time that surrounds us. The latter one reflects a deep faith in worldly wisdom of a correspondence principle, which reads: every new and more precise theory comprises in the utmost case the old and less exact theory. Thus, Newton's classical mechanics was not wrong it simply turned out to be approximate theory. It is easy to check, that all of its formulae can be obtained from the corresponding expressions in the STR at passage to the limit $c \to \infty$. When describing relativistic phenomena ($v \sim c$), the correct answer can be given by the STR only. Max Plank's explanation of radiation spectrum of blackbody was that very spark to stroke the flames of the second revolution in Physics which resulted in creation of non-relativistic quantum mechanics (NQM). This theory explained a matter structure at atomic and molecular levels with striking success. The Planck constant, the action quantum, plays the fundamental role in the NQM. As soon as any dynamic observables of the system having action dimension (energy$\times$time) are comparable in their value with $\hbar$, classical physics ends in a fiasco and the right description can be obtained only within the NQM. Once again we can check, that familiar formulae of classical physics follow from their counterparts of the NQM at passage to the limit $\hbar \to 0$.

In 1915 Einstein's general theory of relativity (GTR) succeeded classical theory of gravitation. The GTR made a revolution both in methods and in the very contents of theoretical physics. Einstein discovered not only the new physical laws, but a new method of establishing the new laws as well. The GTR was based on the equivalent principle which suggested no difference between gravitation and acceleration in a small spatial domain. Unlike Newton's theory of gravitation, the GTR is not only a theory of gravitational interaction, but also a theory of space-time, consequently, a theory of the Universe in general. Non-stationary models of the Universe, obtained by Friedmann on the basis of the GTR equation solutions at that time seemed to be mere fantastic. However, as soon as 1929 astronomical observations by A. Hubble proved the theory of the expanding Universe to be right. Thus, the Universe of Plato and Pifagor is succeeded by the Universe which has a starting point in time and neither beginning nor end in space.

Merging of the STR and the NQM resulted in foundation of the quantum field theory, which constitutes the theoretical base of elementary particle physics. In the beginning it seemed that particle physics consisted of intuitive assumptions and set of recipes taken from ceiling. Each kind of fundamental interactions was studied separately, almost independently from other ones. The existence of divergences in the series of the perturbation theory was the only common trait, unifying all interactions. Only the theory of electromagnetic electron-positron interaction, the quantum electrodynamics, was a pleasant exception from such a dull landscape. The situation, however, changed abruptly by the beginning of the seventies in the XXth century. Velvet gauge revolution started gauge era in physics of microworld. The use of gauge group $SU(2)_{EW}\otimes U(1)_{EW}$ together with the hypothesis of spontaneous symmetry violation allows to unify electromagnetic and weak interactions. Adding the theory of strong interactions based on color gauge group $SU(3)_c$ to this scheme leads to the creation of so called standard model (SM). The SM perfectly explains not only the events in the microworld but also many cosmological phenomena, for example, Big Bang theory. However, nowadays there are a few experiments (oscillations of solar and atmospheric neutrinos, reports about registration of neutrinoless double beta decay etc.) results of which demand some light reconstruction of the SM, namely its electroweak sector.

The next step is to unify strong and electroweak interactions, that is, to build up the grand unification theory (GUT). We have all the reasons to believe that once again the solution will be obtained while searching for a gauge group involving as a subgroup the $SU(3)_c\otimes SU(2)_{EW}\otimes U(1)_{EW}$ gauge group of the SM. There is no doubt, that gauge symmetries will play an important role in creation of the unified field theory, which comprises both the GUT and gravitational interaction theory.

The aim of the textbook is to present the ideas evolution in particle physics, the modern state of this physics division and to display the role this science plays in explaining the processes occurring in the Universe. The textbook is primarily meant for Physics Department students. It will be also useful for technological institute students, for students of the institutes which are specialized in a background of teachers, engineers, researchers and for all who wants to know how the world in which we live really works.

A reader should only know the fundamentals of non-relativistic quantum mechanics and special theory of relativity. Under reading the textbook it is useful to address to other books devoted to elementary particle physics the list of which is placed at the end of the textbook.

Notations

We mark three-dimensional indices by Latin letters and four-dimensional ones (running over 0,1,2,3) by Greek letters. All components of four-vectors are real numbers. We introduce two kinds of four-dimensional tensors. Thus by definition for 4-coordinates we have

$$x^\mu = (x^0, x^1, x^2, x^3) = (ct, x, y, z),$$

$$x_\mu = (x_0, x_1, x_2, x_3) = (ct, -x, -y, -z).$$

Four-vectors with upper (low) index we name contravariant and (covariant) vectors. In the same way the difference is made for covariant and contravariant tensors with the rank higher than one. Let us define the metric tensor

$$g^{\mu\nu} = \begin{pmatrix} 1 & 0 & 0 & 0 \\ 0 & -1 & 0 & 0 \\ 0 & 0 & -1 & 0 \\ 0 & 0 & 0 & -1 \end{pmatrix}.$$

Since determinant of this matrix is not equal to 0, there is its inverse matrix $g_{\mu\nu}$ satisfying the relation

$$g_{\mu\nu} = g^{\mu\nu}.$$

To raise and lower indices the metric tensor is employed. Thus, for example,

$$x_\mu = g_{\mu\nu}x^\nu, \qquad T^{\mu\nu} = g^{\mu\lambda}g^{\nu\sigma}T_{\lambda\sigma}, \qquad \text{etc..}$$

We call twice repeated indices dummy ones and shall mean the summarizing on them. Note, that only spatial components change its sign under transition from covariant to contravariant four-vectors. The product of two four-vectors a_μ b^μ is defined as follows:

$$a_\mu b^\mu = a^0 b^0 - \mathbf{ab},$$

where

$$\mathbf{ab} = a^k b^k = a_k b_k = a^1 b^1 + a^2 b^2 + a^3 b^3.$$

Four-dimensional vector of energy-momentum has the following form

$$p^\mu = (E/c, \mathbf{p})$$

and it satisfies the following relation

$$p^\mu p_\mu = m^2 c^2.$$

Four-dimensional generalization for the Nabla operator is given by the expression

$$\partial_\mu \equiv \frac{\partial}{\partial x^\mu} = (\partial_0, \nabla).$$

Symbol $\Box$ is used for D'Alembert operator

$$\Box = \partial_\mu \partial^\mu = \frac{1}{c^2}\frac{\partial^2}{\partial t^2} - \triangle,$$

where $\triangle$ is the Laplace operator. Quantity ε_{ijk} is a completely antisymmetric tensor

$$\varepsilon_{ijk} = \begin{cases} 1, & n - \text{even} \\ -1, & n - \text{odd} \\ 0, & \text{two and more indices coincide} \end{cases}$$

where n is the number of transpositions which leads indices i, j, k to the sequence 123. Symbol $\varepsilon_{\mu\nu\lambda\sigma}$ denotes four-dimensional generalization of tensor ε^{ijk} with $\varepsilon^{0123} = 1$ (while $\varepsilon_{0123} = -1$)

Upper signs $*$, T and $\dagger$ mean operations of complex conjugation, transposition and Hermitian conjugation respectively. A continuous line above spinors indicates operation of Dirac conjugation

$$\overline{u} = u^{\dagger}\gamma_4,$$

where γ_μ are Dirac matrices.

For basic vectors of representations and state vectors we use Dirac *bra* ($|\ldots>$) and *ket* ($<\ldots|$) vectors. Thus, for example

$$\Psi(\mathbf{p}, s_3) \equiv |\mathbf{p}, s_3>, \qquad \Psi^{\dagger}(\mathbf{p}, s_3) \equiv <\mathbf{p}, s_3|.$$

We mark three-dimensional radius vector by $\mathbf{r}$, and its module by r, where $r = \sqrt{x_1^2 + x_2^2 + x_3^2}$.

In the book Heaviside system of units is used, in which $e^2/(4\pi\hbar c) = \alpha_{em}$. This very normalization of electric charge is adopted in periodic literature on quantum field theory. In Gauss system of units the normalization $e^2/(\hbar c) = \alpha_{em}$ is used (the value of electric charge corresponding to it is usually given in tables of physical constants). In Heaviside system of units the equations of electromagnetic fields have more convenient form, since the multiplier 4π does not enter there. Coulomb law, however, in this system has the following form

$$F_c = \frac{q_1 q_2}{4\pi r^2}.$$

In contrast to this field equations in Gauss system of units contain the factor 4π and Coulomb law has the simple form $F_c = q_1 q_2/r^2$. It is obvious, that the value α_{em} is the same in all systems of units, while the magnitude of the elementary charge e takes different values.

Chapter 1

Interactions and Ways of Their Unification

1.1. Strong Interaction

The main purpose of Physics is to explain all the nature phenomena by means of few simple fundamental principles. Since all the matter consists of particles, then elementary particles physics must give the final answer. By this one of the main subjects is the question about the nature of interactions. It appears that despite of diversity of the world surrounding us, all interactions are reduced to four fundamental types, which differ greatly in intensity of their proceeding. Such a classification is not free from a some conventionality due to the fact, that relative role of different interactions is changed while energy of interacting particles grows. It means that separation of interaction in classes, based on comparison of processes intensity, can be made reliably only for not very high energies.

Let us consider types of fundamental interaction in order of decreasing their intensity. Strong interaction opens our list. This is an overwhelming type of interactions in nuclear physics of high energies. Particles, which participate in strong, weak, and gravitation interactions, are called hadrons. Hadrons with half-integral spin are called baryons, while hadrons with integral spin are called mesons. In addition, charged hadrons participate in electromagnetic interactions. Just strong interaction causes couplings between protons and neutrons in atomic nuclei and ensures the exclusive nuclei strength lying at the heart of matter stability in Earth conditions. Strong interaction is also responsible for the confinement of quarks inside hadrons. Strong interaction may be manifested not only as ordinary attraction in nucleus, but as a force, which causes instability of some elementary particles (particles, which decay is caused by strong interaction, are called resonances). Due to its great value, strong interaction is a source of huge energy. In particular, the main part of heat in the sun is produced by strong interaction, when made by weak interaction deuterium nuclei together with protons are synthesized into helium nuclei. Strong interaction is a short-range interaction with radius of 10^{-13} cm.

Only after quark structure of hadrons had been discovered, strong interaction theory ceased to resemble the plots of Russian popular tales. This theory, called quantum chromodynamics (QCD), resembles quantum electrodynamics (QED) in construction. However,

QED has a local gauge symmetry with respect to group $U(1)_{em}$[1], while a local gauge symmetry group of QCD is the $SU(2)_c$ group. Note, that the both symmetries are internal ones, that is, they are connected with a system symmetry not in the ordinary space-time but in abstract spaces[2]. The lower index c in QCD symmetry group is caused by the fact, that quarks, beside ordinary quantum numbers, have three addition degrees of freedom, for which we use a conventional term "color" or color charge R (red), G (green) and B (blue). For both QCD and QED internal symmetries are exact.

The subject of internal symmetry violation is of the great importance in quantum field theory. There are two mechanisms of symmetry violations, namely, explicit and spontaneous. Under explicit violation the Lagrangian contains terms, which are not invariant with respect to a symmetry group. The value of these terms characterizes the degree of corresponding symmetry violation. Thus, for example, the Lagrangian of strong interaction is variant under isotopic transformations, but the total Lagrangian also contains electromagnetic and weak interactions, which explicitly violate isotopic symmetry. For this reason the complete theory does not possess the exact isotopic invariance.

Under spontaneous symmetry violation the Lagrangian possesses the invariance with respect to the transformations of the internal symmetry group, vacuum (vacuum is the state with the minimum energy), however, looses this invariance. Vacuum non-invariance reveals itself through the fact, that one or more components of quantized field (as a rule these components correspond to scalar particles) acquire the non-zero vacuum averages $< 0|\varphi_i|0 >$ (or vacuum expectation values), which define various energy scales of the theory. Under the spontaneously violated local symmetry the corresponding gauge bosons, which are interaction carriers, turn out to be massive particles, while under exact symmetry these gauge bosons are massless. Thus, carriers of strong interactions between quarks, which we call gluons, are massless particles.

In all observable hadrons the color charges of quarks are compensated, i.e hadrons are colorless (white) formations. Hadron colorlessness can result either from mixing of three main colors (true for baryons) or by mixture of color and anticolor (true for mesons). In strong interactions the color charges of quarks play the same role as the electric charges of particles do in electromagnetic interaction. Color charge is the source of gluon field. As this takes place gluon carries on its both color and anticolor charges, that is, its color composition is the product of color and anticolor. When quark emits gluon its color changes, depending on a gluon color. For instances, red quark, emitting red-antiblue gluon, turns blue. Analogously, blue quark, absorbing red-antiblue gluon, turns red, etc. A total of $3\times3=9$ "color-anticolor" combinations are possible. Among them there is one corresponding to colorless state

$$g_0 = R\overline{R} + B\overline{B} + G\overline{G},$$

which does not change the quark state under emitting or absorbing by quark and, consequently, can not play the role of gluon transferring interaction between quarks. Then only eight gluons are left. Gluons are electrically neutral, have zero-mass and spin equal to 1. All this makes them to be similar to photons. But unlike photons, gluons have a "charge" of the field whose interaction they transfer. Gluon can emit or absorb other gluons, changing

[1]The term "local" means that the transformation parameters are functions of coordinates.

[2]Such symmetries are also called geometric ones.

its own color in so doing. That is, gluons create new gluon field around themselves, not depending on quarks. Photons are deprived of such a property, they have no electric charge and no new electric field is created around them. Electromagnetic field is the most intensive near the charge, which causes the field and further away it is dispersed in space and weakened. Charged gluons produce around themselves new gluons, which produce new gluons and so on. The result is, that gluon field is not decreasing, but increasing further away from the quark creating this field. In other words, effective color charges of quarks and gluons are increasing as the distance grows. At the distances of hadron size order ($\sim 10^{-13}$ cm) color interaction becomes really strong. Perturbation theory, the main mathematical apparatus in microworld physics, is not applicable in this domain, so there are no reliable calculations. However, one could expect on qualitative grounds, that strengthening the interaction with distance must result in impossibility to bring isolated quarks at large separations. To put this another way, it results in imprisonment for life of quarks in hadron prison. This phenomenon is called confinement.

Baryon consists of three differently colored quarks. The quarks are constantly exchanging gluons and changing their own color. These changes, however, are not arbitrary. Mathematical apparatus of QCD restricts severely this play of colors. At any moment of time the summarized color of three quarks must represent the sum $R+G+B$. Mesons consist of quark-antiquark pairs, every pair is colorless. Then, no matter what gluons quark-antiquark pairs are exchanging, the mesons also remain white formations. So, from QCD standpoint, strong interaction is nothing else but a tendency to maintain $SU(3)_c$-symmetry, resulting in conservation of white color of hadrons, while their components change their colors.

Strong interaction intensity is characterized by so called QCD running coupling constant:

$$\alpha_s(q) = \frac{g_s^2}{4\pi},$$

where g_s is a gauge constant of $SU(3)_c$ group. The term "running" reflects dependence of α_s from distance or from transferred momentum q. We remind, that in the microworld to estimate a quantity order one may use Heisenberg uncertainty relation. Then a transition to short (large) distances means a transition to large (small) values of transferred momentum $q \sim \hbar/r$.

Evolution of the running coupling constant of QCD is governed by the equation

$$\alpha_s(q) = \frac{12\pi}{(33-2n_f)\ln(q^2/\Lambda_{QCD}^2)}, \tag{1.1}$$

where n_f is the number of quark kinds (at given color) or the number of quark flavors, and Λ_{QCD} is a scale parameter of QCD. Derivation of this equation is based on the use of the perturbation theory apparatus and the structure of the total Lagrangian describing the theory. At $q^2 \gg \Lambda_{QCD}^2$ the effective constant $\alpha_s(q)$ is small and consequently, the perturbation theory describes behavior of weakly interacting quarks and gluons successfully. At $q^2 \sim \Lambda_{QCD}^2$ it is impossible to use the perturbation theory while strongly interacting gluons and quarks start to form coupled systems — hadrons. Obviously, parameter Λ_{QCD} defines the border between the world of quasifree quarks and gluons and the world of real hadrons, below which confinement becomes substantial. The value of Λ_{QCD} is not predicted by the theory. It is a free parameter which is determined from experiments. Nowadays, despite of joint efforts of experimentalists and theorists, the exact value of Λ_{QCD} remains unknown (its approximate value lays between 100 and 200 MeV). Equation (1.1) leads to decrease of effective interaction as momentum grows and in asymptotic ultraviolet limit effective interaction tends to zero. Then the fields, participating in interaction, become free. The phenomenon of self-switching off interaction at short distances which is a reverse side of confinement, is called asymptotic freedom.

At distances, bigger than hadron size, there is no strong interaction between hadrons at all, that is, hadrons are neutral with respect to color. It is similar to the absence of electromagnetic forces between atoms at big distances, since they are electrically neutral. However, when two or more atoms approach at a distance, when their electron clouds are overlapping, so called Van der Waals forces, or chemical forces come into action. Their radius of action is of the atom size order. Molecular bond is caused by these forces. Its mechanism is based on the exchange of electrons between atoms, i.e. the molecular bond is a complicated manifestation of fundamental electromagnetic interaction between two volume-distributed charged systems. Analogously, hadron interaction can be also viewed as a complicated manifestation of fundamental strong interaction between color quarks, which becomes observable only under approaching the quark cores of hadrons.

1.2. Electromagnetic Interaction

Electromagnetic interaction keeps electrons within atoms and binds atoms in molecules and crystals. This interaction lies at the basis of nearly all the phenomena around us, chemical, physical and biological. Elementary particles with electric charge take part in electromagnetic interaction. Neutral particles could also interact with electromagnetic field due to multipole moments (dipole, quadrupole, anapole, etc.). However, only the particles with composite structure can have such moments. Thus, electromagnetic interaction is not as universal as gravitation one. It is easy to observe electric and magnetic forces, acting between macroscopic bodies. These forces just as the gravitation forces are subjected to inverse square law, that is, they are long-range forces. For instance the Earth magnetic field extends far into cosmic space, and the Sun magnetic field fills all the Solar system. Unification of electric and magnetic forces into the classical theory of electromagnetic field, made by Maxwell in 50s of 18th century, is an example of the first unified field theory. The idea of unification can be illustrated by the example of Lorentz force

$$\mathbf{F}=q\{\mathbf{E}+[\frac{\mathbf{v}}{c}\times\mathbf{H}]\}. \qquad (1.2)$$

Let us assume, that $|\mathbf{E}|\sim|\mathbf{H}|$. Then from Eq.(1.2) it follows, that at $|\mathbf{v}|\ll c$ magnetic forces are very small compared to electric ones and approach in their value to them only at $|\mathbf{v}|\to c$. Thus, the relative intensity of forces is defined by particle velocity, that is, there is the scale on which the unification of electric and magnetic fields takes place and this scale is determined by the light velocity. Since energy is also growing at $|\mathbf{v}|\to c$, we can state, that unification occurs in the region of ultrarelativistic energies of particles.

Quantum theory of electromagnetic interaction of electrons and positrons, called QED, had been built up at the beginning of 50s of 20th century. QED is the most exact of all physical theories. Here the electromagnetic interaction is exhibited in its pure form. Unprecedented accuracy of calculations in QED is caused by usage of apparatus of perturbation theory on small dimensionless parameter

$$\alpha_{em}(q) = \frac{e^2}{4\pi\hbar c},$$

which is called a fine structure constant. $\alpha_{em}(q)$ is also a function of distance or transferred momentum. Its macroscopic value which is defined at $q = m_e c$ equals to $1/137.0359895(61)$. The exact symmetry of QED with respect to local gauge group $U(1)_{em}$, which gauge constant is equal to electron charge, leads to zero-mass of electromagnetic interaction carrier being called photon.

The QED perfectly describes not only electrons, but electromagnetic properties of other charged leptons as well. Contrary to this, electromagnetic properties of hadrons are not amenable to calculation because hadrons are basically controlled by strong interactions. We stress, that the QED is not only the first model of a quantum field theory (QFT), but it is also the simplest and the most extensively studied version of the QFT. Within the framework of the QED many fundamental concepts of the QFT were discovered and formulated. All this allows to build up more complicated quantum field theories in the image and similarity of the QED.

In the QED the phenomenon of vacuum polarization results in screening of electron charge by vacuum positrons. Polarizing vacuum, the electron attracts virtual positrons and repulses virtual electrons. As a result, electron charge is partly screened, if one sees it from a large distance. If one penetrate deep inside of a cloud of virtual pairs, then screening would decrease and effective charge of electron would increase. In other words contrary to the QCD running constant which is increasing with distance, the QED running constant is decreasing as the distance grows. The calculations, made within the QCD scope, define the evolution of $\alpha_{em}(q)$ by means of equation

$$\alpha_{em}(q) = \frac{3\pi\alpha_{em}(m_e c)}{3\pi - \alpha_{em}(m_e c)\ln\left[q^2/(4m_e^2c^2)\right]}. \tag{1.3}$$

From Eq. (1.3) it follows that at $q \sim 80$ GeV$/c$ the value of α_{em} is approximately equal to $\sim 1/128$.

1.3. Weak Interaction

Weak interaction is destructive in its character because it is not able to create stable states of matter in the way, as, for example, gravitation force maintains the existence of the Solar system or electromagnetic interaction ensures atom stability. In other words, the main destination of weak interaction is to regulate life-time of inanimate matter. It is responsible for nuclear $\beta^{\pm}$-decays, for decays of particles not belonging to resonance class (we call such particles "stable"). Only a few stable particles, for example, π^0-meson, η^0-meson and Σ^0-hyperon decay due to electromagnetic interaction.

If electromagnetic multipole moments of neutrino are equal to zero, then all the processes with the neutrino participation are caused by weak interaction only. Weak interaction is also responsible for nuclear and atomic processes going with parity violation. The particles, participating in weak interactions and with the availability of electric charge in electromagnetic interactions but not participating in strong ones, are called leptons.

In some cases weak interaction also influences macroscopic objects. For example, it plays a key role in the Sun energy release, because deuterium nucleus production from two protons is caused by just this interaction

$$p+p\rightarrow {}^2D+e^{+}+\nu_e.$$

Neutrino emission during weak interactions defines stars evolution, especially at their final stages, initiates supernova explosions and pulsar production. If it were possible to switch off weak interaction, then the matter around us would acquire quite another structure. It would contain all the particles, which decay due to weak interaction (muons, $\pi^{\pm}$-mesons, K-mesons, etc.).

Intensity of weak interaction is defined by Fermi constant

$$G_F = 1.16639(1)\times 10^{-5}(\hbar^3c^3)\ \mathrm{GeV}^{-2},$$

which is dimensional as we see. In the reference frame, where a particle rests the probability of the decay Γ due the weak interaction turns out to be proportional to $G_F^2m^5$ (m is the mass of a decaying particle). In virtue of Heisenberg uncertainty relation, elementary particle lifetime τ is inversely proportional to Γ. For particles, decaying due to weak interactions, the value of τ is quite large in microworld scales and lies in the interval 10^3 — 10^{-10} s. The lifetime of a particle decreases as the intensity of interaction, causing decay, grows. For particles, which instability is caused by electromagnetic interaction, τ is of the order $\sim 10^{-16}$ s, while for the particles decaying because of strong interaction, τ is of the order 10^{-23} — 10^{-24} s.

Notwithstanding the fact that the first process, caused by weak interaction, the radioactive β^{-}-decay of nucleus, had been discovered by A. Becquerel in 1896, the attempts of constructing the weak interaction theory was crowned with success only in 60s of XXth century. For the construction of this theory Glashow, Salam and Weinberg were awarded Nobel prize in 1979. In this theory both electromagnetic and weak interactions are the manifestations of one and the same interaction which is called electroweak (EW) interaction. A local gauge symmetry $SU(2)_{EW}\otimes U(1)_{EW}$ makes the base of the theory. In this case, there are two peculiarities, which make EW interaction different from both the QED and QCD.

First, local gauge symmetry of the EW interaction is spontaneously violated up to local gauge symmetry of the QED

$$SU(2)_{EW}\bigotimes U(1)_{EW}\rightarrow U(1)_{em}.$$

Second, from the very beginning the theory is not invariant with respect to operation of the space inversion.

Inviolate local symmetry $SU(2)_{EW}\otimes U(1)_{EW}$ demands the existence of four massless particles with spin 1, two of which are neutral and remaining two are charged. It was known from experiments, that action radius of weak interaction R_W is extremely small $\sim 10^{-16}$ cm. Consequently, carriers of this interaction must have masses of the order $\sim \hbar/(R_Wc)$. To give the mass to the gauge bosons of weak interaction, a doublet of massless scalar fields (Higgs bosons), consisting of neutral and charged components is introduced into the theory. In this case the neutral Higgs boson is not proper neutral. Due to spontaneous symmetry violation (non-zero vacuum average from neutral component of Higgs doublet is chosen) three of gauge bosons of a group $SU(2)_{EW} \otimes U(1)_{EW}$ acquire masses, while the forth one remains massless. Massive gauge bosons $W^{\pm}$ and Z are identified with gauge bosons of weak interaction and massless gauge boson γ is identified with a photon. Out of four massless scalar fields, one neutral fields acquires mass and the remaining three leave physical sector, as if they were eaten by gauge bosons while they are gaining their masses. From massless vector field with two spin states and massless scalar field a massive vector particle with three spin projections is produced, so, the number of degrees of freedom is conserved. The mass production of the gauge field due to spontaneous local symmetry violation is called Higgs mechanism.

By now lots of data have been accumulated, which prove, that experiments fit the theory perfectly. However, the main problem in the EW interaction theory is not solved yet, namely, the mechanism of violation of initial $SU(2)_{EW} \otimes U(1)_{EW}$-symmetry is not established. The most real way to solve this problem is experimental searching for the Higgs boson. Since the theory does not predict its mass m_H, then the range of researching for is rather wide.

The fact, that in the world surrounding us, we discriminate electromagnetic and weak interactions, only means that their unification scale or the boundary of spontaneous symmetry violation in the EW interaction theory lies on a higher energy scale $\sim m_W c^2 = 80.4$ GeV, that corresponds to distances of the order 10^{-16} cm.

To compare different interactions it is convenient to use dimensionless quantities. For this purpose we introduce quantity α_2, which characterizes intensity of the weak interaction according to the relations

$$\alpha_2(q) = \frac{g^2}{4\pi}, \qquad \frac{g^2}{8m_W^2} = \frac{G_F}{\sqrt{2}(\hbar c)^3}. \tag{1.4}$$

1.4. Gravitation Interaction

Though gravitation interaction is the weakest of all, it possesses a cumulative effect. Thus, gravitation interaction between two bodies is a cumulative sum of interactions between elementary masses which form these bodies. Since in the microworld the contribution of gravitation interaction is very small compared to other interactions, it does not result in measurable effects on subatomic level. However, on macroscopic level gravitation interaction is dominating: it keeps together parts of the terrestrial globe, unifies the Sun and planets into the Solar system, connects stars in galaxies and controls the evolution of all the Universe.

Since gravitation interaction was discovered in the first place, then just with its help the term "force" appeared in Physics. Gravitation interaction is universal, because it is in operation between all the bodies having the mass. It belongs to long-range interactions. Building the non-relativistic gravitation theory was completed by I. Newton in 1687. According to this theory two mass points, having masses m_1, m_2 and lying in a distance r, are attracted with the force which value and direction are given by the expression

$$\mathbf{F} = G_N \frac{m_1 m_2 \mathbf{r}}{r^3},$$

where G_N is a Newton constant, $G_N = 6.67259(85)\times 10^{-8}$ cm^{3}g^{-1}s^{-2}. In this theory the force depends on particles position at a given time only and so the gravitation interaction propagates instantly.

At an arbitrary mass distribution the gravitational force, operating on any point mass m_0 in the given spatial point, can be expressed as a product of m_0 on a vector $\mathbf{E}_g$ which is called the gravitation field strength. In the Newton theory the superposition principle is valid for a gravitation field. As this field is potential it is possible to introduce by the usual fashion the gravitation potential

$$\mathbf{E}_g = -\text{grad}\ \varphi_g.$$

The potential of a continuous distribution of a matter density $\rho(\mathbf{r})$ satisfies the Poisson equation

$$\Delta\varphi_g = -4\pi G_N\rho(\mathbf{r}). \tag{1.5}$$

The Newton gravitation theory has allowed to describe with a great precision an extensive range of phenomena, including the motion of natural and artificial bodies in the Solar system, the motion of celestial bodies in other systems: in binary stars, in stellar clusters, in galaxies. On the basis of this theory the existence of the planet Neptune and the satellite of Sirius has been predicted. In the modern astronomy the gravitation law of Newton is the foundation on the basis of which the motions, structure of celestial bodies, their masses and evolution are calculated. The precise definition of the Earth gravitational field allows to establish a mass distribution under its surface and, hence, immediately to solve the important applied problems.

As the Newton theory assumes the instantaneous propagation of gravitation it cannot be made consistent with a special relativity theory (SRT), stating that the interaction propagation velocity can not exceed c. It means, that this theory can not be used when gravitational fields are so strong, that they accelerate bodies, moving in them, up to the velocities of the order c. The velocity v, up to which the body falling freely from infinity ($v|_{t=0}\approx 0$) up to some point with a gravitational potential $\varphi_g(\mathbf{r})$ has been accelerated, can be found from the relation

$$\frac{mv^2}{2} = m\varphi_g(\mathbf{r})$$

(we have put $\varphi(\infty)=0$). Hence, the Newton theory of gravitation is applicable only in the case when

$$|\varphi_g| \ll c^2.$$

For the gravitational fields of usual celestial bodies this requirement is fulfilled. For example, on the Sun surface we have $|\varphi_g|/c^2 \approx 4\times 10^{-6}$, and on the surface of the white dwarfs $|\varphi_g|$ is about 10^{-3}.

Besides the Newton theory likewise is inapplicable under calculation of the particles motion even in a weak gravitational field with $|\varphi_g| \ll c^2$ if the particles flying near the massive bodies, had the velocity $v\sim c$ already far from them. Hence, it will lead to the improper answer under calculation of the light trajectory in a gravitational field. The Newton theory also is not used under investigation of the variable gravitational fields created by moving bodies (for example, binary stars) on distances $r > c\tau$, where τ is the period of revolution in a system of a binary star). Really, the Newton theory, based on the instantaneous propagation of interaction, is unable to take into account the retardation effect which appears to be essential in this case.

Relativistic theory of gravitation, that is, general theory of relativity (GTR) was built by Einstein in 1915. It changed drastically the understanding of gravitation in classical, Newtonian, physics. In the Einstein theory gravitation is not a force, but a manifestation of curvature of space-time. Flat metric of Minkowski $g_{\mu\nu} = \mathrm{diag}(1,-1,-1,-1)$ in the space of the GRT is deformed into metric

$$\eta_{\mu\nu}(x) = g_{\mu\nu} + h_{\mu\nu}(x).$$

The two postulates make the foundation of the GTR. The first one defines the form of the Lagrangian density $\mathcal{L}_g$, describing a propagation and a self-action of a gravitation field. On the basis of the second postulate, namely, equivalence principle, gravitation interaction is introduced by means of substitution $g_{\mu\nu} \to \eta_{\mu\nu}(x)$ into the Lagrangians of all the existing fields, i. e. into $\mathcal{L}_{QCD} + \mathcal{L}_{EW}$. Variation of the total Lagrangian $\mathcal{L}_g + \mathcal{L}_{QCD} + \mathcal{L}_{EW}$ with respect to the gravitation potentials $\eta_{\mu\nu}(x)$ leads to the Einstein gravitation equations

$$R_{\mu\nu}(\eta) - \frac{1}{2}R(\eta)\eta_{\mu\nu}(x) = \frac{8\pi G_N}{c^4}T_{\mu\nu}(\eta), \tag{1.6}$$

where $R_{\mu\nu}(\eta)$ is the Ricci tensor

$$R_{\mu\nu}(\eta) = \partial_\alpha\Gamma^\alpha_{\mu\nu} - \partial_\nu\Gamma^\alpha_{\mu\alpha} + \Gamma^\beta_{\alpha\beta}\Gamma^\alpha_{\mu\nu} - \Gamma^\beta_{\nu\alpha}\Gamma^\alpha_{\mu\beta},$$

$\Gamma^\lambda_{\mu\nu}$ are the Christoffel symbols which play the role of the gravitation field strength

$$\Gamma^\lambda_{\mu\nu} = \frac{1}{2}\eta^{\lambda\sigma}(\partial_\mu\eta_{\nu\sigma} + \partial_\nu\eta_{\mu\sigma} - \partial_\sigma\eta_{\mu\nu}) \tag{1.7}$$

$R(\eta) = R_{\mu\nu}(\eta)\eta^{\mu\nu}$ and $T_{\mu\nu}$ is the symmetric energy-momentum tensor of matter.

Outwardly the Einstein equations (1.6) are similar to Eq. (1.5) for the Newtonian potential. In both cases the quantities characterizing the field stand in the left-hand side and the quantities characterizing matter which creates this field do in the right-hand side. However between these equations there is a number of essential differences. Eq. (1.5) is linear and, consequently, satisfies to the superposition principle. It allows to compute a gravitational potential for any distribution of masses moving arbitrarily. The Newton gravitation field does not depend on the masses motion, therefore Eq. (1.5) in itself does not define immediately their motion. To describe the mass motion we must invoke the second Newton law. In the Einstein theory a pattern is absolutely other. As Eq. (1.6) are non-linear, the superposition principle does not work any more. Further in this theory it is impossible to set arbitrarily a right-hand part of Eq. (1.6) (i.e. $T_{\mu\nu}$), depending on matter motion, and then to calculate the gravitational field $\eta_{\mu\nu}$. The solution of the Einstein equations leads both to the definition of the motion of matter generating the field and to the evaluation of the field itself. In so doing it is essential, that the gravitation field equations also contain the masses motion equations in the gravitation field. From the physical point of view it is equivalent to the fact, that in the Einstein theory the matter creates the space-time curvature which, in its turn, influences on motion of matter originating this curvature. In the GTR all the particles move along extremal lines, called geodesic curves. In flat space-time geodesic curves degenerate into straight lines. Notice, that in the ordinary field theory operating flat space-time, the motion equations are also obtained by means of extremum condition, however, this condition is imposed on the system action. The stronger the gravitation field, the more appreciable is the curvature of the space-time. Thus, non relativistic gravitation theory is not applicable, when gravitation fields are very strong, as it occurs near collapsing objects like neutron stars or black holes. On the other hand in weak fields one may be restricted by the calculation of a small corrections to the Newton equations. The effects corresponding to these corrections, allow to test the GTR experimentally in the ordinary gravitational fields as well.

For today the experimental status of the basic statements of the Einstein theory is as follows. To check the principle of equivalence of the gravitational and the inert masses is carried out with the precision 10^{-12}. The theoretical formula for changing the light frequency (red shift) in the gravitational field which also is a consequence of the equivalence principle, is verified with the precision 2×10^{-4}. The invariance with time of G_N being postulated by the theory was tested by radar observations of the motions both of planets (Mercury, Venus) and spaceships, by measuring the Moon motion with the help of the laser, by observations of the motion of the neutron star, namely, pulsar PSR 1913+16 which enter into the composition of double star-shaped system. All the collection of the experimental data confirms the invariance G_N with the precision

$$\left|\frac{1}{G_N}\frac{dG_N}{dt}\right| < 10^{-11}\ \text{years}^{-1}.$$

The GTR predicts the bending light ray when it is passing near the heavy mass. The analogous bending follows from the Newton theory as well, however in the Einstein theory this effect is twice more. Numerous observations of this effect being done under passage of light coming from the stars near the Sun (during the complete solar eclipse) have confirmed the GTR predictions with the precision up to $\sim 11\%$. The much more precision ($\sim 0.3\%$) has been already reached under observation of the extra-terrestrial point radiation sources.

The Einstein theory also predicts the slow rotation of the elliptic orbits of the planets spinning around the Sun. It should be emphasized that this rotation is not explained by gravitational fields of other planets. The effect has the greatest magnitude for the Mercury orbit — $43''$ in a century. At present the verification precision of this prediction (precession of the Mercury perihelion) reaches 0.5%.

The GTR effects should be rather considerable when the stars are moving in a tight double system. With the greatest precision the motion of the pulsar PSR 1913+16 entering into the composition of the binary star is explored. Here the orbit rotation due to the GTR effects attains 4.2% in one year, and for 14 observations years (1975 — 1989) has given $\sim 60^0$.

One more the GTR effect is the prediction, that the bodies, moving with variable acceleration, will radiate gravity waves. Despite of numerous attempts it has been not possible to register gravity waves as far. However, there are the serious grounds in support of their existence already now. For example, the observations of the pulsar PSR 1913+16 have confirmed an energy loss of the double system due to the radiation of the gravity waves. As a consequence of the effect the period of the star revolution should decrease with the time. The observations confirm the GTR prediction with the precision 1%.

So, since all the GTR predictions prove to be true and there is no facts contradicting to it the GTR is that base on which the modern cosmological model, called the Big Bang model, has been built.

However, the quantization of the GTR faces serious difficulties. So, it follows from the Einstein field equation, that gravitation field theory does not belong to the class of the renormalizable theories. Let us explain what we mean talking about a renormalization. As we know, the mathematical apparatus of the quantum theory is mainly based on the usage of perturbation theory series. In the four-dimensional field theories these series contain infinitely large quantities, which one must be removed in one way or another. Normally that is reached by means of redefinition of the finite number of physical parameters, such as the mass, the charge, etc. This procedure is called the renormalization and the theories, in which it eliminates divergences, are called the renormalizable theories. For non-renormalizable theories there is no procedure to ensure convergence of perturbation theory series.

The presence of dimensional interaction constant makes the ordinary renormalization procedure impossible. To eliminate divergences in the theory, we must summarize all the terms in corresponding series of the perturbation theory. As a result, some divergences are reduced, and the remaining infinities are eliminated by the renormalization of the physical parameters of the theory. However, if the interaction constant is dimensional, then the terms in the perturbation theory series have the different dimensions and their summation has no sense. In the GTR under expanding the metric tensor $\eta_{\mu\nu}$ in a power series near the flat space with the metric $g_{\mu\nu}$, the interaction constant κ appears

$$\eta_{\mu\nu} \approx g_{\mu\nu} + \kappa h'_{\mu\nu}, \tag{1.8}$$

where $\kappa \sim \sqrt{G_N}$. As we see, κ proves to be a dimensional quantity and the theory of the renormalization does not work.

Another difficulty connecting with the quantization of the GTR has the experimental nature. A particle, creating the gravitation field, a graviton, has not been yet discovered. The theory predicts for it zero mass, zero electric charge, and spin being equal to 2.

Just as for other interactions it is possible to introduce nondimensional intensity of gravitation interaction $\alpha_g(q)$. It is defined as follows

$$\alpha_g(q) = \frac{G_N m^2}{4\pi\hbar c}. \tag{1.9}$$

The main difference in the above-enumerated interactions is the strength of their manifestation in nature. There are different ways to compare interactions intensity.

One of them is based on values of corresponding energy effects. Thus, for example, electromagnetic interaction can be characterized by binding energy of electron in ground state of hydrogen atom: $E_{em} \approx 10$ eV, and energy effect of strong interaction can be defined by binding energy of nucleons in nucleus: $E_s \approx 10$ MeV.

Another way is to compare running coupling constants, which describe different interactions. However, since these quantities are energy functions, we must point out the energy value, at which the comparison takes place. One should remember, that running constants of groups $SU(2)_{EW}$ and $U(1)_{EW}$ (α_2 and $\alpha_1 = g'^2/4\pi$) can not be identified with running constants of weak and electromagnetic interactions. The operation is legal only at energies much less then the energy, at which spontaneous violation of local symmetry of electroweak interaction takes place. At the scale 1 GeV running coupling constants of strong, electromagnetic and weak interactions are connected by the relation

$$\alpha_s : \alpha_{em} : \alpha_W \approx 1 : 10^{-2} : 10^{-6}.$$

As soon as the gravitation interaction is switched on, a confusing indefiniteness appears. What elementary particle should be taken as a standard? Now, the mass is the "charge" of gravitation interaction, but the mass spectrum of elementary particles is continuous. So, for example, the ratio of Coulomb and gravitation forces has the form

$$\frac{F_c}{F_g} \approx 10^{36} \tag{1.10}$$

for protons and

$$\frac{F_c}{F_g} \approx 10^{43} \tag{1.11}$$

for electrons. Using Eqs.(1.10) and (1.11) we arrive at two different intensity hierarchies

$$\left.\begin{array}{l} \alpha_s : \alpha_{em} : \alpha_W : \alpha_G \approx 1 : 10^{-2} : 10^{-6} : 10^{-38}, \\ \alpha_s : \alpha_{em} : \alpha_W : \alpha_G \approx 1 : 10^{-2} : 10^{-6} : 10^{-45}. \end{array}\right\} \tag{1.12}$$

1.5. Grand Unified Theory

The extension of the electroweak theory to quarks and inclusion of the QCD resulted in unification of strong and electroweak interactions. Created scheme was called the standard model (SM). The picture of fundamental forces in this model is charmingly simple. Strong, weak and electromagnetic interactions are caused by the existence of the local gauge symmetry group

$$SU(3)_c \bigotimes SU(2)_{EW} \bigotimes U(1)_{EW}$$

with its three gauge constants g_s, g and g' and twelve gauge bosons being the carriers of strong and electroweak interactions. At sufficiently small distances all these forces mainly resemble each other and lead to the potential of Coulomb type $\sim g^2/r$. A scale of short distances for strong interaction represent distances much smaller than hadron size, that is, more smaller than 10^{-13} cm. For electroweak interactions the scale of the small distances is the distance which is much smaller than the Compton wave length of the $W^{\pm}$- and Z-bosons ($\lambda_c = \hbar/mc$), that is, more smaller than 10^{-16} cm. It is obvious, that at such short distances the existence of the mass on the gauge boson is coming inessential.

Since the SM gauge group is the production of three not bound sets of gauge transformations: the groups $SU(3)_c$, $SU(2)_{EW}$ and $U(1)_{EW}$, then three gauge constants of these groups g_s, g and g' are not connected with each other. The gauge constants will bind, if the SM gauge group proves to be embedded in a more wide group of gauge transformations G. Symbolically it is written as follows:

$$G \ni SU(3)_c \bigotimes SU(2)_{EW} \bigotimes U(1)_{EW}.$$

As a result, all the interactions will be described by unified gauge theory, the Grand Unification theory (GUT), with one gauge constant g_{GU}, moreover all the other gauge constants are connected with the latter in unambiguous way, defined by the choice of the group G. The GUT symmetry must be spontaneously violated at supershort distances being many orders smaller than those, at which unification of electromagnetic and weak interactions takes place. In other words, strong interaction with the local $SU(3)_c$-symmetry, described by the QCD, as well as electroweak interaction with the local $SU(2)_{EW} \otimes U(1)_{EW}$-symmetry turn out to be the low energy fragments of the gauge interaction with the group G.

To estimate distance scale, at which Grand Unification takes place, one should turn to equations defining evolution of running constants of the strong and electroweak interactions. In so doing, it is necessary to represent these equations in such a form so that they determine the constants variation not as a function of the transferred momentum q, but as a function of variation of the mass scale μ. The cause of changing the gauge coupling constants is the vacuum polarization, that is, it is stipulated by the processes of creation and consequent destruction of the virtual particles. To take into account these processes in the second order of perturbation theory leads to the sufficiently simple evolution equations for the coupling constants

$$\frac{1}{\alpha_s(M)} = \frac{1}{\alpha_s(\mu)} + \frac{9}{2\pi} \ln\left(\frac{\mu}{M}\right), \tag{1.13}$$

$$\frac{1}{\alpha(M)} = \frac{1}{\alpha(\mu)} - \frac{11}{6\pi} \ln\left(\frac{M}{\mu}\right), \tag{1.14}$$

$$\frac{1}{\alpha_2(M)} = \frac{1}{\alpha_2(\mu)} + \frac{19}{12\pi} \ln\left(\frac{M}{\mu}\right), \tag{1.15}$$

where we have performed the transition from the $U(1)_{EW}$-group running coupling constant α_1 to the electromagnetic interaction running coupling constant α with the help of the relation

$$\alpha = \alpha_1 \cos^2\theta_W,$$

(the absence of the subscript *em* by α underlines the circumstance that the question is the fine structure constant not in the QED, but already in the more precise theory, namely, in the theory of electroweak interactions). Thus, according to the theory, the dependence of $1/\alpha_i$ on $\ln M$ is linear, its slope value defining the polarization effect of relevant vacuum. So, the larger value of the slope of $1/\alpha_s$ compared to the slope of $1/\alpha_W$ is caused by the fact that the number of gluons is larger than the number of carriers of weak interactions ($W^{\pm}$- and Z-bosons) and, as a result, gluons give the bigger anti-screening effect. In $1/\alpha$ the screening effect predominates (tangent of the slope angle is negative by now) and for this reason the value of $1/\alpha$ drops with the growth of M.

Further one may show that in the limit of the exact unified symmetry ($M = M_{GU}$) the following relation is valid

$$\alpha_2(M_{GU}) = \alpha_s(M_{GU}) = 8\alpha(M_{GU})/3 = \alpha_{GU}(M_{GU}). \tag{1.16}$$

Then, from Eqs. (1.13) — (1.16) it is easy to obtain

$$\ln\left(\frac{M_{GU}}{\mu}\right) = \frac{\pi}{11}\left[\frac{1}{\alpha(\mu)} - \frac{8}{3\alpha_s(\mu)}\right]. \tag{1.17}$$

This relation defines the unification scale M_{GU}. Having set the values of μ, $\alpha(\mu)$ and $\alpha_s(\mu)$ it is possible to estimate both the value of M_{GU} under which the relation (1.16) is fulfilled and the value of the unified constant $\alpha_{GU}(M_{GU})$. Having chosen the following parameters values

$$\mu \approx m_W, \qquad \alpha_s^{-1}(m_W) \approx 10, \qquad \alpha^{-1}(m_W) \approx 128,$$

in Fig.1 we present the dependence of the running coupling constants α_s, α and α_2 on M described by Eqs. (1.13) — (1.15).

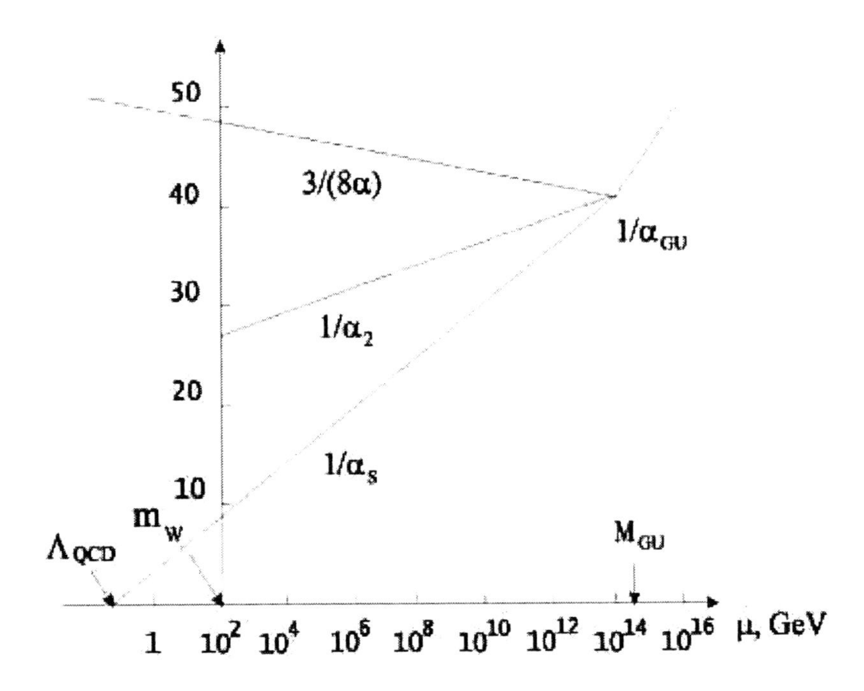

Figure 1. The M-dependence of the running coupling constants α_s, α and α_2.

As it follows from Fig. 1, the Grand Unification takes place at $M_{GU} \approx 2 \times 10^{14}$ GeV/c^2. The analysis based on using, as the definite GUT group, $SU(5)$-group leads to the estimation: $M_{GU} \approx 2 \times 10^{15}\Lambda_{QCD}$. The obtained values correspond to the distances $L_{GU} \approx \hbar c/E_{GU} \sim 10^{-28}$ cm.

At distances, shorter than L_{GU}, the initial symmetry is restored and the interaction is described by a single constant α_{GU}, which evolution law is defined by the structure of the G-group.

At this point it is natural to pose a question: "whether it is possible to trust calculations, altogether ignoring gravitation interaction".

Gravitation effects are of the order 1, when masses of interacting particles have such values, that potential gravitation energy is comparable to the particle rest energy

$$\frac{G_N M_P^2}{r} \approx M_P c^2. \tag{1.18}$$

If we set the distance between masses equal to the Compton wave length, that is to quantity $\hbar/M_P c$, then condition (1.18) is realized at

$$M_P = \left(\frac{\hbar c}{G_N}\right)^{1/2} = 1.22 \times 10^{19}\ \text{GeV}/c^2. \tag{1.19}$$

The obtained mass value is called the Plank mass. The time and the length, corresponding to it

$$t_P = \sqrt{\frac{\hbar G_N}{c^5}} \approx 5.4 \times 10^{-44}\ \text{s}, \qquad L_P = \sqrt{\frac{\hbar G_N}{c^3}} \approx 1.6 \times 10^{-33}\ \text{cm}, \tag{1.20}$$

are called the Plank time and length. Since $L_{GU} \gg L_P$, then on the Grand Unification scale we have the right to neglect gravitation effects.

Probably, in laboratory conditions we shall never be able to produce energies, corresponding to the Unification scale, consequently, experimental check of the GUT is a very complicated task. Among the GUT consequences being available for observations we note the predictions of such effects as proton instability and neutron-antineutron oscillations (neutron transformation into antineutron in vacuum and the reverse process).

In electroweak interactions the gauge constants g and g' are not bound together and their ratio

$$\frac{g}{g'} = \tan\theta_W,$$

where θ_W is the Weinberg angle, is an experimentally obtained parameter. Opposite to that, the GUT allows to calculate the Weinberg angle. The GUT models explain naturally electric charge quantization, manifested through the fact, that quark charges are multiple of $e/3$, while lepton charges are equal either to e or to 0.

GUT's have some cosmological consequences as well. According to adopted point of view, our Universe is made up approximately 2×10^{10} years ago as a result of the Big Bang and it is still expanding. This expansion is described by the GTR equations. Universe size changed from the value of the Plank length order (10^{-33} cm) to contemporary value which is of the order 10^{28} cm. Compressed in such a small value, substance began its evolution with energy of the Plank order, that is, early Universe is a gigantic laboratory, where the GUT consequences could be checked. Within the GUT frameworks it is possible to get explanation of the fact, that the matter at the moment is prevailing over the antimatter in the Universe (baryon asymmetry). The value of the ratio of the baryons concentration n_B to photons concentration n_γ in cosmic microwave background can be also obtained in the context of GUT.

Alongside with the above mentioned achievements, there are some weak points in the existing GUT. Let us enumerate some of them. The models have a great body of free parameters, which number exceeds the number of those in the SM. It is not possible to make any statement concerning the number of fermion generations within the frameworks of the models. The gravitation is excluded from the unification scheme. Serious difficulties are produced under explanation of difference by twelve orders of the distances scales at which unified symmetry G and electroweak interactions symmetry are broken (the hierarchies problem).

1.6. Supersymmetry

Is there any more "grandiose" unification, unified field theory, which includes both gravitation and the SM? Before we discuss the directions, in which the construction of the theory is going on, let us first get acquainted with one more type of symmetry — supersymmetry.

Up to now we have been considering space-time and internal symmetries. Geometrical translations and rotations do not change the nature of a particle: the photon remains the photon after any space-time transformations. Internal symmetries can change the nature of a particle but not its spin value. So, under action of isotopic rotations proton can turn into neutron, but it can not transform into π^0-meson, for instance.

Unlike the above mentioned symmetries, supersymmetry transformations can change not only space-time coordinates of a particle and its nature, but its spin value as well. In other words, supersymmetry implies the invariance of physical system under fermion-boson transitions, that in its turn, allows us to call it as the Fermi-Bose symmetry. The basis for supersymmetric theories is the space-time extension to superspace, which besides the normal space-time coordinates x^μ includes also the internal space coordinates θ_α. In the most general case the points in superspace are characterized by four even coordinates x^μ and $4N$ odd coordinates θ^j_α, where $j = 1, 2, .., N$ (N-extended supersymmetry). Let us restrict ourselves to the supersymmetry with $N = 1$. In the ordinary four-dimensional space-time there is Poincare transformation group with 10 parameters, while in superspace for the case $N = 1$ the extended Poincare group with 14 parameters, comes into action, where besides rotations and translations in the ordinary four-dimensional space the supertranslations in the internal space have been added

$$\left.\begin{array}{l} x'^\mu = x^\mu + \frac{i}{2}\overline{\varepsilon}\gamma^\mu\theta, \\ \theta' = \theta + \varepsilon \end{array}\right\}, \qquad (1.21)$$

where θ is represented in the form of 4-column consisting of θ_α, and ε is the quantity, defining the supertranslation, is determined by four real parameters ($\overline{\varepsilon} = \varepsilon^\dagger\gamma_0$, γ_μ are the Dirac matrices).

Since every boson is associated with supersymmetric fermion and vise versa, then the number of particles in the theory is doubled. Supersymmetric partners get their names either with prefix "s" for scalar partners of normal fermions (for example, squark, selectron) or with ending "ino" for fermion partners of normal bosons (for example, photino, gravitino).

In supersymmetric theories divergences in perturbation theory series, corresponding to bosons and fermions have the opposite signs and mutually compensate each other. Thus, there is no need for renormalization at all. In other words, supersymmetry allows constructing finite, divergence-free theories. To include the supersymmetry in the SM brought the minimal supersymmetric standard model (MSSM) into being. The discovery of superpartners of the known fundamental particles will experimentally prove the existence of the supersymmetry in nature. By now, the search of the superpartners has not given the positive results.

1.7. Supergravitation

In the 60s of the XXth century some papers appeared in which the GTR was reformulated in the form of gauge gravitation theory. For this purpose the symmetry group of the flat space-time, the 10-parameters Lorentz group, was chosen and its localization was carried out (that is, the transformation parameters became the coordinate functions). As a result, the gauge fields appeared, which were associated with gravitation field of the GTR. The theory obtained was completely identical to the GTR and in fact, produced no new results. The development of gauge interpretation of gravitation was defined mainly by hopes to use it in the coming time for unification of gravitation interactions with the other ones. The star hour of a gauge variant of gravitation came in the 1970s when the supersymmetry theory had appeared. The theory, appearing as a result of merging of two origins, the supersymmetry and the gauge principle, was called supergravitation. The supergravitation geometry is as simple and elegant as that of Einstein's GTR (the latter corresponds to supergravitation with $N = 0$). The basis of supergravitation is a relativity principle, which reads, that "*the form of physical laws does not depend on the choice of a coordinate system in superspace*".

In the simple supergravitation ($N = 1$) the fourteen-parametric Lorentz group, extended by supertranslation transformations is localized. In this variant of supergravitation familiar to us the graviton and its superpartner gravitino (with spin $3/2$) are the carriers of gravitation field. In the simplest extended supergravitation ($N = 2$) a symmetry group with 18 parameters is localized, and consequently, there are more interaction carriers, they are: graviton, two gravitino and graviphoton. The $N = 2$ supergravitation represents the first supergravitation theory, which unifies gravitation with electromagnetism in principle. It becomes possible to unify particles with spin 2 and 1 due to the presence of the intermediate stage, particle with spin 3/2. There are much less divergences in supergravitation compared to quantum gravitation theory. Many viable supersymmetric field theories contain the supergravitation as an important component, which helps to spontaneously violate the supersymmetry. In such models the hierarchy problem finds its solution.

1.8. Kaluza-Klein Theory

To get acquainted with another direction in construction of unified field theory, let us consider T. Kaluza concept later developed in the works by K. Klein. The idea was to unify the GTR and Maxwell electrodynamics on the basis of hypothesis that our Universe is the curved five-dimensional space-time. One of the coordinates denotes time, while the four are spatial coordinates. In such a space the five-interval square dS^2 is defined by the relation

$$dS^2 = g_{AB}(x)dx^A dx^B, \tag{1.22}$$

where A, B=0,1,2,3,4, and $g_{AB}(x)$ is a metric tensor with fifteen independent components. Then ten combinations $g_{AB}(x)$

$$g_{\mu\nu}(x) + \frac{g_{4\mu}(x)g_{4\nu}(x)}{g_{44}(x)}$$

are associated with ten components of metric tensor in the GTR $\eta_{\mu\nu}(x)$. The following four combinations

$$\frac{g_{4\nu}(x)}{\sqrt{-g_{44}(x)}}$$

are connected with four components of electromagnetic potential $A_\mu(x)$. Let us explain why this operation is legal. Remember that Christoffel symbols $\Gamma_{AC,B}$ are the field strengths in the curved space-time. We set one of the indices C equal to 4, A and B equal to 0,1,2,3. Then, using (1.7) we obtain

$$\Gamma_{\mu 4,\nu} = \frac{1}{2}\left(\frac{\partial g_{\mu\nu}(x)}{\partial x^4} + \frac{\partial g_{\nu 4}(x)}{\partial x^\mu} - \frac{\partial g_{\mu 4}(x)}{\partial x^\nu}\right). \tag{1.23}$$

When one assumes, that the fifth coordinate is cyclic, then the expression (1.23) takes the form

$$\Gamma_{\mu 4,\nu} = \frac{1}{2}\left(\frac{\partial g_{\nu 4}(x)}{\partial x^\mu} - \frac{\partial g_{\mu 4}(x)}{\partial x^\nu}\right), \tag{1.24}$$

or, having set

$$F_{\mu\nu}(x) = \frac{c^2}{\sqrt{G_N}}\Gamma_{\mu 4,\nu}, \qquad A_\mu(x) = \frac{c^2}{2\sqrt{G_N}}g_{4\mu}(x), \tag{1.25}$$

for (1.24) we arrive at

$$F_{\mu\nu}(x) = \left(\frac{\partial A_\nu(x)}{\partial x^\mu} - \frac{\partial A_\mu(x)}{\partial x^\nu}\right), \tag{1.26}$$

that is, $g_{4\nu}(x)$ and $\Gamma_{\mu 4,\nu}$ can be really identified with potential and tensor of electromagnetic field, respectively. The equations, governing system evolution, follow from the least action principle

$$\delta(S_m + S_f) = 0, \tag{1.27}$$

where S_m and S_f are actions for matter and field, respectively. Variation of the first term in Eq.(1.27) results in five equations for geodesic lines, four of which coincide with the known four-dimensional equations for charged particles moving in gravitation and electromagnetic fields

$$\frac{d^2x_\nu}{ds^2} = -\Gamma_\nu^{\mu\lambda}\frac{dx_\mu}{ds}\frac{dx_\lambda}{ds} + \frac{e}{m}F_\nu^\mu\frac{dx_\mu}{ds}, \tag{1.28}$$

and the fifth equation

$$\frac{dx_4}{ds} = -\frac{1}{2\sqrt{G_N}}\frac{q}{m} \tag{1.29}$$

shows, that while a body is moving in gravitation and electromagnetic fields, its electric charge is conserved.

Varying only the potentials of five-dimensional space-time, we obtain fifteen equations, which break down into the system out of ten of the ordinary four-dimensional GTR equations (1.6) and the system out of four of the Maxwell equations

$$\frac{\partial F^{\mu\nu}}{\partial x^\nu} = j^\mu, \qquad \frac{\partial F^{\mu\nu}}{\partial x^\sigma} + \frac{\partial F^{\nu\sigma}}{\partial x^\mu} + \frac{\partial F^{\sigma\mu}}{\partial x^\nu} = 0. \tag{1.30}$$

In so doing, the equation for the scalar component $g_{44}(x)$ is out of use again.

Despite of obvious merits, the Kaluza-Klein theory has been leaving two questions to be completely opened. 1) What is the physical meaning of the fifth coordinate x_4 and why it is not observable? 2) Why all the physical quantities are cyclic with respect to x_4? The obvious answer could be that the manifestation region of the additional dimension is beyond the existing experimental technology. At the end of the XXth century such statements become the rule of the good form and the mighty imagination of the theorist-physicists produces a great oasis of exotic phenomena in the energy scale close to Plank energy. However, by 1938 only A. Einstein and P. Bergmann could come up with such an idea. They suggested that the fifth coordinate can change from 0 to some value L, that is, five-dimensional world is confined in a layer with thickness L. The assumption was also made, that any function $\Psi(x)$, related to physics, changes little along x_4 over a length of the layer, so that

$$L\frac{d\Psi(x)}{dx_4} \ll \Psi(x)$$

and in the average $\Psi(x)$ may be considered as a function only of four-dimensional coordinates. Actually, instead of restricting x_4 values to quantity L, it is possible to assume, that the fifth coordinate varies within infinite limits, however, only functions, periodic in x_4 with the period L are under consideration. It means, that it is possible to glue together all the points, being distant from each other along x_4 on interval L, without any harm for generality done. As a result, we arrive at five-dimensional space-time being closed by x_4. The world with such a property we shall call cyclic, closed or compactified in the fifth coordinate. In such a theory there is no need for postulating of cyclic character of x_4, since the gauge principle of switching on electromagnetic interaction, which had been armed by us, resulted in the conclusion, that wave functions of charged particles have the form

$$\Psi(x) = \Psi(x_\nu)\exp\left(\frac{iec}{2\sqrt{G_N\hbar}}x_4\right), \tag{1.31}$$

where $\Psi(x_\nu)$ is a ordinary wave function in four-dimensional space. The expression (1.31) describes cyclic dependence of $\Psi(x)$ on x_4 with a period

$$L = \frac{4\pi\sqrt{G_N\hbar}}{ec} \approx 10^{-31}\ \text{cm}. \tag{1.32}$$

Thus, the cyclic period or the world compactification scale in fifth coordinate is infinitesimally small compared to the scale of phenomena, studied by contemporary physics. Consequently, it is not surprising, that the fifth dimension has been skillfully hiding itself from experimentalists up to now. Of course, extension of space-time dimensions can be generalized on larger dimensions as well.

According to the Big Bang theory, the early Universe was a substance, compressed in a volume with radius of the Plank length order 10^{-33} cm. It is easy to understand that evolution of the Universe is completely defined by the structure of elementary particle physics. Consequently, the idea of closed dimensions must find its place in the Big Bang theory, too. The Universe is thought to have carried out the space-time compactification in additional dimensions at the early stages of evolution when its energy was rather high. At present this hypothesis is the required attribute of the contemporary multidimension theories.

1.9. Superstring Theory

The quantum field theory (QFT) and the GTR contain all human knowledge of fundamental forces of nature. The QFT brilliantly explains all the microworld phenomena up to the distances up to 10^{-15} cm. The GTR, in its turn, is beyond any competition in describing both large-scale events in the Universe, as well as evolution history of the Universe itself. The striking success of these theories is caused by the fact, that taken together, they explain both behavior and structure of matter from subnuclear to cosmological scales. All the attempts to unify these theories, however, have been always facing two main difficulties. The former deals with appearance of divergences, while the latter demands to refuse one or more ideas about the Universe nature, which are highly respected nowadays. The QFT and the GTR are based on few postulates, which appear to be mandatory and natural. The postulates, the abandonment of which helps to unify two theories, are as follows: 1) space-time continuity; 2) causality; 3) theory unitarity (the absence of the states with negative norm); 4) interaction locality; 5) point (structureless) particles. The first four items are rather serious. Against their background the abandonment of the fifth item looks like the children's prank.

Any local theory, operating with structureless objects contains divergences at energies, higher than Plank energy E_P. It is caused, as we know, by the fact, that E_P is that very energy value, when quantum theory of gravitation is needed, while it is not renormalizable. To achieve renormalizability, another model can be introduced, in which at "low" energies (¡E_P) fundamental objects behave as point ones and their composite structure reveals itself only at energies, higher than E_P. If one adds to this hypothesis some latest inventions, concerning unification of all interactions (supersymmetry, Kaluza-Klein theory, etc.), then one more way is being opened to the unified field theory, called the superstring theory.

String theories were used by hadron physics as early as 1960s. Their creation was inspired by the fact, that at the distances of the order 10^{-13} cm gluon fields, binding quarks, are concentrated in space not evenly but along the lines connecting quarks. It resulted in interpretation of hadrons as nonlocal objects, one-dimensional strings, on the ends of which either quarks for baryons (fermion strings) or quarks and antiquarks for mesons (boson strings) are placed. Thus, the infinitely thin tube of qluon field is modeled by relativistic string. Relativistic string theories managed to explain many anomalies of quark Universe. Thus, for example, since string energy is proportional to its length L, and string mass square is $m^2 \sim L^2$, then the string angular moment of the string having the form of linear segment is proportional to L^2. Really, experiments confirm linear dependence between the hadron spin J and its mass $J \approx (m/\mathrm{GeV})^2$. Relativistic string, which binds the quark and antiquark, generates potential, linearly growing with distance, that is, it allows to explain quark confinement. The theory is built such a way, that after string breaks no free quarks appear, since at both newly formed string ends a pair "quark-antiquark" is produced.

However, string relativistic theories have some very serious deficiencies, such as, for example, tachyon existence, massless hadrons with spin 2, etc. It also turned out, that consistent quantum theories could be only formulated in 26-dimensional space-time for boson strings and in 10-dimensional space-time for fermion strings.

The renascence of interest in string theories took place after in 1986 M. Green and J. Schwarz proved, that 10-dimensional gauge superstring theory, based on the internal symmetry group $SO(32)$ or $E_8 \otimes E_8$ [1] (subscript indicates the group rank) can be used to unify all the interactions.

Superstrings are one-dimensional in spatial sense (two-dimensional, if the time is taken into account) objects with the typical length of the order L_P. They are put in n-dimensional ($n \geq 10$) space-time manifold. To turn to the observable space-time dimensionality is achieved by compactifying unnecessary dimensions at distances of the order of the Plank length. The theory contains mechanism, ensuring spontaneous compactification of additional dimensions. The initial symmetry is broken up to a symmetry group, involving supergravitation and supersymmetric GUT with fixed parameters and given particles content. If the gauge group $E_8 \otimes E_8$ is used then one E_8-group contains all the low energy physics, while the other E_8 manifests itself only in gravitation interaction and for this reason it describes a shadow matter in the Universe. Phenomenological properties of superstring theory depend in many respects on compactification mechanism. As an example, let us pay attention to the following circumstances. Since the division into normal space-time dimensions and compactified ones is not very strict, then it is possible, that some Universes with non-conventional dimensions of space-time exist.

The important difference between superstring theory and local field theory is that in the former theory the free superstring is characterized by infinite number of supermultiplets, while in the latter one every field describes particles of only one kind. Superstrings have the same number of fermion and boson degrees of freedom. Superstring excitations, (which are: rotations, vibrations or excitations of internal degrees of freedom) are associated with the observed elementary particles. The particle mass scale is regulated by the superstring tension T with $\sqrt{T} \approx M_P c^2$. The number of states with masses, smaller than Plank mass, is finite. It defines the number of elementary particles existing in Nature. There is also a great number of excitations with masses, bigger than Plank mass. The majority of these modes are unstable, however, there are also stable solutions with exotic characteristics (magnetic charge, for example). It is remarkable, that in particle spectrum, which corresponds to superstring theories solution, one massless state with spin 2 appears, which is described by the GTR equations in low energy limit, that is, it is a graviton.

The strings appear in two topologies: as open string with free ends and as closed loops. Besides this, they can possess internal orientation. Quantum numbers of open strings are located at their ends, while in closed loops quantum numbers are evenly spread along the string. The string interaction has the local character, despite the fact, that they are extend objects. When interacting the strings can scatter, produce new strings, and emit point particles as well.

The development of superstring theory showed that it was a fruitful generalization of a local field theory. However, nowadays the superstring theory is still undergoing its development stage and has got no experimental confirmations yet. Let us note that experimental check of superstring models is very difficult due to many unknown parameters they contain. We would like to believe that the completed superstring theory would contain only two fundamental parameters: tension and superstring interaction constant.

---

[1] E_8 together with the groups G_2, E_4, E_6, and E_7 constitutes the exceptional group class. The rank of the exceptional group is fixed, while the normal (regular) group can have any rank (for instance, $SU(2)$, $SU(3)$, $SU(5)$ and so on).

Chapter 2

Three Steps of Quantum Stairway

2.1. Atomic Theory

Two and a half thousands years ago Greek philosophers laid the foundations of our understanding of matter nature, when they embarked on the first attempts to reduce variety of the world to interaction of few initial components, fundamental particles or elements. In IV B.C. Phales suggested water to be the single primary element, of which all existing world was made up. Later Anaximen from Milet extended the list of primary elements to the four, such as soil, air, fire and water. It is generally agreed that Democritus (460 — 370 B.C.) is the creator of the atom idea, however, the history also mentions his teacher Levkipus in this connection.

A legend tells that once Democritus sat on sea shore with an apple in his hand. His stream of thoughts was as follows: "Suppose, I shall cut this apple in halves, and then every half I shall cut in halves again, so that there will be a quarter of an apple left, then one eighth, one sixteenth, etc. The question is, if I continue to cut the apple to the end, will the cut parts always possess properties of an apple? Or, may be, at a certain moment the remaining part of the apple will have no properties of an apple any more?" Upon serious consideration the philosopher arrived at the conclusion, that there was a limit for such a division and he called the last indivisible part an atom. His conclusions mentioned below, he stated in his book "Great diacosmos".

"The beginning of the Universe is atoms and emptiness, other things exist only in our imagination. There exists infinite number of worlds, and they all have their beginning and end in time. And nothing appears from non-existence, nothing goes into non-existence. Atoms are uncountable both in collection and in variety, whirling like the wind they rush in the Universe and so all composite substances create: fire, water, air, soil. The point is that these substances are compounds of some atoms. Atoms are unaffected by any influence, they are unchangeable due to their hardness".

Democritus could not prove his statements, so he suggested his contemporaries to entrust his words. The majority of contemporaries did not believe him, and Aristotle was among them. Aristotle was the author of the opposite teaching. According to him, the process of the apple partition could be infinitely extended. Natural philosophy is not based on experiments and mathematics, it used the element of faith as the single criterion of truth. So it is not surprising, that both teachings for the ancients seemed to be equally reasonable and acceptable. It is difficult to tell, what has weighed down the weights bowl in favour of Aristotle philosophy. Maybe, it was the gleams of military glory of Alexander the Great, whose teacher Aristotle was?. Anyway, the teaching of Aristotle becomes dominant, while Democritus had been forgotten for many centuries.

In XVII the idea of Democritus about atoms has been restored to live by a French philosopher Gassendet. When spring comes, all violets bloom at once. So was with the atomic hypothesis. After twenty centuries of oblivion all the contemporary advanced scientists believed in atomic theory, including great I. Newton, whose credo was "*not to build any hypothesises*". One of the burning questions of atomism was, undoubtedly, the question, whether the variety of bodies in nature means, according to Democritus, the same variety of atoms? If the answer is positive, then the atomic hypothesis not a jot brings us closer to understanding the world. Luckily, the answer was negative. The variety of substances in nature is caused not by a variety of different types of atoms, but by the variety of different compounds these atoms (nowadays, these compounds are called molecules). In 1808 D. Dalton, upon studying many chemical reactions, precisely formulated the notion of a chemical element: "Chemical element is a substance, consisting of atoms of the same type". It turned out, that there were not so many chemical elements. In 1869 D. Mendeleev could manage to place all of them in one periodic table (at that time only 63 elements were discovered, while now their number is reaching 120).

The work by botanist R. Brown (1827) may be considered as the first experimental proof in support of atomic theory. He observed chaotic motions of flower pollen in water (Brownian motion). The discovery by Brown did not attract scientists attention immediately and for a long time its nature remained unclear. Only seventy eight years later, the atomistic theory of Brownian motion was established in works by A. Einstein and M. Smoluchovski. In 1908 J. Perrin carried out a series of experiments to study Brownian motion. Not only did he proved experimentally the works by Einstein and Smoluchovski but he also measured sizes and masses of atoms. The last and, probably, final proof of atomic matter structure was the work by E. Rutherford and Royds on measuring the number of α-particles in radium. By that time it had been known, that in minerals, containing α-radioactive elements (radium, thorium), helium is accumulated. The task was to determine the number of α-particles emitted by a sample and to measure the volume of helium produced on the sample. In a second 13.6×10^{10} particles are emitted by one gram of radium. Having captured two electrons all these α-particles turn into helium atoms and occupy the volume of 5.32×10^{-9} cm^3. Consequently, 1 cm^3 contains $L=2.56\times10^{19}$ atoms. Let us compare the value obtained with Loschmidt number calculated as early as 1865 on the basis of molecular-kinetic theory. One mole of helium (or any other gas) occupies a volume 2.241×10^{-2} cm^3/mole and contains 6.02×10^{23} atoms, that is, 1 cm^3 contains 2.68×10^{19} atoms. The coincidence is impressive. So, existence of atoms got the final experimental proof. That has completed climbing of physics on the first step of "Quantum Stairway". The picture of the world on this step is very simple: the matter in our Universe consists of indivisible atoms of different types, which make all the elements of Mendeleev periodic table.

The variety of elements and explicit systematization in periodic table by Mendeleev install in us far from being utopian belief that the result obtained is not final yet. Existence of the next step of Quantum Stairway became obvious after a series of experiments, which may be called "Roentgen" of atom according to Rutherford (1909 — 1911).

2.2. Nuclear Model of Atom

The fact, that all atoms contain electrons was the first important piece of information about the internal structure of atom. Electrons have negative electric charge, while atoms are electrically neutral. Consequently, every atom must contain enough of positively charge matter to compensate negative charge of electrons. In 1903 in the book "Electricity and matter" D. D. Thomson introduced a model of a radiating atom, which was satisfying to totality of chemical and spectroscopic experiments existing in that time. Thomson's predecessor in the model was his famous namesake William Thomson (Lord Kelvin), who in 1902 suggested coming back to atomic theory by F. Epinus (1759). Epinus atom was a sphere, being uniformly charged with positive electricity, in the center of which a negative charged corpuscle was located. D. D. Thomson developed this model by assuming that electrons rotated inside of a sphere, the number of electrons and their orbit configuration having depended on atom nature. Electrons are spaced as concentric rings (shell) and perform periodic motions, which are causing the observed atom spectrum. Investigating stability of the electron combination, Thomson gave the physical interpretation of valence. The atom model by Thomson was the first real attempt to explain chemical properties of substance and periodic low by Mendeleev. His model, called a "pudding with raisins" was warmly accepted by the scientific public. One should not be to think that spirit of contradiction, being so characteristic for scientific creativity, has led to the construction of some alternative atom models. So, for example, in 1901 A. Perren published the article titled "Nuclear Planetary Structure of Atom" in scientific popular journal "Scientific Review ". Two yeas later Japanese physicist Ch. Nagaoka (disciple of Bolzman), suggested the model according to which atom consists of positive charged nucleus and of electrons ring rotating around nucleus (atom of Saturn type). First Nagaoka published his results in even less popular among physicists journal "Proceedings of Tokyo Physical-Mathematical Society". Then in 1904 he published the article on the same topic in "Philosophical Magazine". However, physicists paid no attention either to Perren or Nagaoka models, although these models could be regarded as predecessors of the atom nuclear model later suggested by Rutherford.

Nine more years passed before the atom model by Thomson had been brought under serious experimental testing, results of which showed that the model had as low as single drawback, namely, it had nothing in common with physical reality. The creators of quantum theory liked to tell "in order to check, what is inside of a pudding with raisins one must simply put finger in it". In reality Rutherford used the method being rather similar with one described above. α-particles from radioactive sources were used as a probe, and thin foils of different substances with typical thickness about 10^4 atomic layers were used as targets. In Fig. 2, the scheme of these experiments done by G. Geiger and E. Marsden under the guidance of Rutherford is represented.

An ampoule with radioactive source (radium C, Po-214) was placed behind a lead screen with a small opening. α-particle beam was passing through a lead collimator and was directed at a target. After interacting with target atoms, α-particles got into a mobile screen of sulphureous zinc causing scintillations on it, which were registered by means of a microscope.

As we know from optics, to see an object, one must use electromagnetic waves having wavelengths the same order as its size. The same rule takes place in quantum world, only

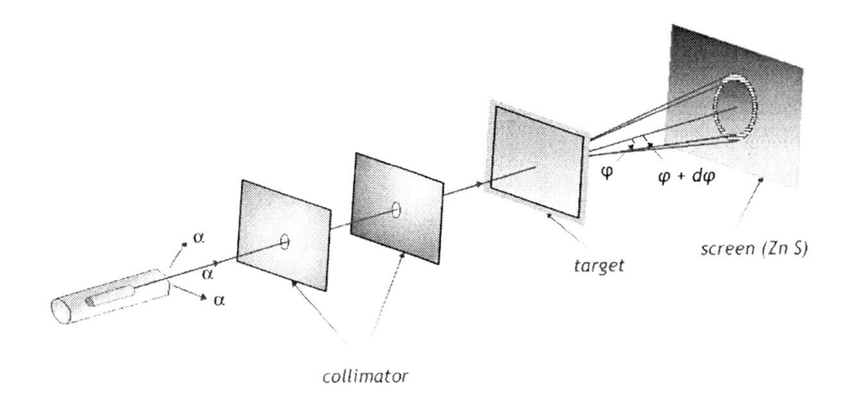

Figure 2. The plan of Rutherford's experiments.

here we are dealing with de Broglie wavelength λ. Of course, in 1909 nobody knew about de Broglie waves, but the Nature once again proved to be favorable to Man, and α-particles with energy 7.7 MeV (typical energy values in Rutherford experiments) had wavelength of $\sim 5\times 10^{-13}$ cm, which was quite enough to "see" a nucleus in an atom.

The Thompson atom model predicted, that the main part of α-particles flew straight through a foil, while the remaining particles is subjected to the small deviation. In reality the behavior of most α-particles fitted in the frames of this model, but parallel with that some particles deviated on big angles. Let us show, quantitatively, that Thompson model is not able to explain the results of these experiments. When, according to Thompson, one assumes that a positive charge Q is spread evenly all over the volume of a gold atom and does not take into account the influence of electrons, then electric field strength inside of an atom at a distance r from its center is given

$$E_+(r) = \frac{Qr}{4\pi R^3},$$

where R is the atom radius. α-particle maximum deviation takes place, if it is slightly touching atom surface where $E_+(r)$ reaches values of the order of 10^{13} V/cm. Thus, the maximum value of the force is determined by the expression

$$F_{max} = \frac{Q|e|}{2\pi R^2}.$$

Since this force rapidly decreases with distance, then for approximate estimation, one can consider interaction on small interval L, which includes the distances before and after contact between α-particle and the atom. We accept $L\sim 2R$, and take the force value being maximum, that is, during time interval $\Delta t = L/v_\alpha \approx 2R/v_\alpha$ α-particle is subjected to F_{max}. In this case, transverse momentum, transferred to α-particle, is equal to

$$\Delta p_\alpha = F_{max}\Delta t \sim \frac{Q|e|}{\pi R v_\alpha}.$$

Consequently, the maximum deviation angle is given by the expression

$$\theta_+ < \frac{Q|e|}{\pi R m_\alpha v_\alpha^2}. \tag{2.1}$$

Using the values 6.6×10^{-27} kg and 2×10^{7} m/s for the mass and the velocity of the α-particle respectively, we obtain in the case of the gold atom ($Q = 79|e|$)

$$\theta_{+} < 0.02^{0}.$$

Now let us take into account the influence of atomic electrons on the α-particles motion. We assume that their initial velocity is equal to zero. Then the momentum transfer to an atomic electron is maximum under the head-on collision. From the conservation laws of momentum and energy (in non-relativistic case) it follows, that after collision an electron has acquired the velocity

$$v_e = \frac{2m_\alpha}{m_\alpha + m_e} v_\alpha \approx 2v_\alpha,$$

where v_α is the initial velocity of the α-particle. Certainly, at such a collision the α-particle does not deviate. At a sliding collision the electron momentum change Δp would be already less $2m_e v_\alpha$. To obtain a value of a maximum possible deviation θ_{-} under scattering off the α-particle on the electron, we assume that the electron after collision flies out at a right angle to the initial direction of the α-particle motion and has momentum being equal to $2m_e v_\alpha$. Then, the result follows

$$\theta_{-} \sim \frac{\Delta p_\alpha}{p_\alpha} < \frac{2m_e}{m_\alpha} \sim 0.02^{0}. \tag{2.2}$$

Notwithstanding the fact that the deviation of the incident α-particle caused by both atomic electron and positive charged sphere is as low as of the order of 0.02^{0}, whether a series of such deviations could give rise to a big scattering angle. Let us suppose, that an average deviation about the angle $\overline{\theta} \sim 0.01^{0}$, caused either by the positive charged sphere or by the electron, occurs under transition through one atomic layer. Using statistic methods to obtain result of a sequence of random deviations, we find out, that after passing through the $N = 10^{4}$ atomic layers, the total average deviation is equal to

$$\overline{\theta}_t = \overline{\theta}\sqrt{N} \sim 1^{0}. \tag{2.3}$$

Really, the experimentally measured average deviation represented about 1^{0}. However, some part of the α-particles was scattered at much bigger angles. For example, one out of 8000 α-particles deviated at angle $\theta \geq 90^{0}$. The probability, that the α-particle is subjected to summarized scattering at the angle larger than θ under average deviation $\overline{\theta}_t$ is

$$P(\geq \theta) = \exp\left(-\frac{\theta^2}{\overline{\theta}_t^2}\right). \tag{2.4}$$

For Thompson atom model $P(\geq 90^{0}) = \exp(-8100) \sim 10^{-3500}$, that is, only one out of 10^{3500} α-particles can be scattered by the angle $\geq 90^{0}$ that contradicts the experimental data.

To explain α-particles scattering results Rutherford suggested the planetary atomic model, which essence was as follows. Practically all the atom mass was concentrated in its nucleus which was located in the center and had the size of 10^{-13} — 10^{-12} cm. Electrons are rotating around a nucleus at a distance of the order of 10^{-8} cm. Soon they found out, that nucleus electric charge exactly equaled the element number in the periodic table by Mendeleev. In the beginning of 1913 this idea was introduced by a Dutch physicist Vander Broek and some months later Rutherford disciple G. Moseley produced its experimental proofs. Moseley performed a set of experiments to measure X-ray spectrum for various elements. It turned out, that the X-ray wave length systematically decreases while the atomic number Z in periodical table increases. Moseley got the conclusion that this regularity is caused by increasing the atomic nucleus charge. The charge increases from atom to atom by one electronic unit and the number of such units coincides with the number of the element position in the Mendeleev table. Since the atom is electrically neutral, it means, that the total number of electrons in the atom is equal to Z as well.

Then, in the Rutherford atom at the nucleus surface the electric field strength is $> 10^{21}$ V/cm, which is almost eight orders higher than that at the atom surface. In Fig. 3 (R is the atom radius) the distribution of the electric field strength $E_+(r)$ in the atomic models by Thomson (Fig. 3a) and Rutherford (Fig. 3b) are showed for comparison.

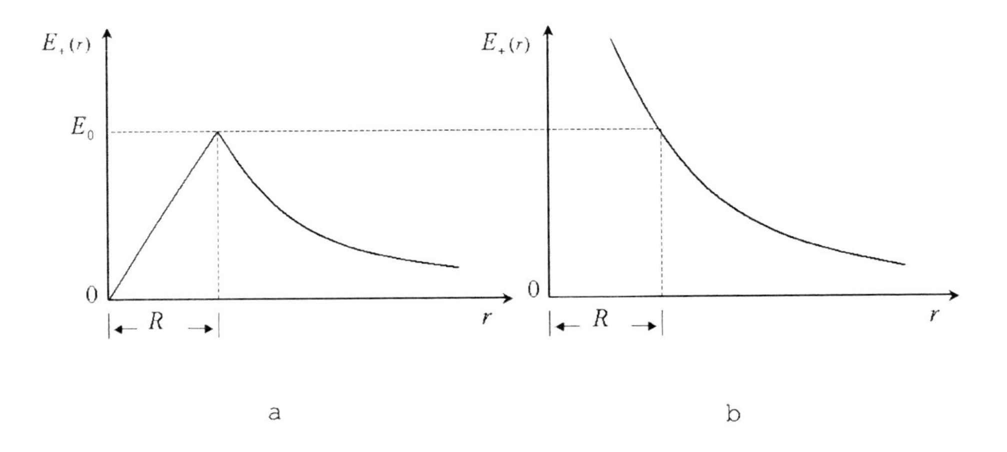

Figure 3. The electric field strength in the Tompson's atom model (a) and in the Rutherford's atom model (b).

It is obvious, that a strong field in the planetary model can cause a strong deviation and even scatter α-particle backwards, when it is flying close to nucleus. However, the secret of the rapid development of physics lies in the fact that for confirming physical hypothesis not only qualitative, but also quantitative coincidences are needed. Let us prove that Rutherford model correctly describes the scattering of α-particles at atoms.

To analyze both elastic and non-elastic collisions either laboratory reference system (LRS) or center of mass system is used (CMS). The LRS corresponds to the standard performance of experiments, namely, a beam of particle of type I strikes on a fixed target, built up from particles of type II. In the CMS, however, the equations, describing scattering processes, are much more simple, since total system momentum equals zero $\mathbf{p}_1 + \mathbf{p}_2 = 0$. In the CMS particle collision is reduced to a motion of one particle with a reduced mass

$$M = \frac{m_1 m_2}{m_1 + m_2}$$

in the field $U(\mathbf{r})$ of a stationary force center, located in particles inertia center. In the LRS scattering angles θ_1 and θ_2 (the second particle rested before collision) are connected with scattering angle in the CMS by the relations

$$\theta_2 = \frac{\pi - \chi}{2}, \qquad \tan\,\theta_1 = \frac{m_2 \sin\chi}{m_1 + m_2 \cos\chi}. \tag{2.5}$$

Notice that the CMS can be practically realized under performance of experiments with colliding beams.

In classical physics the collision of two particles is completely defined by their speeds and impact parameter ρ. However, during real experiments we deal not with individual deviation of a particle but with scattering of a beam, consisting of identical particles, striking on a scattering center at the same speed. Various particles in a beam have a different impact parameters and, consequently, scatter at different angles χ. Let dN be a number of particles, scattered in a unit of time at angles belonging interval χ and $\chi + d\chi$. Since dN is a function of a falling beam density, that is not convenient to characterize scattering process. For this reason we use quantity

$$d\sigma = \frac{dN}{n},$$

where n is a number of particles passing in a time unit through the unit of area of a beam cross section (we assume beam homogeneity over the whole section). In given interval of angles only those particles scatter, which fly with impact parameters enclosed in the interval between $\rho(\chi)$ and $\rho(\chi) + d\rho(\chi)$. The number of such particles is equal to the product of n and area of a ring between circles with radii $\rho(\chi)$ and $\rho(\chi) + d\rho(\chi)$, that is, $dN = 2\pi\rho(\chi)d\rho(\chi)n$. Thus, the effective cross section of scattering within interval of flat angles $d\chi$ (it is also called differential cross section) is defined by the expression

$$d\sigma = 2\pi\rho(\chi) \mid \frac{d\rho(\chi)}{d\chi} \mid d\chi. \tag{2.6}$$

Bearing in mind that the derivative $d\rho(\chi)/d\chi$ could be also negative we used its absolute value only. Passing to solid angle $d\Omega$, we obtain

$$d\sigma = \frac{\rho(\chi)}{\sin\chi} \mid \frac{d\rho(\chi)}{d\chi} \mid d\Omega. \tag{2.7}$$

To integrate the differential cross section over all values of solid angle produces the total cross section σ. The cross section has the dimension of area: it is the "useful" area of the interacting system, consisting of the incident particle and the target- particle. Thus, for the incident particle a target is like an area, collision with which results in interaction. Effective cross sections can be smaller or bigger than the geometrical cross sections of target particles, and can coincide with them as well. To obtain in the LS the scattering cross sections for the incident beam, one should express χ through θ_1 and θ_2 by means of Eqs. (2.5).

In the scattering quantum theory the statement of the problem on its own is changed since the conceptions of trajectories and impacts parameters have no sense under the motion with definite speed. Here the aim of the theory is to calculate the probability, that in the result of the collision the particles scatter at one or another angle. Once again, we can introduce the conception of the effective cross section, which characterizes the transition probability of a system, consisting of two colliding particles, as a result of their elastic or non-elastic scattering to definite final state.

Differential cross section of scattering within angles interval $d\Omega$ is equal to ratio between probability of such transitions $P_{i\to f}$ per time unit to the incident particles flux j_0

$$d\sigma = \frac{W_{i\to f}}{j_0} d\Omega,$$

where $W_{i\to f} = P_{i\to f}/\Delta t$. Integration over the entire interval of solid angle variation gives the total cross section

$$\sigma = \int_0^{4\pi} \frac{W_{i\to f}}{j_0} d\Omega.$$

In the CGS system of units the square centimeter cm^2 is the unit of the effective cross section. This unit, however, is very large for microworld and we use a unit having the order of a geometrical cross section of a nucleus 1 barn=10^{-26}cm^2.

Let us calculate the differential cross section of the α-particles scattering in Born approximation. In the SCM Born formula has the following form

$$d\sigma = \frac{M^2}{4\pi^2\hbar^4} \mid \int U(\mathbf{r}) \exp(\frac{i}{\hbar}\mathbf{qr}) d\mathbf{r} \mid^2 d\Omega, \qquad (2.8)$$

where $d\Omega = 2\pi \sin\chi d\chi$, $U(\mathbf{r}) = Ze^2/2\pi r$, Z is the target atomic number, $\mathbf{q} = \mathbf{p} - \mathbf{p}'$, $\mathbf{p}$ and $\mathbf{p}'$ are the momenta of the particles before and after collision. Using Poisson equation for potential of point charge, located in the beginning of coordinate system

$$\Delta\left(\frac{e}{r}\right) = -4\pi e \delta(\mathbf{r}),$$

and Fourier transformation for delta function

$$\delta(\mathbf{r}) = \frac{1}{(2\pi)^3} \int \exp(i\mathbf{kr}) d\mathbf{k},$$

we obtain the following expression for Fourier transform of potential energy

$$U(\mathbf{q}) = Ze^2 \int \frac{1}{2\pi r} \exp[\frac{i}{\hbar}(\mathbf{qr})] d\mathbf{r} = \frac{Ze^2}{4\pi^3} \int \frac{d\mathbf{k}}{\mathbf{k}^2} \int \exp\{i[\frac{(\mathbf{p}-\mathbf{p}')}{\hbar} + \mathbf{k}]\mathbf{r}\} d\mathbf{r}$$

$$= \frac{Ze^2(2\pi)^3}{4\pi^3} \int \frac{1}{\mathbf{k}^2} \delta(\frac{\mathbf{p}-\mathbf{p}'}{\hbar} + \mathbf{k}) d\mathbf{k} = \frac{2Ze^2\hbar^2}{\mid \mathbf{p}-\mathbf{p}' \mid^2}.$$

Taking into account that

$$\mid \mathbf{p} - \mathbf{p}' \mid = 2p \sin\frac{\chi}{2},$$

where $\mid \mathbf{p} \mid = \mid \mathbf{p}' \mid = p$, the expression for the differential cross section takes the form

$$d\sigma = \left(\frac{Ze^2 M}{4\pi p^2}\right)^2 \frac{d\Omega}{\sin^4(\chi/2)}. \qquad (2.9)$$

Notice some interesting peculiarities of formula (2.9), which is called Rutherford formula. Solving the task exactly (no relativistic effects taken into consideration) we arrive at the expression (2.9) as well. The solution obtained does not depend on the potential energy sign, that is, the solution is the same for both attracting and repulsing force centers. Since the differential cross section does not contain Plank constant, then one can analyze Rutherford scattering by classical or quantum mechanical method with the same success. The latter fact follows from the rule, which reads, that "*if forces of interaction between particles depend on distance as r^n, than the cross section of scattering of such particles at each other is proportional to $\hbar^{4+2n}$*".

If we try to integrate the expression (2.9) over scattering angle, then we obtain infinity. It is caused by long-range character of electrostatic forces. For this reason the particles are scattered, no matter how far away from a scattering center they fly. To obtain a reasonable, that is, finite result, we must take into consideration a screening effect of electron shell. This is achieved under use of the potential

$$U(r) = \frac{Ze^2}{2\pi r}\exp\left(-\frac{r}{a}\right),$$

where a has the value of the order of the atom radius.

Now we must take the last step, namely, to compare theoretical formula (2.9) with experimental results. Let us turn to the LSR and consider the cross section for incident α-particles, taking into account that $m_N \gg m_\alpha$ (m_N is nuclear mass). In this case $\theta \equiv \theta_1 \sim \chi$ and $M \sim m_\alpha$, so that

$$d\sigma = \left(\frac{Ze^2}{4\pi m_\alpha v_\alpha^2}\right)^2 \frac{d\Omega}{\sin^4\frac{\theta}{2}}. \tag{2.10}$$

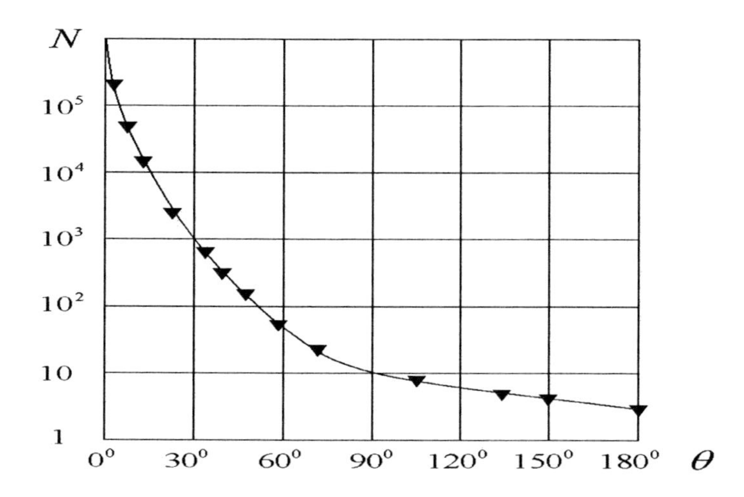

Figure 4. The θ angle dependence of the number of scattered α-particles.

In Fig. 4 the results of calculations according to formula (2.10) are plotted, which define the scattered particles number in relation to the angle θ at $Z = 79$. The experimental data obtained by Geiger and Marsden are displayed by small triangles. Excellent agreement between theoretical and experimental results testifies that the internal structure of the atom indeed includes hard and massive core, which is nucleus.

It should be stressed that the formula (2.10) is valid under fulfillment of the following conditions.

1) The atom nucleus must have the mass exceeding the α-particle mass to such a degree that the recoil energy of nucleus may be safely neglected.

2) Nucleus potential consists of a Coulomb part and the part, which is responsible for nuclear attraction forces φ_N. For the α-particle not to be influenced by nuclear forces the collision diameter d, defining the minimal distance on which the incoming particle can approach to the scattering center, must greatly exceed the action radius of the potential φ_N. In this case the collision diameter is determined by the help of the energy conservation law

$$\frac{m_\alpha v_\alpha^2}{2} = \frac{Ze^2}{2\pi d}. \qquad (2.11)$$

By a lucky chance these both conditions have proved to be satisfied in experiments with gold targets. However, in later experiments with aluminum targets ($Z = 13$) deviations from Rutherford formula were especially noticeable in the region of the large angles.

Using Rutherford formula, we can approximately estimate the size of the atom nucleus. Let us assume, that geometrical cross section of nucleus is of the same order as differential scattering cross-section of the α-particles deflected through the angle being greater than 90^0. Then for $Z = 79$ and $E_\alpha = 7.68$ MeV we have $d\sigma/d\Omega = 6.87 \times 10^{-28}$ m^2, which produces the wholly plausible result $R = 1.5 \times 10^{-14}$ m. If the core were absent in the atom center, that is, Thomson model were true, then decreasing the α-particles deflected through big angles θ would take place along the shortest curve, which is a straight line (see Fig. 4). Thus, the excess of the differential cross section over the straight line crossing the horizontal axis provides direct confirmation of the fact that atoms are not indivisible elements of matter, but represent in themselves a composite structures, consisting of positive charged nucleus and electrons.

Thus, according to Rutherford, the atom is similar to the Solar system. The character of resemblance to the Solar system represents not only qualitative but also quantitative, that is very strange and has not been found an exhausting explanation till now. If one takes the ratio between the diameters of the Sun and the Solar system then this ratio proves to be approximately equal to one between the diameters of the nucleus and the atom. Further on, according to the quantum theory, the electrons in atom are not located at arbitrary distances from nucleus. Their orbital radii are defined by the relation

$$r_n = \frac{4\pi\hbar^2}{me^2} n^2, \qquad n = 1, 2, 3...$$

It turns out, that the planets have analogous behavior, namely, distances between the planets and the Sun are not changed in a random manner but are subjected to the definite law. This fact was known to J. Kepler, but it was first mathematically formulated by D. Titius in 1772.

Later on Bode made some corrections and the law was given the title Titius-Bode law. If the distance from the Sun to Mercury is adopted as 0.4 arbitrary unit, then formula for the planet radii takes the form

$$R_n = 0.3n + 0.4,$$

where n=0 (Mercury), 1 (Venus), 2 (Earth), 3 (Mars), 5 (Jupiter), 6 (Saturn), 7 (Uranus), 8 (Neptune), 9 (Pluto). It is remarkable, that in the scheme a planet with number n=4, which should be spaced between Mars and Jupiter, is missing. Instead of it exactly in this place asteroid belt is located, which represents in astronomers view the fragments of the planet Phaeton to have existed in former time.

As a result of triumphal success of Rutherford atom model the traditional picture of this model with its precise figure of the electron orbits becomes universally recognized emblem of the passed XXth century. It has wandered in books covers, exhibitions and conferences booklets, stamped papers of institutes and universities. Although nowadays we know that in reality there are no electron orbits in classical understanding, these figures have been remaining the deserved tribute of respect to Rutherford atom model.

In the experiments, described above Rutherford has discovered not only the atomic nucleus he has also established the main type of experiment in microworld physics, which is the particle scattering on each other. By the example of Rutherford experiments we see, that scattering experiments are characterized by three time intervals. During the first interval the systems are prepared and brought into contact. During the second interval the interaction takes place. During the third interval the systems appeared in the result of interaction (reaction products) move towards the measuring devices (detectors). Since during the first and the third intervals the systems are isolated from each other, then the systems investigations are reduced to analysis of free motion equations. The description of the second interval demands deriving the solutions of the interacting systems equations and as a result this is one of the most important and fundamental physics problems.

2.3. Proton-Neutron Model of Atom Nucleus

The discovery of the first isotopes in 1919 by F. Aston and establishing the integral numbers rule (all atomic or molecular masses are integral numbers within the limits of observation precision) set in turn the question about the composite structure of nucleus. Long ago forgotten Proud hypothesis (1815), stating that the hydrogen is the part of all atoms, became popular once again. Analogously to "protil", introduced by Proud, Rutherford suggests to use a term "proton" for the hydrogen atomic nucleus (1920). As far back as years of the first World War Rutherford started to investigate the collisions of the α-particles with nuclei of such light elements as hydrogen, nitrogen, oxygen, and air. At central collisions nuclei emitted particles, causing scintillations at a screen made of sulphurous zinc. At that time a deep belief existed, that only nucleus components can be knocked out of a nucleus. Due to lucky coincidence it really took place in experiments on the α-particles scattering, because α-particle energy E_α was not enough to produce even the lightest hadron π^0-meson ($E_\alpha < 135$ MeV). Basing on studying behavior of knocked out particles in magnetic field, in 1924 E. Rutherford and D. Chedvick presented in their paper the final proof that those particles were protons. They also suggested, that the α-particle was stuck in a nucleus. Using Wilson camera, in 1925 P. Blackett made a photograph of the reaction, which showed, that the α-particle was captured by nucleus and its track was ended by the typical "fork", in which the short fat track belonged to the residual nucleus while the long thick track did to the knocked out proton. The reaction, in which one was shown for the first time that the proton enters into the composition of the nucleus

$$^{4}_{2}He+^{14}_{7}N\rightarrow ^{17}_{8}O+p,$$

was defined as an example of the first two-particle reaction.

Since nuclei masses are always larger than Z proton masses, then it is obvious, that some more particles must enter into the composition of the nucleus. It is natural to assume that a nucleus contains electrons and that protons exceeding nuclear electrons cause total positive nuclear charge. However, this hypothesis, despite its charming simplicity (no need to introduce new particles), ran into sequence of serious contradictions. Let us enumerate them.

1) Protons and electrons are fermions having the spin $\frac{1}{2}$. Then, as follows from the rule of summing up the spin moments vectors, nuclei with even number of protons n_p and electrons n_e must have integer spin, while nuclei with odd n_p and n_e must have half-integer spin. For example, for deuteron (nucleus of deuterium atom) spin should equal either $3/2$ or $1/2$ depending on orientation of electron and proton spins. In reality the observed deuteron spin is equal to 1 which is in no way reconciled with hypothesis of nuclear electrons.

2) If electrons were present in nucleus then nuclear magnetic moment would be of the Bohr magneton order ($\mu_B = e\hbar/2m_ec$). However, experiments exhibit, that nuclear magnetic moments are of the nuclear magneton order ($\mu_N = e\hbar/2m_pc$) which is 1000 times less than μ_B.

3) From Heisenberg uncertainty relation

$$\Delta p\cdot\Delta x\geq\hbar/2$$

it follows that in nucleus ($\Delta x\approx 10^{-14}$ m) Δp for the electron has of the order of $\hbar/\Delta x\approx 10^{-20}$ kg $\cdot$ m/sec. Since the momentum must be no less than this value then for energy of such an electron we obtain the value ~ 20 MeV. Maximum energy of electrons emitted in β-decay of various nuclides is in the interval 0.011 — 6.609 MeV that is much less than the energy of electrons inside of a nucleus, if electrons were inside of it.

In 1930 W. Bote and H. Becker obtained deeply penetrating radiation, which arouse under α-particles bombardment of light elements nuclei $^{9}_{4}Be$, $^{11}_{5}B$, and $^{19}_{9}F$. Since this radiation went through lead layer several centimeters thick and did not deflect by magnetic or electric fields, for several years unsuccessful attempts had been made to identify it with high energetic photons. February, 27, 1932 I. Chadwick published the article in journal "Nature", where he showed that Bote-Becker penetrating radiation was a flux of neutral particles having the masses close to the proton mass (nowadays these particles are called neutrons n). After appearing Chadwick's article, the events developed at striking quickness. In the same journal D. Ivanenko published the letter, in which he proposed a hypothesis of proton-neutron nuclear structure. Soon the proton- neutron model was also introduced by W. Heisenberg in his article, published in journal "Zeitschrift fur Physik" (1932, June). Due to numerous similarity of proton and neutron (neutron spin equals $1/2$ and $m_n\approx m_p$) a common name "nucleon" is used for both of them.

In the nucleus nucleons form not such a hard lattice as atoms in crystal do, but rather a liquid structure in which nucleons can move like molecules in liquid. Successful explanation of structure elements in Mendeleev periodic table by means of only three types of particles, namely, protons and neutrons contained in nucleus, which electrons surrounded, provided all the reasons (at that time) to consider that e^-, p and n were structureless matter blocks, of which all the Universe was built. Establishing this level of the matter structure was the climbing on the third step of the "Quantum Stairway". The list of the "elementary" particles, known at that time, was rather short: electron, proton, neutron, photon, and positron (electron antiparticle which was discovered in 1932 by K. Anderson in cosmic rays). So, the world used to be described by a fascinating simple scheme. One could think, that the third step, reached with such trouble, was the last one on the way to Creator. However, it was nothing else, but illusion of understanding which, like a cat that walking in herself, calls on us whenever it wants.

Chapter 3

Leptons and Hadrons

In 1936 muons $\mu^{\pm}$, discovered by K. Anderson and S. Niedermayer in cosmic rays, were added to the list of elementary particles. Before 1953, when the first accelerator was constructed (Brookhaven proton synchrotron with maximum energy 3 Gev), the elementary particle investigations were intimately connected with cosmic ray investigation. In 1947 the group by S. Powell discovered π^+ and π^--mesons. The situation with elementary particles already became not entirely simple. Electrons and nucleons are necessary to build atoms. Photons and pions play a role of the carriers of electromagnetic and nuclear forces respectively. Electron antiparticle, positron, can be viewed as a delicate hint (for the experienced mind) on the existence of antimatter searching of which should be continued until antinucleons are found. However, what shall we do with muons, which do not find their place in this world scheme?

In late 1940s — early 1950s a real demographic explosion has occurred in the elementary particle world. A whole zoo of new particles, called "strange" particles, has been discovered. A main peculiarity of those particles and the ones, discovered later, is that they are not the component of matter observed. They live for a very short time and decay into stable particles (protons, electrons, photons and neutrinos). The first particles from this group, K^+- and K^--mesons, Λ-hyperons were discovered in cosmic rays, the next ones — in accelerators. From the early 1950s the accelerators become the main tool to investigate matter microparticles. The accelerators energy is growing and the tendency for increasing the number of fundamental particles becomes more and more apparent. Nowadays the list of elementary particles has become tremendously large ~ 400. The properties of discovered elementary particles prove to be unusual in many respects. To describe them, the characteristics taken from classical physics, such as electric charge, mass, momentum, angular moment, magnetic moment, proved to be not sufficient. It was necessary to introduce many new quantum numbers, having no classical analogs, which we call *internal quantum numbers*. The first reason for their introduction deals with additional degrees of freedom of elementary particles. The second reason is caused by striving to explain the non-observation of some "acceptable" reactions[1] by means of existence of internal symmetry which leads to conservation of corresponding charge. The latter circumstance allows to carry out the gen-

[1]By the "acceptable" reactions are meant the reactions which are allowed the conservation laws of energy, momentum, angular moment and electric charge.

erous dispensation of the conserved charges to some groups of particles, according to the following principle: dynamic symmetry corresponds to a closed channel of an "acceptable" reaction. It is possible, that a reader, being experienced by any sort direct and inverse theorems about the existence and uniqueness of the solutions, will feel no deep satisfaction. As an quieting reason, one may call attention to aesthetic attractiveness of symmetric approach in all spheres of our life (it is hardly probable that the statue of Venus from Milos being deprived of symmetry could find the place within the walls of Louvre).

Let us proceed to classification of the known particles. Elementary particles are divided in three categories. 1) Hadrons, which participate in strong, weak, gravitation interactions. Being electrically charged they also participate in electromagnetic interaction. 2) Leptons, which do not participate in strong interactions. 3) Field quanta, which carry strong, electromagnetic and weak interactions. Hadrons are divided in baryons with half-integer spin and mesons with integer spin. Maximum spin values of discovered by now hadrons reach the value 6 for mesons $a_6(2450)$ and $f_6(2510)$ and the value $11/2$ for baryons $N(2600)$ and $\Delta(2420)$, where in brackets the hadrons masses are given in MeV/c^2. Electron (e), muon (μ), tay-lepton (τ) and corresponding to them neutrinos (ν_e, ν_μ, ν_τ) belong to lepton class. They all have spin $1/2$. If one subtracts from the total number of discovered particles 12 interaction carriers (8 gluons, $W^\pm$, Z and photon) and 6 leptons, then the total number of hadrons is obtained.

For division of particles with spin $1/2$ into leptons and baryons to have sense, transitions between these particles kinds must be impossible. For example, neutron must not decay into electron-positron pair and neutrino

$$n \to e^- + e^+ + \nu_e. \tag{3.1}$$

In reality, this decay has been never observed. Let us introduce two quantum numbers, namely, baryon B and lepton L charges connecting them with dynamic symmetries with respect to global [1] gauge transformations. For baryons $B = 1$, for antibaryons $B = -1$, for non-baryons $B = 0$. In case of leptons, it is accepted to speak of not lepton charge, but lepton flavor. One discriminates total lepton flavor L and individual lepton flavors L_e, L_μ and L_τ. For e^- (e^+) — $L_e = 1$ ($L_e = -1$), for μ^- (μ^+) — $L_\mu = 1$ ($L_\mu = -1$), for τ^- (τ^+) — $L_\tau = 1$ ($L_\tau = -1$) and

$$L = \sum_{i=e,\mu,\tau} L_i.$$

All non-lepton particles have $L = 0$. By now no reactions with violation of either total or individual lepton flavors have been observed. However, there are no serious reasons in support of the lepton flavor conservation law. Consequently, many vital electroweak interaction theories predict the existence of processes, in which either total or individual lepton flavor is not conserved. Scientists are intensively searching for reactions, which can help to establish upper limits on their cross sections. Some lepton decays going with violation of L_i are given below

[1] When the transformation parameters do not depend on coordinates the transformation is called global.

$$\mu^- \to e^- + \gamma \qquad (4.9 \times 10^{-11}), \tag{3.2}$$

$$\mu^- \to e^- + e^+ + e^- \qquad (1.0 \times 10^{-12}), \tag{3.3}$$

$$\tau^- \to \mu^- + \gamma \qquad (3.0 \times 10^{-6}), \tag{3.4}$$

$$\tau^- \to e^- + \pi^0 \qquad (3.7 \times 10^{-6}), \tag{3.5}$$

where in brackets the upper bounds on their branchings [1] have been pointed. The comparison of theoretical expressions for partial decay width with experimental ones results in establishing the bounds on parameters of the theories in which these decays are allowed.

Despite the fact, that baryon charge conservation law ensures matter stability, its correctness is subjected to question too. Within framework of some GUT's B is not a conserved quantum number and that leads to proton instability. Some decay channels with fixed upper limit on life-time with respect to the given channel are given below

$$p \to e^+ + \pi^0 \qquad (5.5 \times 10^{32} \text{ yrs}), \tag{3.6}$$

$$p \to e^+ + \nu_l + \nu_{l'} \qquad (1.1 \times 10^{31} \text{ yrs}), \tag{3.7}$$

$$p \to e^- + \mu^+ + \mu^+ \qquad (6 \times 10^{30} \text{ yrs}) \tag{3.8}$$

Since the age of our University is as short as 10^9 years, there are no special reasons for inconsolable grief over possible proton instability.

There is an important difference between electric charge conservation and internal quantum numbers conservation. Electric charge is not a simple number, similar, for example, to baryon charge, which is ascribed to various particles. Electric charge governs the system dynamics and is a source of electromagnetic field in itself. Interaction between charged particles is carried out by means of electromagnetic fields whose quanta are nothing but the massless photons. Since with the help of the corresponding devices the electric field is easily measured, then it is possible to measure the object electric charge from a large distance, that is, without close contact with this object. Nothing similar takes place with baryon charge. Baryon does not influence upon space around it, since there exists no baryon field, similar to electromagnetic one. For this reason it is impossible to measure baryon number of an object some distance away.

In consistent description of interacting systems, the above mentioned difference between electric charge and "charges", not being the sources of physical fields, is in different

[1] Branching is the ratio between the width of the given decay channel (partial decay width) and the total decay width.

character of gauge transformations, responsible for conservation laws. For a conserved quantity to be a field source, the theory must be invariant under the local gauge transformations. Thus, the electric charge conservation corresponds to the invariance of the theory under the following transformations

$$U_{em}(1) = \exp\left[ie\alpha(x)\right]. \tag{3.9}$$

The conservation of the charges, not producing physical fields, is connected with the invariance of the theory under a global gauge transformations (transformation phase is not a function of coordinates)

$$U_N(1) = \exp\left[iN\alpha_N\right], \tag{3.10}$$

where $N = B, L_e, L_\mu, L_\tau, \ldots.$

Let us proceed to introduction of inexact internal quantum numbers, that is, numbers which are already conserved not in all interactions. Above we mentioned the so-called strange particles which are produced by pairs (one or more) under colliding π-mesons with nucleons. Since a production of these particles had been caused by strong interaction the probability of their birth was large. However, they decayed into ordinary hadrons or leptons at the expense of only weak interaction and as a result the probability of their decays was very small. The behavior uncommonness of these particles once again reminded to physicists the famous phrase by F. Bacon: " The perfect beauty without touch of strangeness is not available in the world". When in 1954 at Brookhaven cosmotron these particles were obtained for the first time, among the other processes the following one was observed

$$\pi^- + p \to \Lambda + K^0.$$

The large value of its cross section indicated, that it is going on exclusively due to the strong interaction. On the other hand, long life times ($\sim 10^{-10}$ s) of particles Λ and K^0 with respect to decays

$$\Lambda \to p + \pi^-, \qquad K^0 \to \pi^+ + \pi^-$$

testified that these decays are caused by weak interaction. For some reasons Λ and K^0 decays into lighter hadrons are forbidden due to strong interaction and as a result, they live long time. M. Gell-Mann and K. Nishijima independently from each other introduced a new additive quantum number, strangeness s. They postulated its conservation in strong and electromagnetic interactions and non-conservation in processes, caused by weak interaction (for weak interaction $|\Delta s| = 1$). For already known hadrons s takes the values -3, -2, -1, 0, 1. Further extension of hadron sector demanded introduction of such quantum numbers as charm (c) and beauty (b). They are also additive numbers and are conserved in strong and electromagnetic interactions. In hadron decays due to weak interactions they vary according to the rule

$$|\Delta| = |\Delta b| = 1.$$

Using characteristics introduced by us, we can divide known baryons and mesons into the following families: I. normal ($s = \; = b = 0$) hadrons (nucleons, π-mesons); II. strange ($= b = 0, s \neq 0$) hadrons (Λ-, Σ-, Ξ- Ω^--hyperons, K-mesons); III. charmed ($s = b = 0, \neq 0$) hadrons (Λ_c^+- and $\Sigma_c^{\pm,0}$-hyperons, D-mesons); IV. beautiful ($s = \; = 0, b \neq 0$) hadrons (Λ_b-baryons and B-mesons). There are also mixtures of the last three families: V. strange and charmed ($b = 0, s \neq 0, \neq 0$) hadrons (Ξ_c^+- and Ω_c^0-baryons, D_s-mesons); VI. strange and beautiful ($= 0, s \neq 0, b \neq 0$) hadrons (Ξ_b-baryons, B_s^+-mesons); VII. charmed and beautiful ($s = 0, \neq 0, b \neq 0$) hadrons (B_c-mesons).

However, looking at this megapolis of hadrons, we understand, that the above mentioned division is nothing else, but a scheme of streets and squares which does not a jot bring us closer to understanding the idea of the architect creating this elementary particles Babylon. This picture is also far from that, so the human mind could experience reverential delight under a sight on it. Let us remind the guiding thread, which brought us from the first step of Quantum Stairway to the third one. Mendeleev periodic table was built on the basis of accidentally discovered periodic alteration of chemical properties of elements alongside with increasing their atomic masses (nucleus electric charge, to be exact). Sixty three years later with the help of this table we managed to construct hundreds of elements from only three fundamental (as then it seemed) particles p, n, and e^-. Let us concentrate our efforts to find a similar table for hadrons.

Chapter 4

Periodic Table of Hadrons

4.1. Yukawa Hypothesis

In language of quantum field theory, the force field influence on a particle is interpreted as emission and absorption of the quanta of this field. For example, electromagnetic interaction between two electrons appears due to photons exchange. Photon masslessness reveals itself in long-range action of Coulomb forces, that is, Coulomb potential decreases with distance as $1/r$. It is natural to expect, that the massive carrier of interaction will produce a short-range potential. In analogy with exchange nature of electromagnetic forces, in 1935 H. Yukawa introduced a hypothesis about massive quantum exchange, between nucleons in a nucleus. The idea was, that interaction between this massive quantum (π-meson) and nucleon N appears as a result of combination of virtual process

$$N \leftrightarrow N' + \pi. \tag{4.1}$$

In other words, Hamiltonian, causing interaction between nucleons in a nucleus H_{int} must be of the form

$$H_{int} \sim \overline{N}(x)N(x)\pi(x). \tag{4.2}$$

Now we should more carefully consider the concept of virtual states. Physical laws describe only experimentally measurable quantities. In quantum theory uncertainty relations for dynamical observables, represented by non-commutating operators, impose restrictions on precision of simultaneous measurements of these quantities. Thus, Heisenberg uncertainty relations impose precision limits on simultaneous measurements of particle coordinates and momenta

$$\Delta p_i \Delta x_i \geq \frac{\hbar}{2}, \qquad (i = 1,2,3), \tag{4.3}$$

as well as its energy and time

$$\Delta E \Delta t \geq \frac{\hbar}{2}. \tag{4.4}$$

The uncertainty relation (4.4) can be interpreted by the way which is somewhat unusual from the view point of the classical physics. In microworld the processes, going on with the violation of the energy conservation law on the value of ΔE are allowed, provided that this violation lasts no longer than $\Delta t \sim \hbar/\Delta E$. In this case the violation cannot be registered by any physical device. Analogously, the momentum conservation law could be violated on the value Δp in the region $\Delta x \sim \hbar/\Delta p$. We emphasize, that uncertainty relations reflect the inner nature of microworld and have nothing to do with imperfection of our measuring devices.

When considering interactions between particles by means of field quanta exchange, one naturally thinks of ΔE to denote brought in or taken away quantum energy and Δt to denote the exchange duration or the lifetime of this quantum. The particles states with such lifetimes are called as the virtual states. All the elementary particles can occur both in real and in virtual states. Inherent in real particles connection between energy E, momentum $\mathbf{p}$ and mass m

$$E^2 = \mathbf{p}^2c^2 + m^2c^4 \tag{4.5}$$

is violated in case of virtual particles due to appearance of ΔE. Keeping in mind the violation of this equality they say the virtual particles to lay beyond the mass surface. So, according to Yukawa hypothesis, an interaction between nucleons in nucleus is carried out by means of virtual particle exchange. Emitting and absorbing the virtual π-mesons, protons and neutrons turn into each other. It is easy to guess, that for $p \leftrightarrow n$ coupling to exist, π^+ and π^--mesons are necessary. Reasoning from experimentally determined principle of nuclear forces charge independence

$$V_N = V_{pp} = V_{nn} = V_{np}, \tag{4.6}$$

where V_N is nucleons interaction potential, one might be concluded that a neutral π^0-meson is needed to describe $p \leftrightarrow p$ and $n \leftrightarrow n$ interactions. Of course, $p \leftrightarrow p$ and $n \leftrightarrow n$ interactions can take place due to two charged π-mesons as well. Thus, for example, interaction between two protons takes place as follows. Both protons emit one π^+-meson each, in so doing they turn into neutrons. Then, these neutrons absorb π^+-mesons and turn into protons again. A corresponding mutual conversion chain for each proton has the form

$$p \to \pi^+ + n, \qquad n + \pi^+ \to p, \tag{4.7}$$

and $p \leftrightarrow p$ interaction on its own is displayed by the reaction

$$p + p \to n + \pi^+ + n + \pi^+ \to p + p. \tag{4.8}$$

Quite obvious, that emission of two π-mesons necessitates larger value of ΔE and consequently, lessens Δt. In shorter lifetime π-mesons cover shorter distance and that reduces (approximately twice) the action radius of nuclear forces between identical nucleons, and, as a result, the condition (4.6) will be violated.

Let us estimate the π-meson mass by means of uncertainty relation (4.4). We consider interaction between a proton and a neutron. The proton emits the π^+-meson and turns into the neutron, while the initial neutron having absorbed the π^+-meson becomes the proton

$$p \to n + \pi^+, \qquad n + \pi^+ \to p. \tag{4.9}$$

The chain of the reactions (4.9) can be represented as follows

$$p + n \xrightarrow{\pi^+} p + n. \tag{4.10}$$

If one neglects the proton recoil momentum, then the emission of the virtual π^+-meson leads to the energy violation by the value $\sim m_\pi c^2$ at the minimum. The virtual meson exists during the time $\Delta t \sim \hbar/\Delta E \sim \hbar/(m_\pi c^2)$. In Δt time the virtual Δt-meson, even if it moves at a maximum possible speed c, covers the distance, which defines maximum value of the force field action radius

$$R \sim c\Delta t \sim \frac{\hbar c}{m_\pi c^2} \sim \frac{\hbar}{m_\pi c}. \tag{4.11}$$

Substituting in Eq.(4.11) the experimentally obtained value of nuclear force action radius $R_n \sim 10^{-13}$, we obtain $m_\pi \approx 280\, m_e$ (m_e is the electron mass).

In reality, Yukawa predicted, that the mass of nuclear interaction carrier might be equal to 206 m_e. That was caused by using the improper data concerning the value of R_n. Being guided by this number, experimentalists began to search and have found the particle with the mass 207 m_e which was called a muon (μ). However the muons turned out to be weakly interacting particles. Consequently, they are not suitable for the role of a nuclear field quantum. As later as 1947 in interactions between cosmic rays and upper atmospheric layers the predicted by Yukawa π-mesons were discovered.

The notion about the interaction law, caused by π-meson exchange, is given by the Yukawa potential, that is, the potential energy which is derived under assumption that the interacting particles are immovable (particles masses are so large that one may precisely fix their positions and neglects their recoils under emitting and absorbing quants). It could be shown that the interaction potential of two nucleons separated by the distance r is determined by the equation

$$V_N(\mathbf{r}) = -\frac{C}{4\pi|\mathbf{r}|}\exp(-k_0|\mathbf{r}|), \tag{4.12}$$

where C is the constant connected with the "nuclear" charge of nucleon and $k_0 = m_\pi c/\hbar = 1/\lambda_\pi$ (λ_π is the Compton wave length of the π-meson). From the expression for $V_N(\mathbf{r})$ follows the same result as from uncertainty relation: nuclear forces have the finite action radius, approximately equal to the Compton wave length of the π-meson. Notice, that the negative sign in the expression for the Yukawa potential points to the attraction character of the nuclear forces.

The process of emitting and absorbing the π-meson lasts no longer than 10^{-23} s. In all modern experiments such a process may be considered as an instant one. Roughly speaking, protons spend one part of their life in nucleus being protons, while during the second part they are neutrons. So, it is natural to consider proton and neutron as two different states of the same particle given the title nucleon. All nucleons consist of identical cores, surrounded by a cloud of virtual π-mesons. The only difference between p and n lies in the character of such a cloud. When two such clouds are approaching each other at the distance of the order of the π-meson Compton wave length, the π-meson exchange between clouds takes place[1]. So the π-mesons are constantly scurrying between interacting nucleons. The situation reminds a little bit covalent coupling carried out by electrons in molecular ion H_2^+. Since in this case the electron can transfer from one to another proton, the exchange forces appear and they are added to the ordinary Coulomb forces.

[1]At smaller distances exchanging the heavier particles (vector mesons ρ , φ, ω etc.) begins to be more substantial.

Obviously, the π-meson clouds (fur-coat), surrounding neutrons and protons, contribute to nucleon magnetic moments. Since such virtual conversion chains

$$p \to \pi^+ + n, \qquad n \to \pi^- + p$$

are possible for proton and neutron, then anomalous parts of the particle magnetic moments, caused by the π-meson field, must be approximately equal in value and opposite in sign.

Let us consider the process of originating the neutron magnetic moment due to its virtual dissociation in π^- and p. Since π-meson has a zero-spin, it also has a zero intrinsic magnetic moment. Then only the π-meson with non-zero orbital moment, for example, in p-state ($l = 1$), contributes to neutron magnetic moment. For the moment momentum conservation law in this virtual process to be fulfilled, the following demands must be met. a) The direction of orbital moment of the virtual π-meson being equal to 1, must coincide with the neutron spin direction. b) The virtual proton spin must be directed opposite to neutron spin. Since the π^--meson is negatively charged, then the neutron magnetic moment induced by the π^--meson, is negative as well.

To estimate the neutron anomalous magnetic moment value, we must know, how long neutron exists in dissociated state or, what amounts to the same thing, the probability of transition into this state. Since $m_p \approx 6.72 m_\pi$, the magnetic moment of the system $p + \pi^-$ is in value order equal to $(-6.72 + 1)\mu_N$ ($\mu_N = e\hbar/(2m_p c)$). Then the neutron magnetic moment observed is

$$\mu_n = W_0 \mu_n^s - 5.72(1 - W_0)\mu_N, \tag{4.13}$$

where μ_n^s is the neutron intrinsic magnetic moment, that is, the quantity equals zero, W_0 is the probability to find a neutron in a naked neutron state, $(1 - W_0)$ is the probability to find a neutron in a dissociated $p + \pi^-$-state. The experimental value of the neutron magnetic moment $\mu_n = -1.913\,\mu_N$ is obtained if one sets $W_0 \approx 0.665$. . Analogously, for the proton anomalous magnetic moment we have

$$\mu_p = W_0 \mu_N + 6.72(1 - W_0)\mu_N, \tag{4.14}$$

where we took into consideration that the intrinsic magnetic moment of the proton is equal to μ_N and assumed that the probabilities of the proton and neutron virtual dissociations are equal. Then the total proton magnetic moment is pretty close to its experimentally measured value $\mu_p = 2.793\,\mu_N$.

Experimentally proved Yukawa theory of nuclear forces is a milestone in the development of Physics. It has finally strengthened the assurance that quantum interpretation of interaction as exchange of virtual quanta is a correct one. Such interaction interpretation underlies the foundation of modern physical theories. Taking into consideration virtual particles changes our concepts of physical vacuum under transition from classical to quantum theory. The example of electrodynamics is very significant in this case. Electromagnetic field in a classical theory is defined by the values of the strengths of the electric and magnetic fields (**E** and **H**) given in all points of space and in all moments of time. Under transition to quantum electrodynamics in places of those strengths, the operators appear which, in particular, do not commute with operators, defining the number of photons in a given state. However, only the physical quantities, to which commuting operators correspond, can simultaneously have definite values. If operators do not commute, the more precisely the quantity corresponding to one of these operators is defined, the less information can be obtained for the second quantity. At exact definition of **E** and/or **H** the number of photons is absolutely undefined. In the same way, if the number of photons is exactly defined, then fields strengths are not defined.

In the quantum theory we determine a vacuum as a state without real particles or as a state with the least energy. Then, since for the photon vacuum (electromagnetic vacuum) the number of particles is zero, i.e. is exactly defined, the fields strengths are not defined, and this fact, in particular, does not allow to accept these strengths being equal to zero. The impossibility to simultaneously set both the fields strengths and the photons number equal to zero make us to consider the vacuum state in the quantum theory not as the field absence, but as one of the possible field states having the definite properties which are displayed in real physical processes.

Virtual production and absorption of the photons should be viewed as manifestation of photon vacuum, or, in other words, as taking account of photon vacuum effects. The concept of vacuum being the lowest field energy state can be analogously introduced also for other particles. Considering interacting fields, the lowest energy state of all the system could be called a vacuum state. If sufficient energy is supplied to a field in a vacuum state, then the field is exited, that is, the field quantum is produced. Thus, a particle production can be described as a transition from "unobserved" vacuum state to a real one. Without real particles and external fields vacuum, as a rule, does not reveal itself through any phenomena due to its isotropy. A presence of real particles and/or external fields leads to vacuum isotropy violation, since a production of virtual particles and their subsequent absorption results in changes of the state of the real physical system. Virtual particle production and absorption of particles is limited by conservation laws, the electric charge conservation law among them. For this reason a virtual production of a charged particle is impossible without a charge change of real particles (if there are any). If real particles charges are invariable, then in virtual processes the charged particles are created and destroyed in pairs only (*particle-antiparticle*). Thus, in the case of the charged particles one can speak only of *particles-antiparticles* vacuum: *electron-positron* vacuum, *proton-antiproton* vacuum, etc. From this it also follows, that since, for example positrons and electrons can be produced only in pairs, one can not speak of electrons as of isolated and solitary type of matter, just as one impossibly draws a demarcation line between electric and magnetic field.

Electron and positron fields make up a unified electron-positron field, and this circumstance remains imperceptible, provided the processes of pairs productions and pairs destructions may be neglected. Analogously, as in the case of *photon* vacuum, *electron-positron* vacuum or any other *particles- antiparticles* vacuums will lead to observable effects, one of which is a change in physical properties of a particles. The above mentioned effects of charge screening and appearance of nucleons anomalous magnetic moments can serve as example.

So, in quantum theory every particle is enclosed by the fur coat consisting of cloud of virtual quanta, produced and subsequently absorbed by the particle. Quanta can belong to any field (electromagnetic, electron-positron, meson, etc.), with which the particle is interacting. The fur coat contains many layers with different density. For example, since meson interactions of nucleons are hundred times more intensive as electromagnetic ones, then the meson fur coat of the proton should be several orders thicker than the electromagnetic fur coat. The fur coat is not something hardened, since quanta, its components, are continuously produced and destroyed. One can say, that in quantum theory a particle is suffering from striptease- mania, since one part of its lifetime it spends in the dressed state, while during the rest of the time it is naked.

4.2. Feynman Diagrams

Although in the quantum field theory typical non-classical objects are considered, it does not hinder us to give the visual graphical representation for the processes of interactions and conversions of particles.

Such a kind of graphs was first introduced by R. Feynman and now bears his name. Diagrams, or Feynman graphs outwardly resemble the displays of trajectories of all the particles taking part in interaction. Classical description, however, is not applicable in this case, so that interpretation of graphs as classical trajectories is incorrect, one can speak only of outer similarity.

By the example of non-relativistic quantum mechanics we became aware of the fact, that only in the case of most simple potentials we manage to obtain exact solutions for quantum systems. For this reason, perturbation theory method is the main mathematical method in quantum theory. The essence of perturbation theory method is most transparent in the Green function formalism. So, if a system wave function in initial state $\psi(\mathbf{r}_i,t_i)$ is known, then its value at an arbitrary moment of time $t=t_f$ in the point with coordinates $\mathbf{r}_f$ is given by expression

$$\psi(\mathbf{r}_f,t_f)=\int G(\mathbf{r}_f-\mathbf{r}_i,t_f-t_i)\psi(\mathbf{r}_i,t_i)d^4x_i. \tag{4.15}$$

The Green function in non-relativistic quantum mechanics is represented by series on perturbation $H_{int}(\mathbf{r},t)$

$$G(\mathbf{r}_f-\mathbf{r}_i,t_f-t_i)=G^{(0)}(\mathbf{r}_f-\mathbf{r}_i,t_f-t_i)-$$

$$-\frac{i}{\hbar}\int G^{(0)}(\mathbf{r}_f-\mathbf{r}_1,t_f-t_1)H_{int}(\mathbf{r}_1,t_1)G^{(0)}(\mathbf{r}_1-\mathbf{r}_i,t_1-t_i)d^4x_1+\left(-\frac{i}{\hbar}\right)^2\times$$

$$\times\int d^4x_1\int G^{(0)}(\mathbf{r}_f-\mathbf{r}_1,t_f-t_1)H_{int}(\mathbf{r}_1,t_1)G^{(0)}(\mathbf{r}_1-\mathbf{r}_2,t_1-t_2)H_{int}(\mathbf{r}_2,t_2)\times$$

$$\times G^{(0)}(\mathbf{r}_2-\mathbf{r}_i,t_2-t_i)d^4x_2+\ldots.., \tag{4.16}$$

where $G^{(0)}(\mathbf{r},t)$ is the Green function in zero approximation. Expression (4.16) brings us to the idea, that the evolution of a quantum system can be plotted as a chain of transitions from the initial state to the final one through the totality of points (vertices) corresponding to interaction acts. In so doing the number of vertices corresponds to perturbation theory order. These are non-relativistic prerequisites for appearance of Feynman diagram method. This method is also unambiguously connected with taking account of every order of perturbation theory by steps. Diagrams topology is completely defined by the form of interaction Lagrangian. By the appearance of a Feynman diagram, it is possible immediately to write down the corresponding expression for scattering amplitude, i.e. there is a set of rules, so-called Feynman rules connecting any diagram element with particle characteristics. To depict a free particle one or another line is introduced (which is, of course, only a graphic symbol of particles propagation), while lines knot (vertex) corresponds to particles interaction. External lines describe real initial and final particles and the internal ones describe virtual particles. Since particles in initial and final states are considered to be free (it is guaranteed by experiment conditions), then real particles correspond to wave functions, which are solutions of free equations of corresponding fields. Virtual particles, describing interaction, are put in correspondence of the Green functions of the equations, which they would satisfy, if they were real. A coupling constant, characterizing the given interaction, must be in every diagram vertex. In the case of quantum electrodynamics it equals $\sqrt{e^2/(4\pi\hbar c)}$. Moreover, in every vertex the momentum conservation law must be taken into consideration (interaction, taking place at a vertex, can take place at any space point, so that $\Delta x_i = \infty$, and it means, that momentum is precisely defined). In every vertex of "charges" (electric, baryon, lepton, strange, etc.) conservation laws which are valid for the given interaction, must be fulfilled. The Feynman diagrams can be plotted in coordinate and momentum space, in the latter case four-momentum of a corresponding particle is ascribed to every line. The content of the Feynman rules is determined by the structure of the interaction Hamiltonian H_{int}. In the QED this quantity is given by the expression

$$H_{int} = e\overline{\psi}(x)\gamma_\mu\psi(x)A^\mu(x), \tag{4.17}$$

that is, H_{int} has the trilinear structure. It means that in every vertex three lines are encountered, namely two fermion and one photon lines. As this takes place, the electric charge conservation law demands direction invariability of the fermion line over its length. Let us display photon with a dashed line, fermion with a solid line. Sometimes a symbol of a particle will be also mentioned near every line. We agree to direct time axes from left to right in all the diagrams. In fermion lines we put arrows which denote particle propagation direction. The arrows, directed to the time axis denote particle, while antiparticles are indicated by the arrows being opposite to the time axis direction.

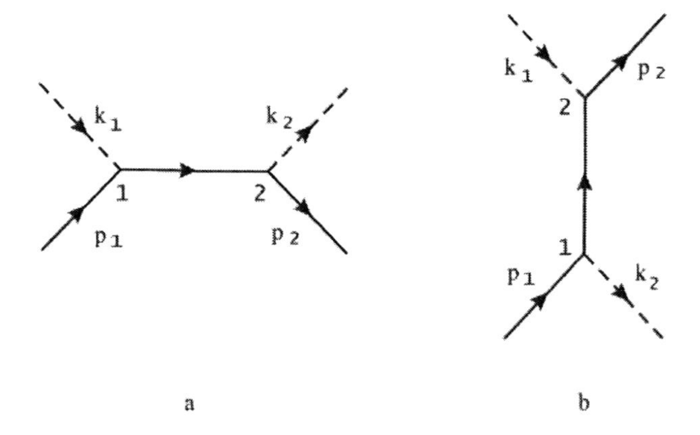

Figure 5. The Feynman diagrams corresponding to the Compton effect in the second order of perturbation theory.

Let us illustrate diagram method by some examples from the QED. In Fig. 5a,b the

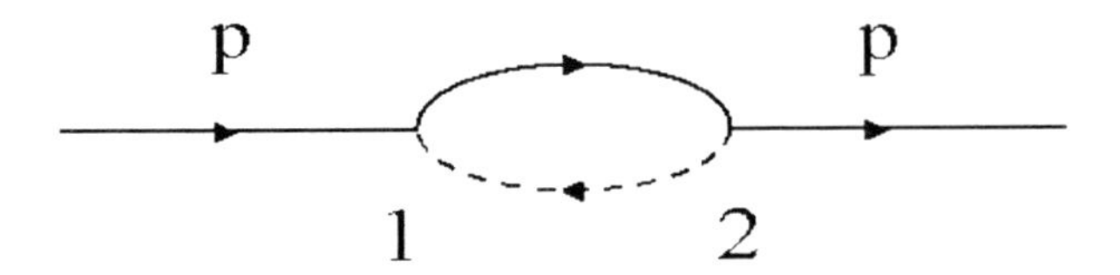

Figure 6. The electron vacuum loop.

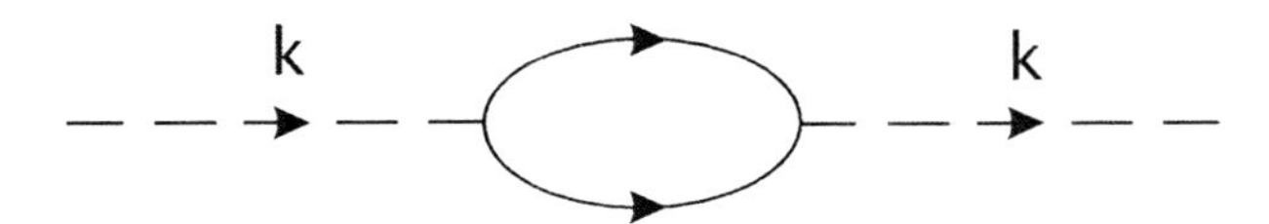

Figure 7. The diagram of the photon self-energy in the second order of perturbation theory.

diagrams for the so-called Compton effect — the elastic scattering of photon off electron

$$e^- + \gamma \rightarrow e^- + \gamma,$$

are displayed. Fig. 5a should be read as follows. In the initial state one photon and one electron are present. In point 1 they annihilate into a virtual electron, in point 2 the virtual electron turns into the real photon and electron. The second order of perturbation theory (in electromagnetic interaction constant) corresponds to two vertices of the diagrams. In Fig. 6 the following process is plotted. In point 1 an electron emits a virtual photon and turns into a virtual electron. In point 2 the virtual electron and photon turn into a real electron. Thus, the diagram in Fig. 6 describes electron interaction with the virtual particles fields, namely, with photon and electron-positron vacuums. As a result of this interaction the electron energy is changed. For this reason the corresponding diagram is called the diagram of the self-energy electron, or the electron vacuum loop. Analogous photon vacuum loop, describing the self-energy photon in the second order of the perturbation theory, is represented in Fig.7. It is obvious that every of the above mentioned diagrams can be complicated by adding new vertices. Thus, for example, the electron vacuum loop can be inserted in the internal line of the diagram in Fig. 5a (Fig. 8). In the diagram obtained there are four vertices, that is, the corresponding amplitude is proportional to e^4 and makes up one of the terms, describing the Compton effect in the fourth order of the perturbation theory. The total number of diagrams, corresponding to the definite order of the perturbation theory is defined by the interaction Hamiltonian structure. Let us calculate the number of all the possible diagrams of the fourth

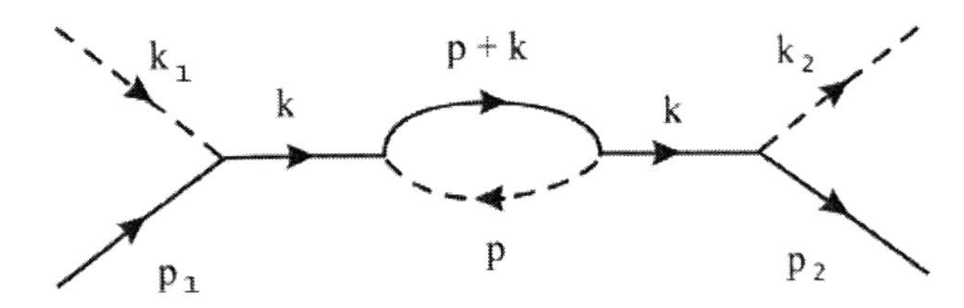

Figure 8. One of diagrams describing the Compton effect in the fourth order of perturbation theory.

order for the Compton effect. The electron vacuum loop can be inserted to the diagram in Fig. 5a in the following ways: 1) into the inner line, 2) into the two external electron lines. The photon vacuum loop can be inserted into any of the two external photon lines. Besides, a virtual photon line can connect the initial and finite electrons, while a virtual electron line connects the initial and finite photon. So, the total number of the diagrams of the fourth order for the Compton effect in the QED is equal to 14.

4.3. Elements of the Group Theory

The concept of symmetry is the main concept of Physics. Its role in Physics has been already perceived after the classical Noether theorem which has connected the physical system symmetry with conservation laws. With the help of symmetry one manages to reveal the basic structures in variety of the physical phenomena and to reduce this variety to a few tens of fundamental formulas. The group theory is the natural mathematical language for describing the symmetry. To study isotopic and unitary hadron families and subsequently to get over to quark models, we should get acquainted with foundations of the group theory. We start our digression to the group theory with studying the three-dimensional Euclidean space transformations. Let us consider the transformations

$$\mathbf{x}' = R\mathbf{x}, \tag{4.18}$$

leaving a scalar product $(\mathbf{x}\cdot\mathbf{y})$ invariant, that is, for which the following is true

$$(\mathbf{x}'\cdot\mathbf{y}') = (\mathbf{x}\cdot\mathbf{y}). \tag{4.19}$$

As it follows from (4.19), these transformations are carried out through orthogonal matrices

$$RR^T = R^T R = I.$$

Since

$$\det(RR^T) = \det R^T \det R = (\det R)^2 = 1,$$

then $\det R = \pm 1$. In case $\det R = +1$ we have proper orthogonal transformations or rotations, while when $\det R = -1$ we obtain improper orthogonal transformations. The latter is exemplified by a spatial inversion, which is represented by the matrix

$$R_- = \begin{pmatrix} -1 & 0 & 0 \\ 0 & -1 & 0 \\ 0 & 0 & -1 \end{pmatrix}. \tag{4.20}$$

Any rotation can be represented as a product of three consecutive rotations around orthogonal axes Ox_1, Ox_2, x_3Q by angles θ_1, θ_2, θ_3.

While rotating around axis Ox_1 coordinates $\mathbf{x}$ transform in the following way

$$x_1' = x_1 + 0\times x_2 + 0\times x_3,$$

$$x_2' = 0\times x_1 + \cos\theta_1 x_2 + \sin\theta_1 x_3,$$

$$x_1' = 0\times x_1 - \sin\theta_1 x_2 + \cos\theta_1 x_3.$$

In matrix form it looks as follows

$$\begin{pmatrix} x'_1 \\ x'_2 \\ x'_3 \end{pmatrix} = R_{\theta_1}\begin{pmatrix} x_1 \\ x_2 \\ x_3 \end{pmatrix} = \begin{pmatrix} 1 & 0 & 0 \\ 0 & \cos\theta_1 & \sin\theta_1 \\ 0 & -\sin\theta_1 & \cos\theta_1 \end{pmatrix}\begin{pmatrix} x_1 \\ x_2 \\ x_3 \end{pmatrix}.$$

In the same way we can represent transformations of rotations around axis Ox_2 Ox_3

$$R_{\theta_2} = \begin{pmatrix} \cos\theta_2 & 0 & \sin\theta_2 \\ 0 & 1 & 0 \\ -\sin\theta_2 & 0 & \cos\theta_2 \end{pmatrix}, \qquad R_{\theta_3} = \begin{pmatrix} \cos\theta_3 & \sin\theta_3 & 0 \\ -\sin\theta_3 & \cos\theta_3 & 0 \\ 0 & 0 & 1 \end{pmatrix}.$$

The matrix of a rotation around an arbitrary axis is defined by the product of matrices $R_{\theta_1}R_{\theta_2}R_{\theta_3}$. We pay attention to the fact, that the final result depends on the order of multiplication of R_θ matrices, that is, rotation transformations are not commutative. Totality of all the orthogonal transformations in the three-dimensional Euclidean space makes up a group, namely, an orthogonal group $O(3)$ (in the name of a group it is indicated by letter O). Transformations with the determinant being equal to 1 are called either "unimodular" or "special" (the letter S). Thus, the rotation group is the group $SO(3)$. The groups, in which the elements are commutative, are called Abelian groups while the groups with noncommutative elements are called non-Abelian groups. The group $SO(3)$ is a non-Abelian group. The group $SO(3)$, supplemented with reflections, makes up the group $O(3)$. It means, that $SO(3)$ is a subgroup of $O(3)$. For matrices in n-dimensions the group $O(n)$ is orthogonal and its dimension d is defined by the formula $d = n(n-1)/2$. At this point it is high time to introduce the rules of the game, adopted in the group theory.

A group is a set of elements G, which meets the following demands:

1. On the set G the group action is defined (call it "multiplication" conventionally), which puts in correspondence to every pair of elements f and g a certain h element from the same set. The element h is called the productions of elements f and g, i. e $h = fg$. In the most general case $fg \neq gf$.

2. The set contains such a unit element e, for which the relation $ef = fe = f$ is true, where f is any element from the set G.

3. Alongside with any element f, the set G contains an inverse element f^{-1}, which possesses the following property

$$ff^{-1} = f^{-1}f = e.$$

The group is finite (infinite) when the number of its elements is finite (infinite). The three-dimensional rotation group is infinite while the reflection group is finite.

The group is called discrete (continuous) when its elements take discrete (continuous) values. The three-dimensional rotation group is continuous, while the reflection group is discrete. The order of group is a number of independent parameters, which defined a group. Three-dimensional rotation group is a group of the third order (its independent parameters are θ_1, θ_2, θ_3). The order of reflection group is the same.

The order of transformation matrix[1] (two-row, three-row, etc.) defines dimension of the group. The dimension of three-dimensional rotation group and of reflection group equals

[1] It should be square matrix.

three. Let us note, that the order of group differs from its dimension. In the case of the orthogonal group this coincidence is accidental. Continuous groups of the finite order are Lie groups. Three-dimensional rotation group is an example of the Lie group.

The next goal is to define all representations of orthogonal group. In general, a representation of a certain G group is a mapping, which compares every element g of G with linear operator U_g, acting in some vector space V. In such a mapping a multiplication table for a group is maintained, and the unit element e of the group G is represented by identity transformation of I in V

$$U_{g_1}U_{g_2} = U_{g_3} \qquad (g_3 = g_1 g_2), \qquad U_e = I.$$

Subspace V_1 of the space V is called invariant subspace with respect to representation U_g in case if all the vectors v in V_1 are transformed by any U_g into vectors v' belonging again to V_1. A representation is irreducible, if the only subspace V invariant with respect to representation $g \to U_g$ is the whole space and subspace, consisting of one zero vector. Otherwise a representation is reducible. Studies of representations can be limited to studies of irreducible representations. Moreover, only inequivalent irreducible representations are of main interest for us. Two representations U_g and U'_g are equivalent, if there is a unitary operator M, which ensures

$$U'_g = MU_gM^{-1} \qquad v' = Mv,$$

where v and v' are vectors of representation space. Thus, two equivalent representations could be seen as realizations of one and the same representation in terms of two different bases in vector space. All irreducible representations are finite-dimensional and any representation is a direct sum of irreducible finite-dimensional representations. The direct sum of two square matrices D_1 and D_2 is by definition a square matrix D_3 for which

$$D_3 = \begin{pmatrix} D_1 & 0 \\ 0 & D_2 \end{pmatrix}.$$

is true. Symbolically it is written down as follows $D_3 = D_1 \bigoplus D_2$.

Further on, infinitesimal rotations and matrices, corresponding to them will play a fundamental role. Their importance is caused by the fact, that they produce one-parameter subgroups and any finite rotation can be composed of a sequence of infinitesimal rotations. To obtain matrices for infinitesimal rotations, we expand every element of matrix of finite rotations into Taylor series and constrain ourselves by the terms of the first order

$$A_i = \frac{d}{d\theta_i} R(\theta_i)|_{\theta_i=0}, \qquad (i = 1,2,3) \tag{4.21}$$

The obtained matrices A_i we shall call generators of rotations about ith axis. Their obvious form is the following

$$A_1 = \begin{pmatrix} 0 & 0 & 0 \\ 0 & 0 & 1 \\ 0 & -1 & 0 \end{pmatrix}, \qquad A_2 = \begin{pmatrix} 0 & 0 & -1 \\ 0 & 0 & 0 \\ 1 & 0 & 0 \end{pmatrix},$$

$$A_3 = \begin{pmatrix} 0 & 1 & 0 \\ -1 & 0 & 0 \\ 0 & 0 & 0 \end{pmatrix}. \qquad (4.22)$$

From (4.22) it is clear, that infinitesimal rotations commute with each other. A rotation about a finite angle θ_i can be viewed as a result of n rotations by angle θ_i/n

$$R(\theta_i) = \lim_{n\to\infty}\left(1+\frac{\theta_i}{n}A_i\right)^n = \exp(A_i\theta_i). \qquad (4.23)$$

Thus, the matrices of transformations corresponding to finite values of the transformations parameters are defined with the help of the generators A_i. This generators property is common for any class of transformations. It is possible to check by direct calculations that the generators A_i satisfy the commutation relation

$$[A_i, A_k] = -\varepsilon_{ikl}A_l \qquad (4.24)$$

and commute with the reflection operator

$$[A_i, R_-] = 0. \qquad (4.25)$$

The set of generators A_i constitutes the so-called Lie algebra of a three-dimensional rotation group.

The set of elements N is Lie algebra, corresponding to a certain group, if it meets the following demands:

1. If X and Y are elements of set N, then both the sum of $X+Y$ and the product αX (where α is an arbitrary number) belong to N once again.

2. Commutator of two elements X and Y belonging to set N is again expressed in terms of elements of the set N.

3. Commutators of elements of set N satisfy the relations

$$[X,Y]+[Y,X]=0, \qquad [X,(Y+Z)] = [X,Y]+[X,Z],$$

$$[X,[Y,Z]]+[Y,[Z,X]]+[Z,[X,Y]]=0, \qquad \text{(Jacobi identity)}. \qquad (4.26)$$

The representation of Lie algebra is a correspondence $X \to T(X)$, which confronts every element X with the linear operator $T(X)$ acting in vector space in such a way, that

$$T(X+Y) = T(X)+T(Y), \qquad T(cX) = cT(X),$$

$$T([X,Y]) = [T(X), T(Y)].$$

By the example of rotation group we shall show, that representation of Lie algebra of a group uniquely determines a representation of the group. An infinitesimal rotation by the angle $\delta\theta$ is set down as follows

$$R(\delta\theta) = I + A_i\delta\theta_i + O(\delta\theta^2).$$

The corresponding representation operator has the form

$$U(\delta\theta) = I + M_i\delta\theta_i + O(\delta\theta^2), \qquad (4.27)$$

where M_i constitute a representation of Lie algebra generators and obey the same commutation relations as A_i

$$[M_i, M_k] = -\varepsilon_{ikl} M_l. \tag{4.28}$$

Let us consider two rotations about the finite angles $a\theta$ and $b\theta$. Since rotations about one and the same axis commute then

$$R(a\theta)R(b\theta) = R((a+b)\theta)$$

and as consequence

$$U(a\theta)U(b\theta) = U((a+b)\theta). \tag{4.29}$$

Differentiating (4.29) with respect to a, using (4.27) and taking into consideration that

$$\frac{d}{da} \leftrightarrow \frac{d}{db},$$

with $a = 0$, we obtain the equation

$$\frac{d}{da} U(a\theta)U(b\theta)|_{a=0} = \frac{d}{db} U(b\theta) = (M_i\theta_i)U(b\theta). \tag{4.30}$$

To integrate Eq. (4.30) with the initial condition $U(0) = 1$ produces the expression for operators of representation of a three-dimensional rotation group

$$U(\theta) = \exp(M_i\theta_i). \tag{4.31}$$

Since in unitary representations the operators U are unitary, then the operators M_i are anti-Hermitian. We pass to Hermitian operators $J_i = -iM_i$, which satisfy ordinary permutation relations for angular moment operators

$$[J_k, J_m] = i\varepsilon_{kmn} J_n. \tag{4.32}$$

The problem of finding all the irreducible rotation group representation is reduced to finding all the possible set of matrices J_1, J_2, J_3, complying with the permutation relations (4.32).

Shur lemma has the fundamental meaning in theory of representation of group, realized by complex matrices. It reads: "For representation to be irreducible, it is necessary and sufficient, that the only matrices, which commute with all the representation matrices are the matrices being multiple to the identity matrix". Such matrices being polynomials of generators, are called Casimir operators U_i^K. According to Shur lemma, a representation is irreducible in the case of

$$U_i^K = \lambda I$$

and as a result

$$U_i^K \psi = \lambda \psi,$$

where ψ denotes the field function. Since Casimir operators enter the complete set of commuting operators (which undoubtedly contains the Hamiltonian as well), then conserved physical quantities correspond to them. So, if we have found all the Casimir operators, have chosen one eigenvalue from each operator and built up a representation which acts in space spanned on corresponding eigenfunctions, such a representation is irreducible, according to Shur lemma. In other words, the problem of classifying the irreducible group representations is reduced to finding the spectrum of eigenvalues and eigenfunctions of the Casimir operators.

Three-dimensional rotations group has one Casimir operator

$$\mathbf{J}^2 = J_1^2 + J_2^2 + J_3^2, \tag{4.33}$$

which is nothing else but the squared angular moment operator with eigenvalues $j(j+1)$, where $j = 0, 1/2, 1, 3/2, 2, \ldots$. Consequently, every irreducible representation of three-dimensional rotation group is characterized by the positive integer or half-integer number j, which also defines representation dimension by the formula $2j+1$. From quantum mechanics we know, that squared angular moment operator commutes with projection of angular moment operator onto a certain direction singled out in space, for example, axis x_3. Thus, basic functions of representation are eigenfunctions of operators $\mathbf{J}^2$, J_3 and could be marked by their eigenvalues. Passing to the classification of irreducible representations of orthogonal group we keep in mind that the linear operator U_- corresponding to the operation of spatial inversion R_- commutes with all the operators of the rotation group representation. According to Shur lemma, in every irreducible representation the operator U_- must be multiple to the identity operator. Thus the irreducible representation of orthogonal group is classified by a pair of indices (j, η_p), where the latter is the eigenvalue of U_- corresponding to the given representation. Representations with integer j are called single-valued or tensor representations, in the case of half-integer j representations are two-valued or spinor representations.

a. Tensor representations of the group $SO(3)$. Since at integer values j the relation takes place

$$(U_-)^2\psi = I\psi,$$

then eigenvalues of the reflection operator are $+1$ and -1. Thus, there are two different representations of the orthogonal group, in the former $U_- = I$, and in the latter $U_- = -I$.

At $j = 0$ the representation is one-dimensional, each group element is mapped by the identity operator, and generators are identically equal to zero. Let us call representation {0,1} a scalar and {0,-1} — pseudoscalar. The quantities transformed according to (pseudo)scalar representation are called (pseudo)tensor of zero-rank or (pseudo)scalars.

At $j = 1$ the representation is three-dimensional. One might use matrices of generators of three-dimensional rotation group as matrix representation of generators M_i (representation, constructed directly from generators of transformation group is called regular or associated representation of group). For the representations {1,-1} and {1,1} we use terms vector representation and pseudovector representation, respectively. Three-dimensional quantities transformed with respect to (pseudo)vector representations are called (pseudo)tensor of the first rank or (pseudo)vectors. Very often pseudovectors are called axial vectors and vectors are called polar vectors. All the representations with $j = 0, 1$ are irreducible, while the representations with $j \geq 2$ are reducible.

To summarize the aforesaid, we can give the following definition: three-dimensional tensor (pseudotensor) of the n-rank is a quantity, which transforms under the representation

$\{n,(-1)^n\}$ ($\{n,(-1)^{n+1}\}$) of the group $O(3)$. In more convenient for practical use language, tensor and pseudotensor of the n-rank are the quantities, which components, when rotating, are transformed according to the law

$$\psi'_{\underbrace{ik..m}_{n}}(\mathbf{r}') = R_{ip}R_{kl}...R_{ms}\psi_{\underbrace{pl..s}_{n}}(\mathbf{r}). \tag{4.34}$$

So, under rotations they behave in the same way. However, under reflections we have

$$\psi'_{\underbrace{ik..m}_{n}}(\mathbf{r}') = (-1)^n\psi_{\underbrace{ik..m}_{n}}(\mathbf{r}), \tag{4.35}$$

for a tensor

$$\psi'_{\underbrace{ik..m}_{n}}(\mathbf{r}') = (-1)^{n+1}\psi_{\underbrace{ik..m}_{n}}(\mathbf{r}), \tag{4.36}$$

and for a pseudotensor, respectively.

Ability to expand reducible representations in irreducible ones will make one more element of our mathematical culture. As an example, we consider a field function ψ_{ik} which is a tensor of the second rank with respect to three-dimensional rotation group and consequently it has nine components. A tensor of the second rank is a product of two three-dimensional vectors, which transform with respect to three-dimensional representations. Schematically it can be written down in the form

$$3\bigotimes 3. \tag{4.37}$$

Let us represent ψ_{ik} as a sum of symmetric and antisymmetric tensors

$$\psi^s_{ik} = \frac{1}{2}(\psi_{ik}+\psi_{ki}), \qquad \psi^a_{ik} = \frac{1}{2}(\psi_{ik}-\psi_{ki}).$$

By the direct check one can be sure, that

$$\psi^{s\prime}_{ik} = R^{(1)}_{im}R^{(1)}_{kn}\psi^s_{mn} \qquad \psi^{a\prime}_{ik} = R^{(1)}_{im}R^{(1)}_{kn}\psi^a_{mn},$$

i. e., under rotations the components of the symmetric and antisymmetric tensors are not mixed with each other. In other words, under transformation of the group $SO(3)$ the nine components of the tensor ψ_{ik} fall into two independent totalities: three-dimensional ψ^a_{ik} and six-dimensional ψ^s_{ik}. The symmetric tensor can be also further expanded in two independent totalities

$$\psi^0 = \frac{1}{3}\mathrm{Sp}\{\psi_{mn}\} = \frac{1}{3}\psi_{nn} = \frac{1}{3}(\psi_{11}+\psi_{22}+\psi_{33})$$

scalar (we remember, that scalar product of three-dimensional vectors is invariant under rotations) and five components, which constitute a matrix with a zero-trace

$$\psi^s_{ik} = \psi^s_{ik} - \frac{1}{3}\delta_{ik}\psi_{nn}.$$

So, from the viewpoint of rotation transformations, the second rank tensor ψ_{ik} is the sum of the three independent quantities: the one-dimensional ψ^0, the three-dimensional ψ^a_k

$(k = 1,2,3)$ and the five-dimensional ψ_l^s $(l = 1,2,3,4,5)$. The corresponding representation of the rotation group consists of matrices of reduced dimensions and has the box-diagonal form

$$U\begin{pmatrix}\psi_1^a\\ \psi_2^a\\ \psi_3^a\\ \psi_1^s\\ \psi_2^s\\ \psi_3^s\\ \psi_4^s\\ \psi_5^s\\ \psi^0\end{pmatrix} = \begin{pmatrix}U_{11} & U_{12} & U_{13} & 0 & 0 & 0 & 0 & 0 & 0\\ U_{21} & U_{22} & U_{23} & 0 & 0 & 0 & 0 & 0 & 0\\ U_{31} & U_{32} & U_{33} & 0 & 0 & 0 & 0 & 0 & 0\\ 0 & 0 & 0 & U_{44} & U_{45} & U_{46} & U_{47} & U_{48} & 0\\ 0 & 0 & 0 & U_{54} & U_{55} & U_{56} & U_{57} & U_{58} & 0\\ 0 & 0 & 0 & U_{64} & U_{65} & U_{66} & U_{67} & U_{68} & 0\\ 0 & 0 & 0 & U_{74} & U_{75} & U_{76} & U_{77} & U_{78} & 0\\ 0 & 0 & 0 & U_{84} & U_{85} & U_{86} & U_{87} & U_{88} & 0\\ 0 & 0 & 0 & 0 & 0 & 0 & 0 & 0 & 1\end{pmatrix}\begin{pmatrix}\psi_1^a\\ \psi_2^a\\ \psi_3^a\\ \psi_1^s\\ \psi_2^s\\ \psi_3^s\\ \psi_4^s\\ \psi_5^s\\ \psi^0\end{pmatrix}. \qquad (4.38)$$

If a representation matrix can be written down in the box-diagonal form, then the corresponding representation is reducible (otherwise it is irreducible). Then Eq. (4.38) means

$$3\bigotimes 3 = 9 = 5\bigoplus 3\bigoplus 1. \qquad (4.39)$$

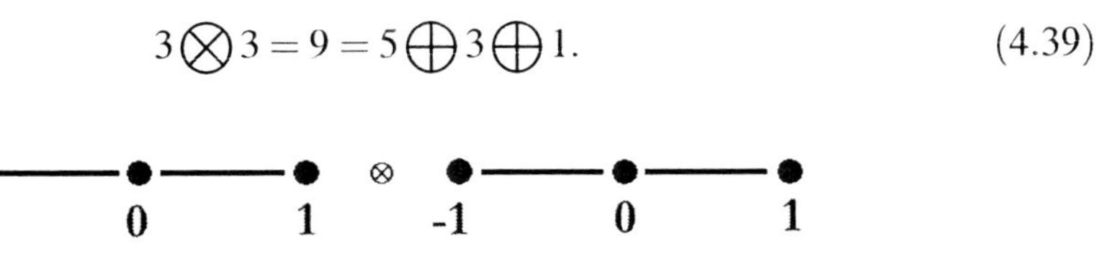

Figure 9. The basic vectors of the representations 3 and 3.

Let us present all the chain of manipulations from (4.37) to (4.39) in a more obvious form. In Fig. 9 we plot basis vectors of representations entering the left-hand side of Eq. (4.39), where the angular momentum projection eigenvalues m_j are denoted with dots. To obtain the totality of the basis vectors, composing representations in the right-hand side of Eq. (4.39), one might use the moment multiplication rule, which is well known from quantum mechanics. In order to get the product of two moments, we combine the center $J_3 = 0$ of the second diagram with each of the three states $J_3 = 1,0,-1$ of the first diagram, then we calculate frequency n_i with which every summation result I_3 appears (in our example $I_3 = \pm 2$ occurs once, $I_3 = \pm 1$ does twice, and $I_3 = 0$ does triply). From the obtained states we single out diagrams with the same value of J. This procedure is displayed in Fig. 10. When one has taken into account that the number of states having the angular moment eigenvalue j is equal to $2j+1$, that is,

$$(2j_1+1)\bigotimes(2j_2+1) \equiv j_1\bigotimes j_2,$$

then the generalization of Eq. (4.39) takes the form being already familiar to us

$$j_1\bigotimes j_2 = \sum_{\oplus} j, \qquad (4.40)$$

where every j representation occurs once, $j = \mid j_1 - j_2 \mid, \mid j_1 - j_2 + 1 \mid, \ldots \mid j_1 + j_2 \mid$ and subscript $\bigoplus$ indicates, that the sum is direct, not normal.

b. Spinor representations of three-dimensional rotation group. At $j = 1/2$ representations of rotations group is two-dimensional, consequently, operators of the wave function

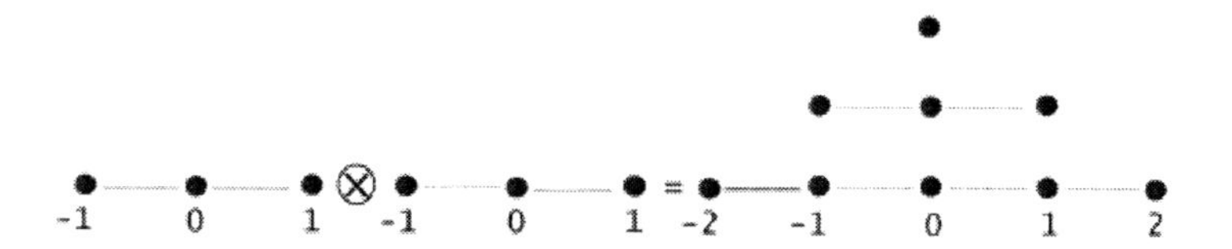

Figure 10. Graphical interpretation of the moments addition rule.

transformations and generators, together with them, constitute 2×2-matrices. Since any 2×2-matrix could be expanded in terms of the Pauli matrices σ_k and the identity matrix I[1], then the generators $J_k^{(1/2)}$ might be chosen in the shape $i\sigma_k/2$, where the Pauli matrices have the form

$$\sigma_1=\begin{pmatrix}0 & 1\\ 1 & 0\end{pmatrix},\qquad \sigma_2=\begin{pmatrix}0 & -i\\ i & 0\end{pmatrix},\qquad \sigma_3=\begin{pmatrix}1 & 0\\ 0 & -1\end{pmatrix},\tag{4.41}$$

and satisfy the relation

$$\sigma_k\sigma_l=\delta_{kl}+i\varepsilon_{klm}\sigma_m.\tag{4.42}$$

The operator, transforming a wave function under rotations by the angle θ about axis with direction unit vector $\mathbf{n}=(n_1,n_2,n_3)$ is written down as follows

$$U^{(1/2)}(\mathbf{n},\theta)=\exp[i(\mathbf{n}\cdot\sigma)\frac{\theta}{2}]=\sum_{k=0}^{\infty}\frac{1}{k!}\left(\frac{i}{2}(\mathbf{n}\cdot\sigma)\theta\right)^k=[1-\frac{1}{2!}\left(\frac{\theta}{2}\right)^2+\frac{1}{4!}\left(\frac{\theta}{2}\right)^2-$$

$$....]+i(\mathbf{n}\cdot\sigma)[\frac{\theta}{2}-\frac{1}{3!}\left(\frac{\theta}{2}\right)^3+\frac{1}{5!}\left(\frac{\theta}{2}\right)^5-...]=I\cos\frac{\theta}{2}+i(\mathbf{n}\cdot\sigma)\sin\frac{\theta}{2}.\tag{4.43}$$

From Eq. (4.43) follows that the rotation matrices are unitary and their determinant equals 1. The objects transformed with respect to representation $j=1/2$ are called spinors of the first rank. As a basis in space of such spinors, one might choose eigenvectors of matrices $(1/4)\sigma^2$ (spin square) and $(1/2)\sigma_3$ (spin projection on the third axis) which have the following form:

$$\omega^+=\begin{pmatrix}1\\ 0\end{pmatrix},\qquad \omega^-=\begin{pmatrix}0\\ 1\end{pmatrix}.$$

Since:

$$\frac{1}{4}\sigma^2\omega^+=\frac{1}{4}\sigma^2\omega^-=\frac{3}{4}$$

and

$$\frac{1}{2}\sigma_3\omega^+=+\frac{1}{2}\omega^+,\qquad \frac{1}{2}\sigma_3\omega^-=-\frac{1}{2}\omega^-,$$

it becomes obvious, that spinors of first rank are used to describe particles with the spin 1/2.

The system rotation by the angle 2π causes quite unexpected result

$$U^{(1/2)}(\mathbf{n},\theta+2\pi)\psi=-U^{(1/2)}(\mathbf{n},\theta)\psi,\tag{4.44}$$

[1]The matrices σ_k and I are linear independent, i. e., they constitute the complete set.

namely, that the system does not return to the initial state. It means that observable quantities can not be represented by a spinor, since spinors have no definite transformation properties under rotations. From the physical point of view, it makes the main difference between spinor and tensor representations. It should be reminded, in tensor representations the observable quantities could be connected with field functions directly (electromagnetic field strengths may serve as an example).

Hermitian conjugation of a spinor is carried out in a ordinary way, that is, it consists of transposition and complex conjugation. Needless to say a bilinear combination of Hermitian conjugated spinors will possess single-valued transformation properties. Thus, for example, quantity $\xi^\dagger\chi$ behaves as scalar under rotations

$$\xi'^\dagger\chi' = \xi^\dagger U^{(1/2)\dagger}(\mathbf{n},\theta)U^{(1/2)}(\mathbf{n},\theta)\chi = \xi^\dagger\chi.$$

It can be shown in the same way, that quantity $\xi^\dagger\sigma\chi$ is a vector and so on.

From Eq. (4.44) follows that representation of weight $j = 1/2$ is two-valued, namely,

$$R(\mathbf{n},\theta) \to \pm U^{(1/2)}(\mathbf{n},\theta). \tag{4.45}$$

It brings us to the idea, that from the point of view of the $SO(3)$-group, the existence of two kinds of spinors is possible.

Let us slightly alter the above mentioned facts in form. In two-dimensional complex space $S^{(2)}$ (spinor space) we introduce basis e_1 and e_2 by which any spinor ξ can be expanded

$$\xi = \xi^1 e_1 + \xi^2 e_2.$$

Under rotation about angle θ around an axis with a direction unity vector $\mathbf{n}$ spinors belonging to $S^{(2)}$ are transformed according to the law

$$\xi'^\alpha = U^{(1/2)\alpha}{}_\beta \xi^\beta, \tag{4.46}$$

where α and β are spinor indices taking the values 1 and 2. The operation of raising and lowering indices is carried out by means of an antisymmetric metric tensor

$$\varepsilon_{\alpha\beta} = \varepsilon^{\alpha\beta} = \begin{pmatrix} 0 & 1 \\ -1 & 0 \end{pmatrix}$$

i. e. $\varepsilon = i\sigma_2$. For any basis (e_1, e_2) of the spinor space $S^{(2)}$ one may build dual basis $(e^{\dot 1}, e^{\dot 2})$ of the space $S^{(\dot 2)}$ by the sole way. In so doing the both bases are connected by the relation:

$$e^{\dot\alpha} e_\beta = \delta^{\dot\alpha}_\beta = \begin{cases} 0, & \dot\alpha \neq \beta \\ 1, & \dot\alpha = \beta, \end{cases}$$

We supplied spinor indices of the space $S^{(\dot 2)}$ with points to emphasize the fact, that the transformation law for spinors in this space differs from that for spinors of space $S^{(2)}$ in these very indices. Arbitrary spinor η from $S^{(\dot 2)}$ is expanded in dual basis according to the relation

$$\eta = \eta_{\dot 1} e^{\dot 1} + \eta_{\dot 2} e^{\dot 2}.$$

Under three-dimensional rotations the spinor transformation law of space $S^{(\dot{2})}$ has the form

$$\eta^{\dot{\alpha}'} = \tilde{U}^{(1/2)\dot{\alpha}}_{\dot{\beta}} \eta^{\dot{\beta}}. \tag{4.47}$$

Matrix of transformation $\tilde{U}^{(1/2)}$ fits the choice of M_k rotation generators in the form $-i\sigma_k^*/2$. It is obvious that such a choice ensures the fulfillment of the commutation relations (4.28) as well.

Passing in Eq. (4.47) to spinors with lower indices with the help of the metric tensor $\varepsilon_{\dot{\alpha}\dot{\beta}} = \varepsilon^{\dot{\alpha}\dot{\beta}} = \varepsilon^{\alpha\beta}$, we obtain

$$\eta'_{\dot{\sigma}} = \varepsilon_{\dot{\sigma}\dot{\alpha}} \tilde{U}^{(1/2)\dot{\alpha}}{}_{\dot{\beta}}(\mathbf{n}, \theta) \varepsilon^{\dot{\beta}\dot{\tau}} \eta_{\dot{\tau}} = U^{(1/2)\dot{\alpha}\dagger}_{\dot{\sigma}}(\mathbf{n}, \theta) \eta_{\dot{\alpha}}. \tag{4.48}$$

When writing this expression we have taken into account

$$\sigma_2 \sigma^* \sigma_2 = -\sigma.$$

Now we adopt the following terminology: the two-component quantity

$$\xi^\alpha = \begin{pmatrix} \xi^1 \\ \xi^2 \end{pmatrix},$$

which under three-dimensional rotations is transformed according to the law (4.46) will be called the contravariant spinor of the first rank, while the one

$$\eta^\alpha = \begin{pmatrix} \eta_1 \\ \eta_2 \end{pmatrix}$$

which are transformed according to the law (4.48) will be called the covariant spinor of the first rank. From now we shall omit dots above indices, but always imply that the transformation law on subscripts is determined by formula (4.48). Now it is possible to obtain bilinear form from covariant and contravariant spinors, which is invariant with respect to three-dimensional rotations

$$(\xi, \eta) = \xi^\alpha \varepsilon_{\alpha\beta} \eta^\beta = \xi^1 \eta_1 + \xi^2 \eta_2.$$

We can interpret it as a scalar product.

Spinors of the higher ranks are built by analogy with tensors theory. The set 2^n of complex numbers $\alpha_1, \alpha_2, \ldots \alpha_n$, which is transformed according to the law

$$\left(\psi^{\alpha'_1 \alpha'_2 .. \alpha'_n}\right)' = U_{\alpha'_1 \alpha_1} U_{\alpha'_2 \alpha_2} \ldots U_{\alpha'_n \alpha_n} \psi^{\alpha_1 \alpha_2 .. \alpha_n}. \tag{4.49}$$

is called contravariant spinor of rank n. Similarly, covariant spinor of rank n is defined through its transformation properties

$$\left(\psi_{\beta'_1 \beta'_2 .. \beta'_n}\right)' = U^{\beta'_1 \beta_1 \dagger} U^{\beta'_2 \beta_2 \dagger} \ldots U^{\beta'_n \beta_n \dagger} \psi_{\beta_1 \beta_2 .. \beta_n}. \tag{4.50}$$

The totality of quantities, which is transformed according to the rule

$$\left(\psi^{\alpha'_1 \alpha'_2 .. \alpha'_n}_{\beta'_1 \beta'_2 .. \beta'_l}\right)' = U_{\alpha'_1 \alpha_1} U_{\alpha'_2 \alpha_2} \ldots U_{\alpha'_n \alpha_n} U^{\beta'_1 \beta_1 \dagger} U^{\beta'_2 \beta_2 \dagger} \ldots U^{\beta'_l \beta_l \dagger} \psi^{\alpha_1 \alpha_2 .. \alpha_n}_{\beta_1 \beta_2 .. \beta_l} \tag{4.51}$$

will be called mixed spinor of rank $n+l$. For the spinors of lower rank to be obtained one should summarize over covariant and contravariant indices of the higher rank spinors.

All representations with $j > 1/2$ are reducible. Let us consider, for example, a field wave function, which is the mixed spinor of the second rank ψ^{α}_{β}, i.e. it is transformed under representation $j = 3/2$. Using the above formulated diagrams rule for expansion into irreducible representations, we obtain

$$2 \bigotimes \bar{2} = 3 \bigoplus 1,$$

where 2 ($\bar{2}$) denotes covariant (contravariant) components of the representation of rotations group. Hereinafter the term "representation of group" is used in two meanings: as an operator, transforming the field wave functions, and as field wave functions transformed under the given representation.

In order to describe the relativistic phenomena in the ordinary space-time we have to assign four-dimensional meaning to the developed classification. However, this generalization is beyond the framework of our book. The three-dimensional classification will be extremely useful for the basic laws of microworld to be understood. With its help we could study properties of tensor and spinors in abstract spaces, such as the spaces of isotopic, unitary and color spins.

4.4. Isotopic Multiplets

The experiments show that nuclear forces, acting between a proton and neutron, a proton and a proton as well as between two neutrons, are practically the same (of course, particles are in the same states). In the other words, peculiarity of interaction between nucleons is that the switching off proton charge liquidates difference between a proton and a neutron. All this allowed us to consider a neutron and a proton as two states of one and the same particle, a nucleon. We had already faced the similar problem in the atomic physics when we took into account the spin of electron. Really, the electron can be in two different states corresponding two possible spin directions. Then for the charge degree of freedom to be described, one may use the mathematical formalism of the spin theory and calls the new degree of freedom as an isospin. Further, except for the isospin operator, we shall have to introduce the whole set of spinlike operators, namely, unitary spin, color spin and weak isospin operators. For the spinlike operators we agree to use the symbol S with the superscript defining the operator class ($S^{(is)}$, $S^{(un)}$, $S^{(col)}$ and S^W). As for the eigenvalues of these operators we shall use for them the notations adopted in "Review of Particle Physics". So, the eigenvalues of the isotopic spin $S^{(is)}$ (ordinary spin S) operator are marked as I (J).

The different values of a projection of isotopic spin operator on the third axis correspond to two possible nucleon states. We agree to ascribe $I_3 = 1/2$ to the proton and $I_3 = -1/2$ to the neutron (I_3 is the eigenvalue of the operator $S_3^{(is)}$). Thus, nucleon wave function can be presented in the form

$$\Psi_N = N \bigotimes \psi(x; J), \tag{4.52}$$

where

$$N = \sum_{i=p,n} a_i \psi_i, \qquad \psi_p = \begin{pmatrix} 1 \\ 0 \end{pmatrix}, \qquad \psi_n = \begin{pmatrix} 0 \\ 1 \end{pmatrix},$$

$|a_p|^2$ and $|a_n|^2$ define probability of nucleon to be found in the proton and neutron state, respectively ($|a_p|^2 + |a_n|^2 = 1$) and $\psi(x;J)$ is a part of wave function which includes coordinate and spin dependence.

At this point we should recollect some facts, related to the ordinary spin. Although the spin describes particle behavior with respect to rotations in ordinary three-dimensional space, it can not be connected with any particle spatial rotations. Spin can be related with indestructible particle rotation only in the internal space. Despite belonging to different kind of spaces, spin and orbital moment of momentum can be summarized, their sum being the total moment of momentum of a particle. A more convenient and quantum streamlined spin definition is simply indicating the existence on particles of the new (spin) degrees of freedom, which are the eigenvalues of spin projection operator S_3, their number being equal to $2J+1$. Isospin is also connected with particle behavior in the internal space, where the third axis is correlated with a charge. The number of isospin degrees of freedom, that is, the number of the possible values of $S_3^{(is)}$, is again equal to $2I+1$. However, unlike the ordinary spin, the isospin does not contribute to the quantities, which define the particle behavior in the ordinary space. It can be checked easily, that with our agreements concerning the eigenvalues of the operator $S_3^{(is)}$ for nucleon, Gell — Mann — Nishijima formula takes place

$$Q = I_3 + \frac{B}{2}, \tag{4.53}$$

where Q is a nucleon charge, expressed in units of $|e|$. From the aesthetic point of view it is attractive to explain the aforesaid by the existence of isospin symmetry, which is the mathematical copy of the spin symmetry. It means that the isospin generators satisfy the same permutation relations as the operators of the ordinary angular moment (4.32). Thus, in the states space there is a $SU(2)$-group of special ($\det U = 1$) unitary ($U^\dagger U = I$) transformations, for which states Φ and $M\Phi$ (M is the transformation matrix of the $SU(2)$-group) describe one and the same phenomenon when only strong interactions are taken into account. A nucleon is the most simple (spinor) representation of the rotation group in the isotropic space. In this case the generators of the $SU(2)$-group themselves form the group representation, that is, in the isospace the transformation matrix of nucleon state at rotations about angle θ looks as follows

$$U(\theta) = \exp[i(\sigma \cdot \theta)/2]. \tag{4.54}$$

Thus, the Pauli matrices, multiplied by $1/2$, play roles of representation generators $S_\alpha^{(is)}$ ($\alpha = 1,2,3$). Since

$$S_3^{(is)} \psi_p = \frac{1}{2} \begin{pmatrix} 1 & 0 \\ 0 & -1 \end{pmatrix} \begin{pmatrix} 1 \\ 0 \end{pmatrix} = \frac{1}{2} \begin{pmatrix} 1 \\ 0 \end{pmatrix} = \frac{1}{2} \psi_p, \tag{4.55}$$

$$S_3^{(is)} \psi_n = \frac{1}{2} \begin{pmatrix} 1 & 0 \\ 0 & -1 \end{pmatrix} \begin{pmatrix} 0 \\ 1 \end{pmatrix} = -\frac{1}{2} \begin{pmatrix} 0 \\ 1 \end{pmatrix} = -\frac{1}{2} \psi_n, \tag{4.56}$$

then the operator $S_3^{(is)}$ really is the isospin projection operator. From $S_1^{(is)}$, $S_2^{(is)}$ the following operators can be composed. They influence the proton and neutron states in the following way

$$S_+^{(is)}\psi_p = \begin{pmatrix} 0 & 1 \\ 0 & 0 \end{pmatrix}\begin{pmatrix} 1 \\ 0 \end{pmatrix} = 0, \qquad S_+^{(is)}\psi_n = \begin{pmatrix} 0 & 1 \\ 0 & 0 \end{pmatrix}\begin{pmatrix} 0 \\ 1 \end{pmatrix} = \psi_p, \tag{4.57}$$

$$S_-^{(is)}\psi_p = \begin{pmatrix} 0 & 0 \\ 1 & 0 \end{pmatrix}\begin{pmatrix} 1 \\ 0 \end{pmatrix} = \psi_n, \qquad S_-^{(is)}\psi_n = \begin{pmatrix} 0 & 0 \\ 1 & 0 \end{pmatrix}\begin{pmatrix} 0 \\ 1 \end{pmatrix} = 0. \tag{4.58}$$

From Eqs. (4.57) and (4.58) it is obvious, that operator $S_+^{(is)}$ ($S_-^{(is)}$) raises (lowers) the isospin projection value by 1 for the nucleon states. In what follows we shall call $S_+^{(is)}$ and $S_-^{(is)}$ as the raising and lowering operators, for short. Their meaning can be understood without resorting to their obvious form (not to be relating to concrete representation), but by using only commutation relations

$$[S_+^{(is)}, S_-^{(is)}] = 2S_3^{(is)}, \qquad [S_3^{(is)}, S_\pm^{(is)}] = \pm S_\pm^{(is)}, \tag{4.59}$$

For this purpose it is enough to find the result of acting the operator $S_3^{(is)}$ on the state $S_\pm^{(is)}\Psi_{I,I_3}$,, where in the wave function Ψ_{I,I_3} for the sake of simplicity the spin and spatial variables are neglected. Eqs. (4.57) and (4.58) are the special cases of relations

$$S_\pm^{(is)}\Psi_{I,I_3} = \sqrt{(I \mp I_3)(I \pm I_3 + 1)}\Psi_{I,I_3\pm 1}. \tag{4.60}$$

Let us prove these relations, using the angular moment operators L_i as the operators $S^{(is)}$. We build the "raising" and "lowering" operators

$$L_\pm = L_x \pm iL_y, \qquad L_k = -i\hbar\varepsilon_{klm}x_l\partial_m,$$

and act by them on the eigenfunctions of the operators $\mathbf{L}^2$ and L_3 which are the spherical functions $Y_l^{m_l}(\theta,\varphi)$

$$\mathbf{L}^2 Y_l^{m_l}(\theta,\varphi) = l(l+1)\hbar^2 Y_l^{m_l}(\theta,\varphi), \qquad L_3 Y_l^{m_l}(\theta,\varphi) = m_l\hbar Y_l^{m_l}(\theta,\varphi).$$

As a result we give the required relation

$$L_\pm Y_l^{m_l}(\theta,\varphi) = \hbar\sqrt{(l \mp m_l)(l \pm m_l + 1)}Y_l^{m_l\pm 1}(\theta,\varphi). \tag{4.61}$$

For antinucleons the scheme of building the wave function must be changed. It is caused by the fact that wave functions of the form $\overline{\psi}_p C$ and $\overline{\psi}_n C$ (C is the charge conjugation matrix) correspond to antinucleons. Thus, relation

$$N' = U(\theta)N, \tag{4.62}$$

means, that

$$\overline{N}' = U^\dagger(\theta)\overline{N}. \tag{4.63}$$

Having used the obvious form of the transformation matrix $U(\theta)$, one can be easily persuaded, that

$$U^*(\theta) = \exp[-i(\sigma^* \cdot \theta)/2] = (i\sigma_2)\exp[i(\sigma \cdot \theta)/2](-i\sigma_2) =$$

$$= (i\sigma_2)U(\theta)(-i\sigma_2). \tag{4.64}$$

Consequently, if we write down for the antinucleon

$$\xi = -i\sigma_2(\overline{N}C) = \begin{pmatrix} -\overline{\psi}_n C \\ \overline{\psi}_p C \end{pmatrix} = \begin{pmatrix} -\psi_{\overline{n}} \\ \psi_{\overline{p}} \end{pmatrix}, \tag{4.65}$$

then we get the transformation law

$$\xi' = U(\theta)\xi. \tag{4.66}$$

Since in isospace transformation properties of ξ and N are the same, then working with ξ and not with antinucleon doublet directly, we do not make any difference between particles and antiparticles. Thereby, there is no need to modify the theory of moments with an eye to extending this theory in case of antiparticles.

Since nucleons are fermions, then, according to Pauli principle, no more than one nucleon can present in one state, in this case the state now is characterized by one more additional number, the eigenvalue of the third component of the isospin operator. Consequently, the wave function of two nucleons must be completely antisymmetric with respect to nucleons transpositions.

Let us consider, as case in point, a two-nucleon system, a deuteron, which is described by a production of two wave nucleon functions ψ_A and ψ_B. Since the isotopic spin is the additive quantum number, then the two-nucleon state can possess the isotopic spin being equal either to 1 or to 0. The isospin operator of a compound system with the wave function Ψ_{AB} is defined by the expression

$$\mathbf{S}^{(is)} = \mathbf{S}_A^{(is)} + \mathbf{S}_B^{(is)}, \tag{4.67}$$

where subscripts indicate, on which a wave function of the subsystem the operator acts. It is convenient to represent the square of the operator $\mathbf{S}^{(is)}$ in the form

$$(\mathbf{S}^{(is)})^2 = (\mathbf{S}_A^{(is)})^2 + (\mathbf{S}_B^{(is)})^2 + 2\mathbf{S}_A^{(is)} \cdot \mathbf{S}_B^{(is)} = (\mathbf{S}_A^{(is)})^2 + (\mathbf{S}_B^{(is)})^2 + (S_{A1}^{(is)} + iS_{A2}^{(is)})(S_{B1}^{(is)} -$$

$$-iS_{B2}^{(is)}) + (S_{A1}^{(is)} - iS_{A2}^{(is)})(S_{B1}^{(is)} + S_{B2}^{(is)}) + 2S_{A3}^{(is)}S_{B3}^{(is)} = (\mathbf{S}_A^{(is)})^2 + (\mathbf{S}_B^{(is)})^2 +$$

$$+ S_{A+}^{(is)}S_{B-}^{(is)} + S_{A-}^{(is)}S_{B+}^{(is)} + 2S_{A3}^{(is)}S_{B3}^{(is)}. \tag{4.68}$$

Two-nucleon function in states with the definite values of I and I_3 is the linear combination of the states

$$\psi_p^{(A)}\psi_p^{(B)}, \qquad \psi_p^{(A)}\psi_n^{(B)}, \qquad \psi_n^{(A)}\psi_p^{(B)}, \qquad \psi_n^{(A)}\psi_n^{(B)}. \tag{4.69}$$

It is absolutely obvious, that the states

$$\psi_p^{(A)}\psi_p^{(B)} \qquad \text{and} \qquad \psi_n^{(A)}\psi_n^{(B)} \tag{4.70}$$

enter to the isotopic triplet ($I = 1$), and the values $I_3 = 1$ and $I_3 = -1$ correspond to them. A certain combination of the remaining states $\psi_p^{(A)}\psi_n^{(B)}$ and $\psi_n^{(A)}\psi_p^{(B)}$ must play the role of the third component of the triplet with $I_3 = 0$. Finally, the other orthogonal combination of the states $\psi_p^{(A)}\psi_n^{(B)}$ and $\psi_n^{(A)}\psi_p^{(B)}$ can only belong to the singlet state with $I = 0$ and $I_3 = 0$. To single out the triplet state, with $I_3 = 0$, the state with $I_3 = 0$ must be acted by lowering operator $S_-^{(is)} = S_{A-}^{(is)} + S_{B-}^{(is)}$, which, as we see, is symmetric with respect to indices A and B. Since the state $\psi_p^{(A)}\psi_p^{(B)}$ is also symmetric with respect to these indices, then the emerging state has the form

$$S_-^{(is)}\psi_p^{(A)}\psi_p^{(B)} = \text{const} \times \left(\psi_p^{(A)}\psi_n^{(B)} + \psi_n^{(A)}\psi_p^{(B)}\right), \tag{4.71}$$

where from the normalization condition follows the value $1/\sqrt{2}$ for the constant const. Because the state with $I = I_3 = 0$ must be orthogonal to that defined by Eq. (4.71), then it is given by the expression

$$\frac{1}{\sqrt{2}}\left(\psi_p^{(A)}\psi_n^{(B)} - \psi_n^{(A)}\psi_p^{(B)}\right). \tag{4.72}$$

Using the eigenvalues equations

$$\left.\begin{aligned} (\mathbf{S}^{(is)})^2\Psi_{AB} &= I(I+1)\Psi_{AB}, \\ (\mathbf{S}_A^{(is)})^2\psi_A &= I_A(I_A+1)\psi_A, \\ (\mathbf{S}_B^{(is)})^2\psi_B &= I_B(I_B+1)\psi_B, \end{aligned}\right\}, \tag{4.73}$$

and the expression for $(\mathbf{S}^{(is)})^2$ (4.68), one can easily check, that the symmetric states

$$\left.\begin{array}{ll} pp, & I_3 = 1 \\ \frac{1}{\sqrt{2}}(pn+np), & I_3 = 0 \\ nn, & I_3 = -1 \end{array}\right\} I = 1, \tag{4.74}$$

really form the isotriplet, and the antisymmetric state

$$\frac{1}{\sqrt{2}}(pn-np) \tag{4.75}$$

is the isosinglet (in formulas (4.74) and (4.75) we switched over to more economic and obvious notations). The deuteron, which is the spatially symmetric two-nucleon state, must have such a isospin, that its total wave function is antisymmetric. Thus, the conclusion is, that the deuteron has the isospin being equal to zero.

For a nucleon-antinucleon system ($N\overline{N}$) the isotriplet and the isosinglet states now have the form

$$\left.\begin{array}{ll} -p\overline{n}, & I_3 = 1 \\ \frac{1}{\sqrt{2}}(p\overline{p} - n\overline{n}), & I_3 = 0 \\ n\overline{p}, & I_3 = -1 \end{array}\right\} I = 1, \tag{4.76}$$

$$\frac{1}{\sqrt{2}}(p\overline{p} + n\overline{n}). \qquad I = 0 \tag{4.77}$$

Isodoublets are also formed by some strange particles, for example, by K-mesons

$$\begin{pmatrix} K^+ \\ K^0 \end{pmatrix} \qquad (s=1), \qquad \begin{pmatrix} \overline{K}^0 \\ K^- \end{pmatrix} \qquad (s=-1),$$

cascade Ξ-baryons

$$\begin{pmatrix} \Xi^0 \\ \Xi^- \end{pmatrix} \qquad (s=-2)$$

and so on. Since the masses of the strange particles, constituting the isodoublet are very close to each other

$$m_{K^\pm} = 493.577 \pm 0.016\,\text{MeV}/c^2 \qquad m_{K^0,\overline{K}^0} = 497.672 \pm 0.031\,\text{MeV}/c^2,$$

$$m_{\Xi^-} = 1321.32 \pm 0.13\,\text{MeV}/c^2 \qquad m_{\Xi^0} = 1314.9 \pm 0.6\,\text{MeV}/c^2,$$

then it is possible to speak of confirmation of conserving the isospin under strong interactions between the strange particles. However, the appearance of a new quantum number, strangeness, forces us to change Gell — Mann — Nishijima formula (4.52) to

$$Q = I_3 + \frac{B+s}{2} = I_3 + \frac{Y}{2}, \tag{4.78}$$

where Y is a particle hypercharge. The formula (4.78) can be viewed as the Nature's delicate hint at the structure of electroweak interaction gauge model, which many years later will be destined to be discovered by Weinberg, Salam and Glashow. Actually, the first term in the right-hand side of Eq. (4.78) is connected with the $SU(2)$-group, while the second one is connected with the $U(1)$-group. Both the isospin and hypercharge symmetries are not exact ones. However, the quantity in the left-hand side of Eq. (4.78), the electric charge, belongs among the exactly conserving quantities. So, if one suggests, that the Creator was using the same rules of the game to produce both strong and electroweak interactions, then electroweak interaction theory must contain the following elements: 1) the weak isospin group $SU(2)_{EW}$ and weak hypercharge group $U(1)_{EW}$; 2) the $SU(2)_{EW}$- and $U(1)_{EW}$-symmetries must be violated to the level of the $U(1)_{em}$-symmetry. In Chapter 6 we shall learn, that these very elements made the foundation of Weinberg — Salam — Glashow model.

According to Yukawa hypothesis, the nuclear forces between nucleons are caused by $\pi^\pm$-, π^0-meson exchanges (although, as we know at present, it is a very approximate statement). Consequently, pion-nucleon interaction must be isotopically invariant as well. From the reaction

$$N \to N' + \pi,$$

which is the basis for Yukawa hypothesis, follows that π-meson must have the isospin equal either to $I=1$ or $I=0$. Since there are three π-mesons having the same spins, parities and almost equal masses, then it is natural to assume, that the Nature choose the possibility with $I=1$. Thus, π-mesons can be viewed as the different charge states of one and the same particle, whose wave function is transformed as a vector in the isospace, that is, has the following form

$$\Psi_\pi = \Pi \bigotimes \Phi(x;J), \tag{4.79}$$

where

$$\Pi=\sum_{i=1}^{3}b_i\psi_i', \qquad \sum_{i=1}^{3}|b_i|^2=1, \qquad \psi_{\pi^+}'=\frac{\alpha_+}{\sqrt{2}}\begin{pmatrix}1\\ i\\ 0\end{pmatrix},$$

$$\psi_{\pi^0}'=\begin{pmatrix}0\\ 0\\ 1\end{pmatrix}, \qquad \psi_{\pi^-}'=\frac{\alpha_-}{\sqrt{2}}\begin{pmatrix}1\\ -i\\ 0\end{pmatrix},$$

and $\alpha_\pm$ are phase factors, which we choose later. Now the matrices

$$S_1^{(is)\prime}=\begin{pmatrix}0&0&0\\ 0&0&-i\\ 0&i&0\end{pmatrix}, \qquad S_2^{(is)\prime}=\begin{pmatrix}0&0&i\\ 0&0&0\\ -i&0&0\end{pmatrix},$$

$$S_3^{(is)\prime}=\begin{pmatrix}0&-i&0\\ i&0&0\\ 0&0&0\end{pmatrix}.$$

play the role of the representation generators. It is easy to check the correctness of the relations

$$S_3^{(is)\prime}\psi_{\pi^+}'=\psi_{\pi^+}', \qquad S_3^{(is)\prime}\psi_{\pi^0}'=0, \qquad S_3^{(is)\prime}\psi_{\pi^-}'=-\psi_{\pi^-}'.$$

Raising and lowering operators in this case are of the form

$$S_+^{(is)\prime}=\begin{pmatrix}0&0&-1\\ 0&0&-i\\ 1&i&0\end{pmatrix}, \qquad S_-^{(is)\prime}=\begin{pmatrix}0&0&1\\ 0&0&-i\\ -1&i&0\end{pmatrix}.$$

Their action on the isotopic parts of the wave functions of the pion triplet is governed by the relations

$$\left.\begin{array}{ll} S_+^{(is)\prime}\psi_{\pi^0}'=-\sqrt{2}(\alpha_+)^{-1}\psi_{\pi^+}', & S_+^{(is)\prime}\psi_{\pi^-}'=\sqrt{2}\alpha_-\psi_{\pi^0}', \\ S_-^{(is)\prime}\psi_{\pi^+}'=-\sqrt{2}\alpha_+\psi_{\pi^0}', & S_-^{(is)\prime}\psi_{\pi^0}'=\sqrt{2}(\alpha_-)^{-1}\psi_{\pi^-}'. \end{array}\right\} \qquad (4.80)$$

The choice of phase factors in the form of

$$\alpha_+=-1, \qquad \alpha_-=1$$

makes the right-hand parts in Eq. (4.80) positive to turn the relations (4.80) in particular case of formula (4.60). However, it is often more convenient to work with the representation, in which the matrix $S_3^{(is)\prime}$ is diagonal. The transition to such a representation is carried out by the transformation

$$\psi_\pi=O\psi_\pi',$$

where matrix O has the form

$$O=\frac{1}{\sqrt{2}}\begin{pmatrix}1&-i&0\\ 0&0&\sqrt{2}\\ 1&i&0\end{pmatrix}.$$

In the new representation the generators are defined according to

$$\left.\begin{array}{c} S_1^{(is)} = \frac{1}{\sqrt{2}}\begin{pmatrix} 0 & -1 & 0 \\ -1 & 0 & 1 \\ 0 & 1 & 0 \end{pmatrix}, \qquad S_2^{(is)} = \frac{1}{\sqrt{2}}\begin{pmatrix} 0 & i & 0 \\ -i & 0 & -i \\ 0 & i & 0 \end{pmatrix}, \\ S_3^{(is)} = \begin{pmatrix} 1 & 0 & 0 \\ 0 & 0 & 0 \\ 0 & 0 & -1 \end{pmatrix}, \end{array}\right\}, \tag{4.81}$$

and the isospin parts of the π-meson wave functions have the form

$$\psi_{\pi^+} = \begin{pmatrix} 1 \\ 0 \\ 0 \end{pmatrix}, \qquad \psi_{\pi^0} = \begin{pmatrix} 0 \\ 1 \\ 0 \end{pmatrix}, \qquad \psi_{\pi^-} = \begin{pmatrix} 0 \\ 0 \\ 1 \end{pmatrix}. \tag{4.82}$$

Besides, there are isomultiplets formed of one particle only, so-called isosinglets. The isotopic parts of the waves functions for such particles are isoscalars, that is, invariants under the isotopic transformations. The Λ-hyperon can serve as an example.

To illustrate the power of the formula (4.60) under obtaining the spin or isospin parts of wave function of composite systems, let us define the isotopic wave function of the pion-nucleon system (ΠN). According to the rule of vector composition, the total isospin of the system can take values $I = 3/2$ or $I = 1/2$. Consequently, there are six states

$$\left.\begin{array}{c} (3/2,3/2), \qquad (3/2,1/2), \qquad (3/2,-1/2), \qquad (3/2,-3/2) \\ (1/2,1/2), \qquad (1/2,-1/2). \end{array}\right\} \tag{4.83}$$

which must be expressed through the π-meson and nucleon states. Obviously, the "highest" state is

$$(3/2,3/2) = \pi^+ p, \tag{4.84}$$

since it is the only one with $I_3 = 3/2$. Now let the operator $S_-^{(is)}$ act on both sides of Eq. (4.84). Then, according to (4.60) in the left-hand side we obtain

$$S_-^{(is)}(3/2,3/2) = \sqrt{3}(3/2,1/2). \tag{4.85}$$

In so doing, the right-hand side of Eq. (4.84) takes the form

$$S_-^{(is)}(\pi^+ p) = S_{\pi-}^{(is)}(\pi^+ p) + S_{N-}^{(is)}(\pi^+ p) = (S_{\pi-}^{(is)}\pi^+ p) + (\pi^+ S_{N-}^{(is)} p) =$$
$$= \sqrt{2}(\pi^0 p) + (\pi^+ n). \tag{4.86}$$

Combining Eqs. (4.85) and (4.86) we get

$$(3/2,1/2) = \sqrt{\frac{2}{3}}(\pi^0 p) + \sqrt{\frac{1}{3}}(\pi^+ n). \tag{4.87}$$

Action of the operator $S_-^{(is)}$ on Eq. (4.87) results in

$$S_-^{(is)}(3/2,1/2) = 2(3/2,-1/2) \tag{4.88}$$

and

$$S_-^{(is)}[\sqrt{\frac{2}{3}}(\pi^0 p)+\sqrt{\frac{1}{3}}(\pi^+ n)]=\sqrt{\frac{2}{3}}[(S_{\pi-}^{(is)}\pi^0 p)+(\pi^0 S_{N-}^{(is)}p)]+$$

$$+\sqrt{\frac{1}{3}}[(S_{\pi-}^{(is)}\pi^+ n)+(\pi^+ S_{N-}^{(is)}n)]=\sqrt{\frac{2}{3}}[\sqrt{2}(\pi^- p)+(\pi^0 n)]+$$

$$+\sqrt{\frac{1}{3}}[\sqrt{2}(\pi^0 n)+0]=\frac{2}{\sqrt{3}}(\pi^- p)+2\sqrt{\frac{2}{3}}(\pi^0 n). \tag{4.89}$$

Eqs. (4.88) —(4.89) produce

$$(3/2,-1/2)=\sqrt{\frac{1}{3}}(\pi^- p)+\sqrt{\frac{2}{3}}(\pi^0 n). \tag{4.90}$$

Now build-up of the state $(3/2,-3/2)$ is possible without using the operator $S_-^{(is)}$, since the only state with $S_3^{(is)}=-3/2$ is the combination

$$(3/2,-3/2)=(\pi^- n). \tag{4.91}$$

The remaining two states with $(1/2,1/2)$ and $(1/2,-1/2)$ can be easily obtained, according to their orthogonality to the states $(3/2,1/2)$ and $(3/2,-1/2)$ respectively, that is, they have the form

$$(1/2,1/2)=\sqrt{\frac{2}{3}}(\pi^+ n)-\sqrt{\frac{1}{3}}(\pi^0 p), \tag{4.92}$$

$$(1/2,-1/2)=\sqrt{\frac{1}{3}}(\pi^0 n)-\sqrt{\frac{2}{3}}(\pi^- p). \tag{4.93}$$

Solving Eqs. (4.87), (4.90), (4.92) and (4.93) we obtain the final answer

$$\left.\begin{aligned}
&(\pi^+ p)=(3/2,3/2), \qquad (\pi^- n)=(3/2,-3/2),\\
&(\pi^0 p)=\sqrt{\tfrac{2}{3}}(3/2,1/2)-\sqrt{\tfrac{1}{3}}(1/2,1/2)\\
&(\pi^+ n)=\sqrt{\tfrac{1}{3}}(3/2,1/2)+\sqrt{\tfrac{2}{3}}(1/2,1/2),\\
&(\pi^- p)=\sqrt{\tfrac{1}{3}}(3/2,-1/2)-\sqrt{\tfrac{2}{3}}(1/2,-1/2),\\
&(\pi^0 n)=\sqrt{\tfrac{2}{3}}(3/2,-1/2)+\sqrt{\tfrac{1}{3}}(1/2,-1/2).
\end{aligned}\right\} \tag{4.94}$$

Thus, we learned the mechanism of defining the isospin (or any spinlike) part of the wave function for any compound system, that is, in this situation we have known the answer on questions "how" , "where ..., and "from". Thereafter it is the high time to become aware on the existence of tabulated values of the numeric coefficients, which appear in the theory of adding the moments (orbital, spinlike, total). By means of these coefficients the isospin function of the pion-nucleon system Φ_{I,I_3} is expressed through the isofunctions of π-meson and nucleon N as follows

$$\Phi_{I,I_3} = \sum_{I''_3,I'_3} C_{I,I_3}^{I',I'_3;I'',I''_3} \Pi_{I',I'_3} N_{I'',I''_3},$$

where $C_{I,I_3}^{I',I'_3;I'',I''_3}$ are Clebsch-Gordan coefficients, which can be found in "Review of Particle Physics" published every two years.

4.5. Resonances

From all the plurality of the elementary particles only eleven are stable, the modern experiments as say us. They are: three neutrino $(\nu_e, \nu_\mu, \nu_\tau)$, three antineutrinos $(\overline{\nu}_e, \overline{\nu}_\mu, \overline{\nu}_\tau)$, the photon, the electron, the positron, the proton and antiproton. Other particles are unstable. Unstable particles can be divided in two classes: metastable particles and resonances. Metastable particles decay due to weak or electromagnetic interactions, that is, they are tolerant to decay caused by strong interaction. Normally these particles are included to the class of stable particles. Particles-resonances decay predominantly due to strong interactions (there may be the channels caused by the electromagnetic and weak interactions, but these channels are greatly suppressed). Typical resonance life-time belongs to the interval $10^{-23} - 10^{-24}$ (this time is necessary for a relativistic particle to cover the distance of the order of the hadron size $\sim 10^{-13}$ cm). Such short life-times do not allow to register resonance traces in track detectors. Resonances are not observed in a free state, they reveal themselves while scattering in the form of quasi-stationary states of two or three strongly interacting particles. They possess such the particle characteristics as the spin, the electric charge and they can be specified by internal quantum numbers, conserved in strong interactions (the isospin, the parity, the hypercharge, etc.). Resonances, however, have no definite mass value, unlike stable particles. They are described by mass spectrum of dispersion type, the maximum of this spectrum is called the resonance mass m. The resonance mass spectrum width Γ supplies information about the probability of resonance decay and must not exceed mc^2.

The nature of an unstable particle is the most transparent when one uses the concept of the quasi-stationary state. Considering the unstable particle f we can write down its total Hamiltonian H in the form

$$H = H_f + H_d,$$

where H_d is the part of Hamiltonian, which is responsible for the decay. When neglecting H_d, the particle becomes stable and, as this takes place, its states are eigenstates of the operator H_f. In the case of the metastable particle, H_d contains the weak and electromagnetic interactions while in the case of the resonances in addition a part of the strong interactions. With H_d, taken into account, the states of the particle f are quasi-stationary.

Let us find out the connection of the quasi-stationary state decay law with the function of energy distribution or, what is one and the same, with the mass spectrum of this state. Let $\Psi(\alpha, t = 0)$ be the initial state of a system. Here α denotes the plurality of variables, according to which the system states are classified. Now we expand $\Psi(\alpha, t = 0) \equiv \Psi(\alpha, 0)$

in terms of the eigenfunctions of the energy operator $\psi(\alpha, E)$ (continuous spectrum)

$$\Psi(\alpha,0) = \int a(E)\psi(\alpha,E)dE. \tag{4.95}$$

Then the system state at the moment of time t is determined by the expression

$$\Psi(\alpha,t) = \int \exp(\frac{iEt}{\hbar})a(E)\psi(\alpha,E)dE. \tag{4.96}$$

From Eqs. (4.95) and (4.96) we obtain the expression for the probability of finding the system in the initial state after a period of time t

$$W(t) = |\int \Psi^{\dagger}(\alpha,0)\Psi(\alpha,t)d\alpha|^2 = |\int \exp(\frac{iEt}{\hbar})|a(E)|^2 dE\,|^2 =$$

$$= |\int \exp(\frac{iEt}{\hbar})w(E)dE|^2, \tag{4.97}$$

where $w(E)dE = |\alpha(E)|^2 dE$ is the function of the energy distribution for the initial $\Psi(\alpha,0)$ and, consequently, for the final $\Psi(\alpha,t))$ state. So, the decay probability of the state $\Psi(\alpha,0)$ is only defined by the function of the energy distribution in this state.

In the case of a rest unstable particle, the energy distribution $dW(E) = w(E)dE$ is nothing else, but a particle mass spectrum. The states $\psi(\alpha, E)$ therewith include the decay products states as well. It becomes obvious from (4.97) that for the quasi-stationary state to decay it is necessary and sufficient, that integral function of the energy distribution is continuous. Thus, the discrete mass spectrum is excluded.

To obtain the radioactive decay law, familiar to us from nuclear physics

$$W(t) = W_0 \exp(-\frac{\Gamma t}{\hbar}), \tag{4.98}$$

where Γ is the total decay width of the unstable particle f, it is enough to assume, that the function $a(E)$ has the form

$$a(E) = \frac{\Gamma/2}{E - E_0 + i\Gamma/2}, \tag{4.99}$$

where $E_0 = m_f c^2$. Really, with the help of the residues theory we have

$$W(t) = |\int \frac{\Gamma^2/4}{(E-E_0)^2 + \Gamma^2/4} \exp(\frac{iEt}{\hbar})dE\,|^2 =$$

$$= |\frac{2\pi i\Gamma}{4} \exp(-\frac{\Gamma t}{2\hbar}) \exp(\frac{iE_0 t}{\hbar})\,|^2 = \frac{1}{4}(\pi\Gamma)^2 \exp(-\frac{\Gamma t}{\hbar}). \tag{4.100}$$

Thus, from (4.99) follows, that the mass distribution of the rest unstable particle has a dispersion character

$$w(mc^2)dm = \frac{(\Gamma/2)^2}{(mc^2 - m_f c^2)^2 + \Gamma^2/4} dm. \tag{4.101}$$

If one neglects the interaction causing decay, the formalism of relativistic quantum theory can be generalized to unstable states. Such a approximation is necessary to describe unstable particles scattering. As a rule, in scattering processes, the particles in initial and final states are not supposed to interact, that can be realized when they are separated at infinitely great distances from each other. In this case, one does not follow to forget that in reality unstable particles have been decayed long before asymptotical division was achieved.

The basic methods of resonances detection are based on the fact that resonances have the mass spectrum of the dispersion type. The first method deals with investigating the maxima in the total scattering cross section. To make it definite, let us assume, that we deal with the reactions

$$a+b \to Y \to a+b, \tag{4.102}$$

$$d+f \to Y \to d+f, \tag{4.103}$$

$$a+b \to Y \to d+f, \tag{4.104}$$

where for Y the both decay channels are possible

$$Y \to a+b, \qquad Y \to d+f. \tag{4.105}$$

Due to fulfillment of Eq. (4.105), these reactions are going through s-channel[1]. The existence of such s-channel diagrams is a necessary condition for the resonance to be observed. Thus resonance peak, connected with Y, appears in all the above mentioned reactions. If one presumes, that the reaction (4.105) is going on only by means of the production of the resonance-particle Y in a virtual state (s-channel is the only one of the reaction), then the total cross section of the elastic df-scattering as function of the energy E near the resonance is defined by Breit-Wigner formula

$$\sigma(E) = \sigma_0 \frac{(\Gamma/2)^2}{(E-E_0)^2+\Gamma^2/4}. \tag{4.106}$$

As we see, this expression coincides with the mass distribution $w(E)$ to an accuracy of kinematic factor. Energy E_0, corresponding to the cross section maximum $\sigma(E) = \sigma_0$ being divided by c^2, defines what we agreed to consider the resonance mass. The maximum width informs us about the resonance decay probability. In this case the particles in the final state appear with retardation $\Delta t \sim \hbar/\Gamma$ as compared to scattering without the resonance production. The main drawback of this method is that it does not allow us to calculate resonance quantum numbers completely.

The next method is the phase analysis method to be more universal since with its help it is possible to define all the resonance characteristics (mass, width, spin, parity, isotropic spin, etc.). The method is based on measurements of elastic scattering differential cross section $d\sigma = 2\pi|f(\theta,E)|^2 \sin\theta d\theta$ (θ is the scattering angle). If the particles having the spin participate in scattering, then the scattering amplitude $f(\theta,E)$ is expanded as a series in the spherical functions $Y^l_m(\theta,\varphi)$. For spinless particles this expansion has the form

$$f(\theta,E) = \sum_l \sqrt{4\pi(2l+1)} f_l(E) Y^0_l(\theta) = \sum_l (2l+1) f_l(E) \mathcal{P}_l(\theta), \tag{4.107}$$

where the coefficients $f_l(E)$ are partial scattering waves with the moment l, which are determined from the experimental data as the complex functions of E. The resonance with

[1]The s-channel diagram is the Feynman diagram, where annihilation of initial particles takes place at one space-time point, while final particles production takes place at the other one.

the spin J=l reveals itself as Breit-Wigner contribution to $f_l(E)$. If the spins of two particles are equal to 0 and $1/2$ respectively, then instead of Eq. (4.107) we have

$$f(\theta,E)=\sum_l\sqrt{4\pi(2l+1)}[\sqrt{\frac{l+1}{2l+1}}f_{l+\frac{1}{2}}(E)Y^{1/2}_{l+\frac{1}{2}}(\theta,\varphi)-$$

$$-\sqrt{\frac{l}{2l+1}}f_{l-\frac{1}{2}}(E)Y^{1/2}_{l-\frac{1}{2}}(\theta,\varphi)]. \tag{4.108}$$

In the case, when there are three or more particles in the final state, then the method of maxima in mass distributions is used to search the resonances. Let us consider the inelastic scattering reaction

$$a+b\to c_1+...+Y\to c_1+...+d+f, \tag{4.109}$$

where the resonance Y decays into the stable particles d and f. During such scattering, momentum and energy of particles a and b are distributed between groups of particles $c_1+...$ and $d+f$ in the final state. If the resonance Y were the stable particle, then in its intrinsic system (rest system) its energy, and the mass$\times c^2$ as well, would have the definite value. But the resonance is unstable and is specified by the distribution function $w(E)$, which in particle rest system is directly connected with the mass spectrum of decay products of particle. In other words, the total set of states in (4.109) consists of the two-particle states $\psi_{df}(E)$ (if the decay channel $Y\to d+f$ is not the only one, and there exists another one, for example $Y\to k+m+l$, then three-particle states $\psi_{kml}(E)$ will enter to Eq. (4.109) and so on). In distribution of the invariant mass square of the particles d and f

$$M^2_{df}c^4=(E_d+E_f)^2-(\mathbf{p}_d+\mathbf{p}_f)^2c^2, \tag{4.110}$$

the brightly expressed maximum will be observed. Thus studying the distribution of the masses in complexes of particles belonging to the final states it is possible to obtain directly the distributions on the masses of unstable particles, which are the resonance states in such complexes. Of course, it does not necessarily mean, that every peak in the masses distributions of the particles being the reaction products can be identified with the resonance, because kinematic peaks also occur, which are inherent in the given reaction only [1]. The difference between the real resonance and the ghost peak is that the energy, corresponding to appearance of the real resonance in different experiments, is always the same, while the energy, connected with the ghost peak, is changed from one experiment to another. The distribution of invariant mass square in the system $\Sigma^+\pi^-$, arising in the reaction

$$\pi^-+p\to\Sigma^++K^0+\pi^- \tag{4.111}$$

at momenta of the incident π^--meson from 2.2 to GeV/ is shown in Fig. 11.

Three peaks are distinguished against the background of uncorrelated events. All of them turn out to be the real resonances.

First resonances (Δ-resonances) were discovered in 1952 by E. Fermi under scattering of π-mesons by protons

$$\pi+p\to\Delta\to\pi+p. \tag{4.112}$$

[1] Such peaks are called ghost ones.

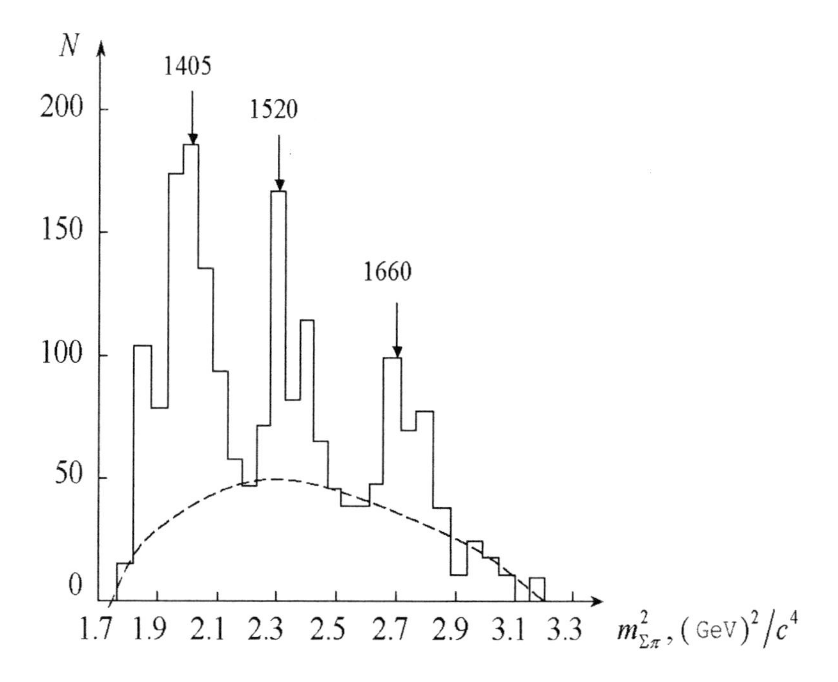

Figure 11. The distribution of the invariant mass square in the $\Sigma^+\pi^-$ system.

The total cross section of the pion-nucleon scattering versus the center of mass energy is displayed in Fig. 12.

The solid line corresponds to σ_{π^+p}, while the dashed one describes σ_{π^-p}. As it follows from Fig. 12, at the π-mesons kinetic energy $T = 195$ MeV the resonance peak appears in both π^+p-scattering (Δ^{++}- resonance) and π^-p-scattering (Δ^0-resonance). From Gell — Mann — Nishidjima formula, it is evident that when Δ^{++}- and Δ^0-resonance enter to one and the same isotopic multiplet they have $S^{(is)} = 3/2$.

Further, by the example of reaction (4.112), we show how to define such resonance characteristics as the mass and isotopic spin from the experimental data. In the laboratory reference system the proton rests, and consequently, three-momentum and energy of Δ-resonance are defined by

$$\mathbf{p}_\Delta = \mathbf{p}_\pi = \frac{1}{c}\sqrt{(T^r_\pi + m_\pi c^2)^2 - m^2_\pi c^4}, \tag{4.113}$$

$$E_\Delta = T^r_\pi + m_\pi c^2 + m_p c^2. \tag{4.114}$$

From Eqs. (4.113) and (4.114) follows

$$m_\Delta = \frac{1}{c^2}\sqrt{E^2_\Delta - \mathbf{p}^2_\Delta c^2} = 1236\,\text{MeV}/c^2$$

(contemporary data lessen down the Δ mass value to 1232 MeV$/c^2$).

To find the isotopic spin value we use isotopic invariance of the cross sections of the reactions (4.112). Since the isotopic spins of nucleon and π-meson are equal to 1/2 and 1 respectively then their sum could be equal either to 3/2 or 1/2. We use the pion-nucleon states

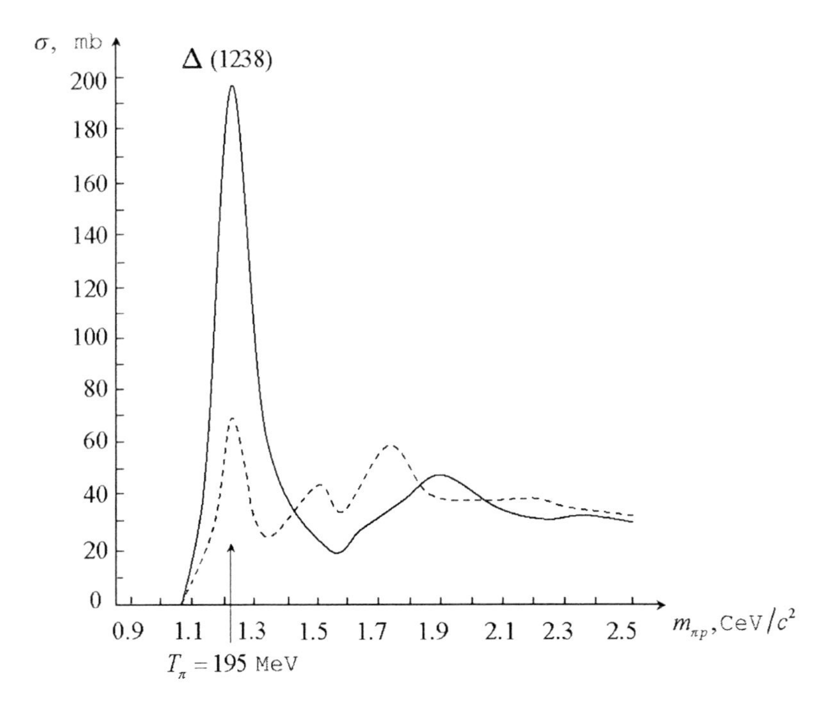

Figure 12. The resonance peak (Δ^0-resonance) in $\pi^- p$ scattering.

expansion in the isotopic spin eigenstates (4.94). Then, from the isotopic spin conservation law it follows that only the following scattering amplitudes exist

$$(\Psi^f_{\pi^+p}, \Psi^i_{\pi^+p}) = (\Psi^f_{3/2,3/2}, \Psi^i_{3/2,3/2}) = A_{3/2}, \qquad (4.115)$$

$$(\Psi^f_{\pi^-p}, \Psi^i_{\pi^-p}) = \frac{1}{3}A_{3/2} + \frac{2}{3}A_{1/2}, \qquad (4.116)$$

$$(\Psi^f_{\pi^0n}, \Psi^i_{\pi^-p}) = \frac{\sqrt{2}}{3}A_{3/2} - \frac{\sqrt{2}}{3}A_{1/2}. \qquad (4.117)$$

The total scattering cross sections in the region, where the multiple π-mesons production is insignificant, are given by the relations

$$\sigma_{\pi^+} = \rho \mid A_{3/2} \mid^2,$$

$$\sigma_{\pi^-} = \sigma_{\pi^-p\to\pi^-p} + \sigma_{\pi^-p\to\pi^0n} = \rho\left(\frac{1}{3} \mid A_{3/2} \mid^2 + \frac{2}{3} \mid A_{1/2} \mid^2\right), \qquad (4.118)$$

where ρ is a kinematic factor (which is constant for all the three processes, if we neglect a mass difference in isomultiplets). If we assume, that the isotopic spin of the Δ-resonance is equal to $3/2$, then the second term in Eq. (4.118) at $E = T^r_\pi$ goes to zero and we arrive at the relation

$$(\sigma_{\pi^+}/\sigma_{\pi^-}) \mid_{E=T^r_\pi} = 3,$$

which is in excellent accord with the experiment.

4.6. Unitary Multiplets

By using isospin symmetry we can reduce a number of independent elementary particles. For example, three π-meson are simply three states of one and the same particle, etc. However, by the 1960s so many isotopic multiplets had been discovered, that it became highly necessary to search for a higher symmetry which could unify particles into more densely populated families. Since such a symmetry has been thoroughly escaping from observations by now, it was obvious, that somehow and somewhere this symmetry has passed a violation stage, i.e. to date it is approximate. Once again, as a violation criteria of this symmetry, we can take mass difference of particles, entering to new multiplets, which we are going to call unitary multiplets. Unitary multiplets must contain isomultiplets, that is, unitary symmetry group contains in itself a subgroup of isotopic transformations. Thus among Casimir operators of the unitary symmetry group, the isotopic spin square is present. What other operators can apply for the role of invariants in the new group? If we agree once again to denote particle states with dots in some abstract space, then the dimension of a new space of states is defined by the number of invariants of the unitary symmetry group. From the physical point of view it is reasonable to demand, that "coordinates" of this space must be dependent (in the same way as space and time are interconnected in the STR). The experiments, which had been carried out by that time, demonstrated, that the strangeness s and the isospin I are not independent. Let us explain what by this is meant. In strong interactions s and I are the conserving quantities, and as a result, the following relation takes place

$$\Delta s = \Delta I_3 = 0. \tag{4.119}$$

Behavior of the isospin and the strangeness is not so faultless as far as strong interaction is concerned. Since the isospin and the strangeness are not defined for leptons, then it makes sense to analyze only semilepton and nonlepton weak interaction. In the former case the final state is formed of both leptons and hadrons, while in the latter case leptons are absent in the final state.

Below we give some examples of typical semilepton decays and indicate the changes of the strangeness and the isotopic spin projection

$$n \to p + e^- + \overline{\nu}_e, \qquad \Delta s = 0, \qquad \Delta I_3 = 1 \tag{4.120}$$

$$\pi^+ \to \pi^0 + e^+ + \nu_e, \qquad \Delta s = 0, \qquad \Delta I_3 = -1 \tag{4.121}$$

$$\Lambda \to p + e^- + \overline{\nu}_e, \qquad \Delta s = 1, \qquad \Delta I_3 = 1/2, \tag{4.122}$$

$$K^+ \to \pi^0 + \mu^+ + \nu_\mu, \qquad \Delta s = -1, \qquad \Delta I_3 = -1/2. \tag{4.123}$$

For nonlepton decays the selection rules according to s and I_3 can be illustrated by the example of the reactions

$$\Lambda \to p + \pi^-, \qquad \Delta s = 1, \qquad \Delta I_3 = -1/2, \tag{4.124}$$

$$K^+ \to \pi^+ + \pi^0, \qquad \Delta s = -1, \qquad \Delta I_3 = 1/2. \tag{4.125}$$

In electromagnetic interaction the strangeness remains a good quantum number. Two typical decays

$$\pi^0 \to 2\gamma, \qquad \eta^0 \to 2\gamma,$$

demonstrate this circumstance

$$\Delta s = 0, \qquad \Delta I_3 = 0. \tag{4.126}$$

From the aforesaid it follows that in all the existing processes for a closed system a change in the strangeness entails a strictly defined change of the isotopic spin projection

$$\left.\begin{array}{lcl} |\Delta s| = 0 & \longrightarrow & |\Delta I_3| = 1, 0, \\ |\Delta s| = 1 & \longrightarrow & |\Delta I_3| = 1/2, \end{array}\right\}. \tag{4.127}$$

Thus, the choice of two quantum numbers, the isospin projection and the strangeness, for dependent coordinates of the unitary spin space is quite well grounded.

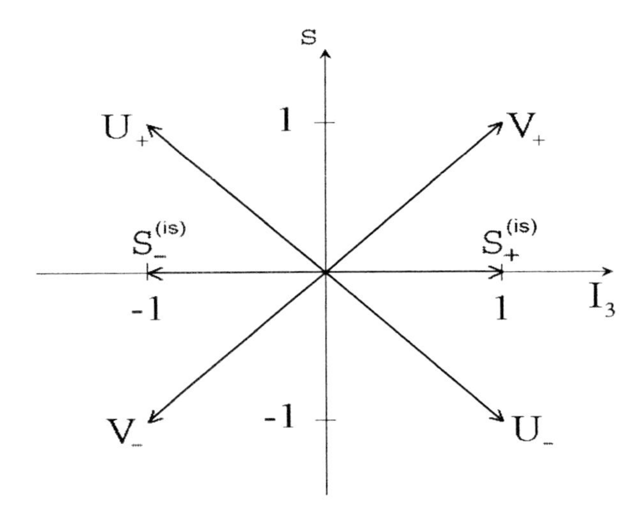

Figure 13. The unitary spin plane.

In Fig. 13 we plotted the plane with the coordinates (s, I_3) and on it we shall specify the states of unitary multiplets. From the mathematical point of view, the transitions between states must be carried out by means of operators, which we agree to denote by vectors. The raising and lowering isospin operators $S_+^{(is)}$ and $S_-^{(is)}$ allow to cross from one state to another inside of one and the same isomultiplet. Since acting $S_\pm^{(is)}$ on a state changes the values of I_3 by ± 1 and in the process $\Delta s = 0$, then the vectors with module 1, parallel to axis I_3, are correlated with these operators. It is obvious, that in order to ensure the transitions between isomultiplets, i.e. the transitions with a strangeness change, the similar raising and lowering operators must exist. It is natural that these operators must be spinlike, i. e. their algebra is determined by the same commutation relations as for the ordinary spin algebra. Raising operator S_+^U corresponds to the selection rules

$$\Delta s = 1, \qquad \Delta I_3 = -1/2,$$

while lowering operator S_-^U describes the situation with

$$\Delta s = -1, \qquad \Delta I_3 = 1/2.$$

The operators S_+^U, S_-^U and S_3^U are the group generators of the so-called U-spin, and for them the commutation relations are valid

$$[S_+^U, S_-^U] = 2S_3^U. \tag{4.128}$$

Obviously, without introduction of one more type of generators it is not possible to obtain on the plane (s, I_3) closed and symmetric (with respect to axis s) figures, which we are going to associate with unitary multiplets. The transitions with the selection rules

$$\Delta s = 1, \qquad \Delta I_3 = 1/2,$$

$$\Delta s = -1, \qquad \Delta I_3 = -1/2, \tag{4.129}$$

are carried out by the operators S_+^V and S_-^V. These operators alongside with S_3^V constitute V-spin group, and for them the ordinary commutation relations of the moments are fulfilled

$$[S_+^V, S_-^V] = 2S_3^V. \tag{4.130}$$

Notice, that in the chosen scale the modules of vectors, which we correlate with the operators $S_\pm^U$ and $S_\pm^V$, are equal to $\sqrt{5}/2$, and tangents of angles, which the operators S_+^U and S_+^V form with the strangeness axis, are equal to 1/2.

Nine generators $S_{1,2,3}^{(is)}$, $S_{1,2,3}^V$ and $S_{1,2,3}^U$ form a closed system and generate a group of the second rank $SU(3)$, the unitary spin group, which was proposed by Y. Neumann and, regardless of him, M. Gell-Mann to classify hadrons. In this scheme, just as in the Mendeleev periodic table, the objects are placed in the order of increasing their masses and a separation into families, unifying elementary particles with similar properties, is used. The proposed theory is very often called the octal way, since, according to its statements, the majority of particles is grouped in 8-plets (octets).

Taking into consideration hadron classification in isotopic multiplets, practically all hadrons which were reliably established by the 1960s, could be unified in four families. These families are characterized by the baryon charge, the spin and the parity (J^P). Let us take the following notations. After a isomultiplet symbol the approximated value of average mass in MeV$/c^2$ will be given in brackets, while the hypercharge value will be indicated under every multiplet. Then these families are as follows:

8 baryons with $1/2^+$ and $B = 1$:

$$\begin{array}{cccc} N(939) & \Lambda(1115) & \Sigma(1193) & \Xi(1317) \\ p, n & \Lambda^0 & \Sigma^+, \Sigma^0, \Sigma^- & \Xi^0, \Xi^- \\ +1 & 0 & 0 & -1 \end{array}, \tag{4.131}$$

8 pseudoscalar mesons and meson resonances with 0^-:

$$\begin{array}{cccc} \pi(137) & K(495) & \overline{K}(495) & \eta(548) \\ \pi^+, \pi^0, \pi^- & K^+, K^0 & \overline{K}^-, \overline{K}^0 & \eta \\ 0 & +1 & -1 & 0 \end{array}, \tag{4.132}$$

8 vector resonances with 1^-:

$$\begin{array}{cccc} \rho(770) & K^*(892) & \overline{K}^*(892) & \omega(782) \\ \rho^+,\rho^0,\rho^- & K^{*+},K^{*0} & \overline{K}^{*-},\overline{K}^{*0} & \omega^0 \\ 0 & +1 & -1 & 0 \end{array}, \qquad (4.133)$$

9 baryon resonances with $\frac{3}{2}^+$ and $B=1$:

$$\begin{array}{ccc} \Delta(1232) & \Sigma^*(1385) & \Xi^*(1530) \\ \Delta^{++},\Delta^+,\Delta^0,\Delta^- & \Sigma^{*+},\Sigma^{*0},\Sigma^{*-} & \Xi^{*0},\Xi^{*-} \\ 1 & 0 & -1 \end{array}. \qquad (4.134)$$

In subsection 4.6 we shall find explicit form of both the operators of the U-, V-spin and corresponding to them the operators of Y_U-, Y_V-hypercharge. We shall be also convinced that the commutation relations (4.128) and (4.130) are correct. Forestalling events we want to point out, that all these operators are in fact definite combinations of the isospin and hypercharge operators. However, at the moment we are not interested in the explicit form of the operators $S^V_\pm$, $S^U_\pm$. Our task is first of all, to be sure that they are able together with operators $S^{(is)}_\pm$ to place all the known by that time hadrons in the superfamilies (4.131) — (4.134).

Let us start from a supermultiplet including the nucleons. For the sake of convenience on the abscissa we shall plot the hypercharge values and not the strangeness values. In the plane (Y, I_3) which we are going to call the unitary spin plane, the point with coordinates $Y=1$, $I_3=1/2$ corresponds to the proton (Fig. 14).

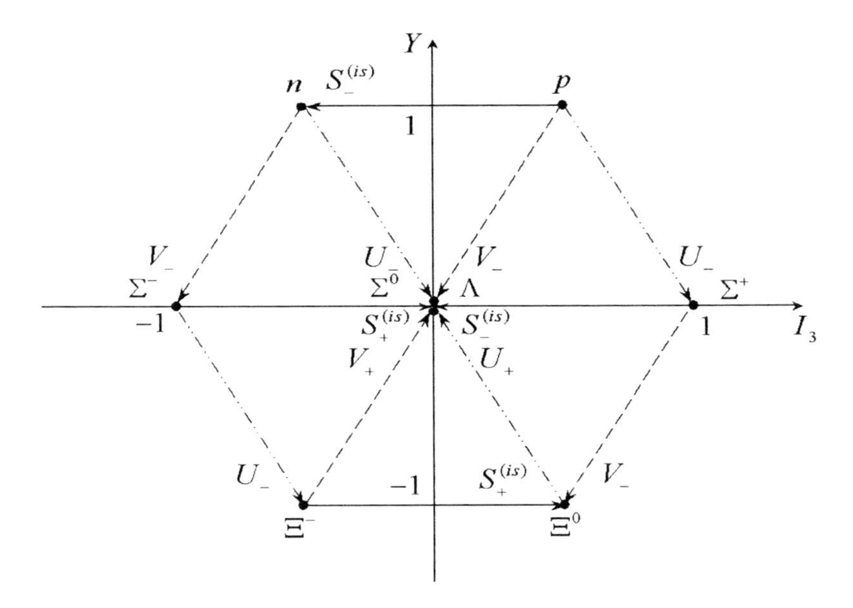

Figure 14. The baryon octet.

Let us assume, that p occupies the highest state, that is,

$$S^{(is)}_+ p = S^V_+ p = S^U_+ p = 0. \qquad (4.135)$$

To obtain other particles, lowering operators must act on proton state. It gives

$$S_-^{(is)}p = n, \qquad (Y = 1, I_3 = -1/2), \tag{4.136}$$

$$S_-^V p = \Lambda, \Sigma^0 \qquad (Y = 0, I_3 = 0), \tag{4.137}$$

$$S_-^U p = \Sigma^+ \qquad (Y = 0, I_3 = 1). \tag{4.138}$$

The action of S_-^U on n brings forth the same result as the action of S_-^V on p, that is, it yields the state with $Y = 0 = I_3 = 0$ which we have already identified with Λ- and Σ^0-hyperons. A new state with the quantum numbers $Y = 0$ and $I_3 = -1$, the Σ^--hyperon, can be obtained, if operator S_-^V acts on n. Important to remember, that our final goal is to obtain a closed symmetrical figure. Consequently, the lowest state must have the quantum numbers with opposite sings compared to the highest state, that is, it has $Y = -1$, $I_3 = -1/2$. Ξ^--hyperon is such a state and transition to it from the Σ^--state is carried out by the S_-^U-operator. Then, by means of the $S_+^{(is)}$-operator we arrive at the last state with $Y = -1$, $I_3 = 1/2$, which is Ξ^0-hyperon. As one can see, particle masses in baryon octet are much more different from each other, then those in isomultiplets. Thus, for example,

$$\frac{m(\Xi) - m(N)}{m(\Xi) + m(N)} \approx 17\%.$$

In other words, the unitary symmetry has been violated stronger than the isotopic one.

In the same way by means of operators $S_\pm^{(is)}$, $S_\pm^U$ and $S_\pm^V$, 0^+-mesons of (4.132) and 1^--meson resonances of (4.133) can be grouped into octets. There are also unitary singlets, for example, $\eta'(957)$-meson forms 0^--singlet. Unlike mesons (where particles and antiparticles enter to one and the same families), antibaryons form individual families which are the same as baryons ones (see, for example, (4.134)).

So, we introduced a new quantum number, the unitary spin, to be a generalization of the isospin, and involve both the isospin and the strangeness. Our world is made in such a way, that strong interactions are approximately invariant under rotations in the unitary spin space. For this reason hadrons are grouped into unitary multiplets. This is an axiom of the unitary symmetry theory. All the particles of such superfamilies can be viewed simply as a set of state of one and the same particle, degenerated in the electric charge and hypercharge. According to concepts of the $SU(3)$-symmetry, baryons with spin $1/2$ must be unified in unitary octet, while baryons with spin $3/2$ must be grouped into unitary decuplet.

Nine baryon resonances with $3/2^+$ (4.134) might be placed in a decuplet with one vacant lower place (Fig. 15).

From Fig. 15 follows, that the masses difference between neighboring isomultiplets is constant and it is approximately equal to 146 MeV$/c^2$. Thus, it is possible to predict the mass, the strangeness and the electric charge of a missing member of the baryon decuplet $3/2^+$, which we are going to denote as Ω^-

$$m_\Omega = 1676\,\text{MeV}/c^2, \qquad s = -3.$$

The strangeness conservation law forbids the decay of this particle through strong interaction. Actually, a decay channel with the least mass of the strange particles

$$\Omega^- \to \Xi^0 + K^- \tag{4.139}$$

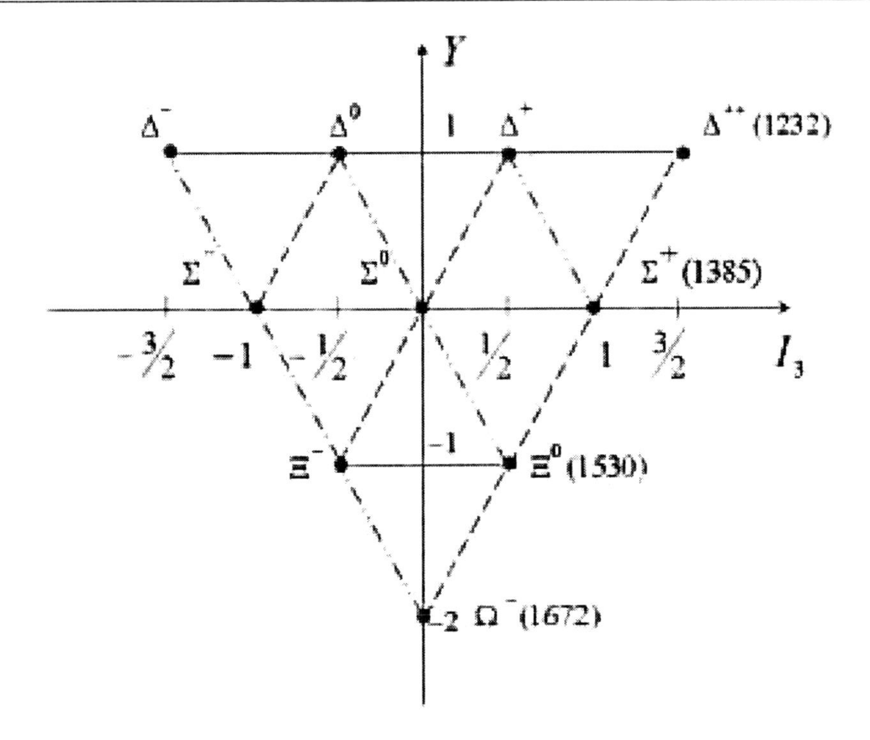

Figure 15. The decuplet of the baryon resonances.

turned out to be closed due to the energy arguments ($m_\Omega < m_{\Xi^0} + m_{K^-}$). Weak interaction with the following decay channels

$$\Omega \to \begin{cases} \Lambda + K^- \\ \Xi^- + \pi^0 \\ \Xi^0 + \pi^- \\ \Xi^- + \pi^+ + \pi^- \\ \Xi^0 + e^- + \overline{\nu}_e \end{cases} \tag{4.140}$$

can give the only chance for the Ω^--hyperon to maintain its status of an unstable particle. Then the calculations of the Ω^--hyperon life-time gave the value of the order of 10^{-10} s. On microworld scale it was a long-lived particle and its track had to be seen in a bubble chamber which has already existed by that time. In 1964 at Brookhaven accelerator the particle was discovered, which characteristics exactly coincided with the predicted ones. It was one of those miraculous enlightenments, which are so rare in the human mind history. The discovery of the Ω^--hyperon was similar to that of the planet Neptune by Leverrie — discovery made on the pen edge. It was regarded to be a brilliant proof of hadrons classification according to the $SU(3)$-symmetry.

4.7. $SU(3)$-Symmetry

A certain isotopic multiplet is described by a wave isotopic function, being one of an irreducible representations of the $SU(2)$-group. In the same way a certain unitary multiplet can be described by a multi-component wave unitary function, which is an irreducible representation of the $SU(3)$-group. Let us study in details the $SU(3)$-group and in particular its Lie-algebra and its irreducible representations.

The number of linearly independent vectors defined in the linear space, to act the transformations matrix, is called the representation dimension. In the case of the internal symmetry the number of particles in the corresponding multiplet is the representation dimension. The most simple representations, from which all the rest of group representations can be built with the help of the multiplication, are named by the fundamental ones. For the $SU(n)$-group those are n-component spinors. Thus the triplet is the fundamental representation of the $SU(3)$-group. To be exact, there are two of these representations: covariant and contravariant triplets, but we discuss it later. Let us consider the transformations, which leave three-component $SU(3)$-spinor invariant

$$\chi^k \longrightarrow \chi'^k = U_{kl}\chi^l, \tag{4.141}$$

where U are arbitrary and unimodular 3×3-matrices. Canonic representation of matrix U has the form

$$U = \exp\left(\frac{i}{2}\alpha_a\lambda_a\right), \tag{4.142}$$

where $a = 1, 2, ..8$ and λ_a are 3×3-matrices satisfying the relations

$$\text{Sp}\,(\lambda_a\lambda_b) = 2\delta_{ab}, \qquad \text{Sp}\,\lambda_a = 0 \qquad \lambda_a^\dagger = \lambda_a. \tag{4.143}$$

We remind that the number of independent parameters and the number of generators of the $SU(n)$-group is equal to $n^2 - 1$. Here the matrices λ_a play the same role as Pauli matrices do in case of the $SU(2)$-symmetry, that is, $\lambda_a/2$ are the generators of fundamental representation of the unitary spin. The standard writing of these matrices, introduced by Gell-Mann is as follows

$$\lambda_1 = \begin{pmatrix} 0 & 1 & 0 \\ 1 & 0 & 0 \\ 0 & 0 & 0 \end{pmatrix}, \qquad \lambda_2 = \begin{pmatrix} 0 & -i & 0 \\ i & 0 & 0 \\ 0 & 0 & 0 \end{pmatrix}, \qquad \lambda_3 = \begin{pmatrix} 1 & 0 & 0 \\ 0 & -1 & 0 \\ 0 & 0 & 0 \end{pmatrix},$$

$$\lambda_4 = \begin{pmatrix} 0 & 0 & 1 \\ 0 & 0 & 0 \\ 1 & 0 & 0 \end{pmatrix}, \qquad \lambda_5 = \begin{pmatrix} 0 & 0 & -i \\ 0 & 0 & 0 \\ i & 0 & 0 \end{pmatrix}, \qquad \lambda_6 = \begin{pmatrix} 0 & 0 & 0 \\ 0 & 0 & 1 \\ 0 & 1 & 0 \end{pmatrix},$$

$$\lambda_7 = \begin{pmatrix} 0 & 0 & 0 \\ 0 & 0 & -i \\ 0 & i & 0 \end{pmatrix}, \qquad \lambda_8 = \frac{1}{\sqrt{3}}\begin{pmatrix} 1 & 0 & 0 \\ 0 & 1 & 0 \\ 0 & 0 & -2 \end{pmatrix}. \tag{4.144}$$

The choice of the generators in the form (4.144) is convenient, because the first three matrices $\lambda_{1,2,3}$ are Pauli matrices (they form Lie algebra of the $SU(2)$-group). What this means is the $SU(3)$-group contains the isotopic spin group as the subgroup. Matrices λ_3 and λ_8 commute with each other, that is, $SU(3)$ is really the group of the second rank. The permutation relations, to characterize the group and to be satisfied by the matrices λ_a, resemble in form those for the matrices σ_l

$$[\lambda_a, \lambda_b] = 2if_{abc}\lambda_c. \tag{4.145}$$

Structural constants f_{abc} are real and antisymmetric with respect to all the indices. They can be determined by means of the relations

$$f_{abc} = \frac{1}{4i}\mathrm{Sp}([\lambda_a, \lambda_b]\lambda_c). \tag{4.146}$$

The components of f_{abc} being different from zero have the following values

$$\left.\begin{array}{c} f_{123} = 1, \qquad f_{147} = f_{246} = f_{345} = f_{257} = -f_{156} = -f_{367} = \frac{1}{2}, \\ f_{458} = f_{678} = \frac{\sqrt{3}}{2}, \end{array}\right\}. \tag{4.147}$$

For the matrices (4.144) antipermutation relations exist as well

$$\{\lambda_a, \lambda_b\} = 2d_{abc}\lambda_c + \frac{4}{3}\delta_{ab}, \tag{4.148}$$

where constants d_{abc} are completely symmetric with respect to indices transpositions. By means of relations

$$d_{abc} = \frac{1}{4}\mathrm{Sp}(\{\lambda_a, \lambda_b\}\lambda_c), \tag{4.149}$$

the non-zero components of d_{abc} can be obtained

$$\left.\begin{array}{l} d_{118} = d_{228} = d_{338} = -d_{888} = \frac{1}{\sqrt{3}}, \qquad d_{448} = d_{558} = \\ = d_{668} = d_{778} = -\frac{1}{2\sqrt{3}}, \qquad d_{146} = d_{157} = -d_{247} = \\ = d_{256} = d_{344} = d_{355} = -d_{366} = -d_{377} = \frac{1}{2} \end{array}\right\}. \tag{4.150}$$

The matrices λ_a are usually called the unitary spin operators. As in case of the normal spin, the sum of the matrices squares is proportional to an identity matrix

$$\lambda_a\lambda_a = \frac{16}{3}\begin{pmatrix} 1 & 0 & 0 \\ 0 & 1 & 0 \\ 0 & 0 & 1 \end{pmatrix} \tag{4.151}$$

and that gives us the right to use the above mentioned term. The λ_a-matrices can be also viewed as components of the eight-dimensional $SU(3)$-vector. The aforesaid is also valid for generators of any representation of the $SU(3)$- group. So, from generators of the unitary spin $S_a^{(un)}$ the operator of unitary spin square can be obtained

$$\mathbf{S}^{(un)2} = S_a^{(un)}S_a^{(un)}, \tag{4.152}$$

whose eigenvalues characterize the given representation. However, it is not the only Casimir operator in the $SU(3)$-group. If we introduce the $SU(3)$-vector

$$D_a = \frac{2}{3}d_{abc}S_b^{(un)}S_c^{(un)} \tag{4.153}$$

it is easy to see that the quantity

$$F = S_a^{(un)}D_a, \tag{4.154}$$

is the second Casimir operator. Although the D_a-vector is made of the $S_a^{(un)}$-generators, however, D_a and $S_a^{(un)}$ are linearly independent. It follows from the fact that after multiplying on $S_a^{(un)}$ both of them form two different Casimir operators of the $SU(3)$-group. These eight-dimensional $SU(3)$ vectors satisfy typical for moments theory the commutation relations

$$[S_a^{(un)}, S_b^{(un)}] = i f_{abc} S_c^{(un)}, \tag{4.155}$$

$$[D_a, S_b^{(un)}] = i f_{abc} D_c. \tag{4.156}$$

Notice, that in case of the $SU(2)$-group the well-known relation

$$[p_k, M_l] = i\varepsilon_{klm} p_m, \tag{4.157}$$

characterizing vector property of $\mathbf{p}$, is analog of (4.156). It is convenient in some cases to use for $S_a^{(un)}$ and D_b their representations in the form of 8×8-matrices

$$(S_a^{(un)})_{bc} = i f_{abc}, \tag{4.158}$$

$$(D_a)_{bc} = d_{abc}. \tag{4.159}$$

Equality (4.158) is proved by substitution of λ_a, λ_b, λ_c in Jacobi identity

$$[[A,B],C] + [[B,C],A] + [[C,A],B] = 0, \tag{4.160}$$

by a consequent multiplication on every λ and by calculating the trace. Further the validity of choice of matrix representation for D_a- operator can be checked by fulfillment of (4.156).

Connection of $S_a^{(un)}$ both with the isospin operators and with the U-, V-spin operators is given by the following expressions

$$\left.\begin{array}{lll} S_\pm^{(is)} = S_1^{(un)} \pm i S_2^{(un)}, & S_\pm^U = S_6^{(un)} \pm i S_7^{(un)}, & \\ S_\pm^V = S_4^{(un)} \pm i S_5^{(un)}, & S_3^{(is)} = S_3^{(un)}, & Y = \frac{2}{\sqrt{3}} S_8^{(un)}. \end{array}\right\} \tag{4.161}$$

Since the hypercharge commutes with the third projection of the isospin, then the hypercharge operator can differ from $S_8^{(un)}$ only in constant factor. If in (4.161) we set this factor equal to $2/\sqrt{3}$, then we arrive at the correct expressions for hadrons hypercharges. Taking into account (4.161) it is easy to check the validity of the following commutation relations

$$[S_+^U, S_-^U] = \frac{3}{2} Y - S_3^{(is)} \equiv 2 S_3^U, \tag{4.162}$$

$$[S_+^V, S_-^V] = \frac{3}{2} Y + S_3^{(is)} \equiv 2 S_3^V. \tag{4.163}$$

The relations (4.162) and (4.163) prove, that $SU(3)$-group besides the isospin subgroup $SU(2)$ really contains two more $SU(2)$-subgroups, the subgroups of the U- and V-spin. Since the charge operator

$$Q = S_3^{(un)} + \frac{1}{\sqrt{3}} S_8^{(un)}, \tag{4.164}$$

satisfies the commutation relations

$$[Q, S^U_{\pm}] = [Q, S^U_3] = 0, \qquad (4.165)$$

then the operator $Y_U = \text{const} \times Q$ plays the role of the hypercharge operator for the U-spin. It appears, that to obtain the correct values of hadron quantum numbers, the constant in definition of the U-hypercharge must be set to -1. Similar considerations concerning the operator Y_V give the following expression

$$Y_V = S_3^{(is)} - \frac{1}{2}Y. \qquad (4.166)$$

The next step is to find irreducible representations of the unitary spin group. The unitary scalar is the most simple irreducible representation. It describes particles forming unitary singlets. Thereupon the fundamental representation of the $SU(3)$-group (the unitary triplet), about which we told above, follows. Let us add some mathematical details to the aforesaid.

In space E_1, transformed with respect to the fundamental representation, we introduce the orthonormalized basis vectors $\mathbf{e}_k$ ($k = 1, 2, 3$), which can be chosen in the form

$$\mathbf{e}_1 = \begin{pmatrix} 1 \\ 0 \\ 0 \end{pmatrix}, \qquad \mathbf{e}_2 = \begin{pmatrix} 0 \\ 1 \\ 0 \end{pmatrix}, \qquad \mathbf{e}_3 = \begin{pmatrix} 0 \\ 0 \\ 1 \end{pmatrix}. \qquad (4.167)$$

For objects, defined in unitary spin spaces we are going to use the term "vector". Concrete nature of a vector is decoded in writing down its components. Thus, vectors in the space E_1 are spinors of the first rank and arbitrary spinor Ψ is defined by the formula

$$\Psi = \Psi^k \mathbf{e}_k. \qquad (4.168)$$

Under the $SU(3)$-transformations the components of the spinor Ψ^k are transformed according to the law

$$\Psi'^k = U_{kl}\Psi^l. \qquad (4.169)$$

Just as in case of other groups, we continue to call spinors transforming with respect to the fundamental representation contravariant spinors of the first rank. Thus in the space $\mathcal{E}_1$ the such spinors are defined.

Using the obvious form of the fundamental representation generators, it is easy to check the validity of the following relations

$$\left.\begin{array}{ll} S_3^{(is)}\mathbf{e}_1 = \frac{1}{2}\mathbf{e}_1, & Y\mathbf{e}_1 = \frac{1}{3}\mathbf{e}_1, \\ S_3^{(is)}\mathbf{e}_2 = -\frac{1}{2}\mathbf{e}_2, & Y\mathbf{e}_2 = \frac{1}{3}\mathbf{e}_2, \\ S_3^{(is)}\mathbf{e}_3 = 0, & Y\mathbf{e}_3 = -\frac{2}{3}\mathbf{e}_3, \end{array}\right\}. \qquad (4.170)$$

It is convenient to display the basis vectors of the space $\mathcal{E}_1$ on a unitary spin plane, what is done in Fig. 16.

Obviously, according to our agreements, the unitary contravariant spinor χ with the components $\chi^1 = \chi^2 = \chi^3 = 1$ describes the unitary triplet of particles, in which the first two elements form the isodoublet ($I = 1/2$) and the third one forms the isosinglet (I=0). Let us agree to call it as the fundamental triplet and display it as the triangle in the plane (I_3, Y).

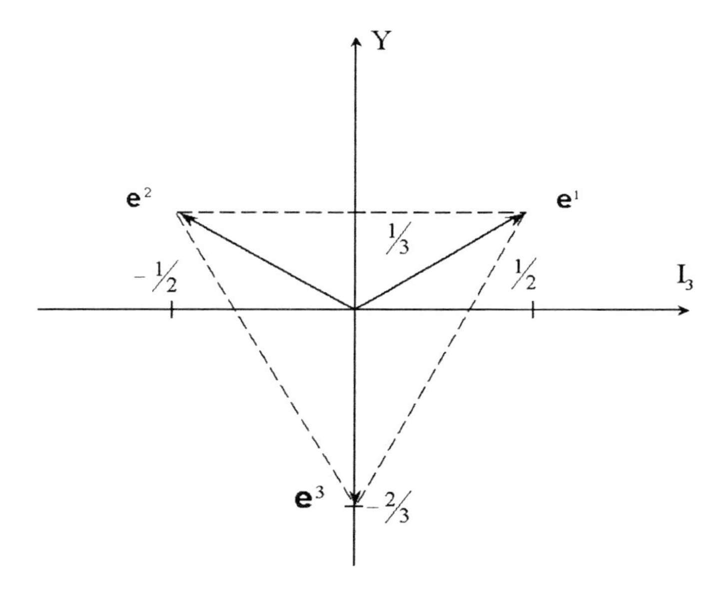

Figure 16. The basic vectors in the E_1 space.

The space $E_{\overline{1}}$, transforming with respect to the representation which is conjugate (contragradient) relative to the fundamental representation of the $SU(3)$-group, is also irreducible. In this space, just as in the previous case, arbitrary spinor Φ can be presented in the form

$$\Phi = \Phi_k \mathbf{e}^k, \tag{4.171}$$

where the Φ_k-spinor components are transformed according to the following rules

$$\Phi'_k = U^*_{kl}\Phi_l = \Phi_l U^\dagger_{lk}. \tag{4.172}$$

The spinors, defined in the space $E_{\overline{1}}$ shall be called the covariant spinors of the first rank. Thus under transition from contravariant to covariant spinors, in the transformation matrix U the change takes place

$$\lambda_a \to -\lambda^*_a. \tag{4.173}$$

Inserting (4.173) in the expression for the operators $S_3^{(is)}$, Y and choosing the basis vectors of the space $E_{\overline{1}}$ also in the form (4.167), we obtain the following eigenvalues equations

$$\left.\begin{array}{ll} S_3^{(is)}\mathbf{e}^1 = -\frac{1}{2}\mathbf{e}^1, & Y\mathbf{e}^1 = -\frac{1}{3}\mathbf{e}^1, \\ S_3^{(is)}\mathbf{e}^2 = \frac{1}{2}\mathbf{e}^2, & Y\mathbf{e}^2 = -\frac{1}{3}\mathbf{e}^2, \\ S_3^{(is)}\mathbf{e}^3 = 0, & Y\mathbf{e}^3 = \frac{2}{3}\mathbf{e}^3. \end{array}\right\} \tag{4.174}$$

The basis vectors of space $E_{\overline{1}}$ on the unitary spin plane are displayed in Fig. 17.

The components of the Hermitian conjugate covariant spinor $\Psi^\dagger_k$ are transformed as the components of the contravariant spinor Ψ^k. Since Hermitian conjugate wave function is connected with antiparticles, then the covariant spinor η with the components $\eta_1 = \eta_2 = \eta_3 = 1$ describes the triplet of antiparticles. In what follows we are going to call it as the antitriplet and depict it as the triangle on the unitary spin plane (Fig. 17).

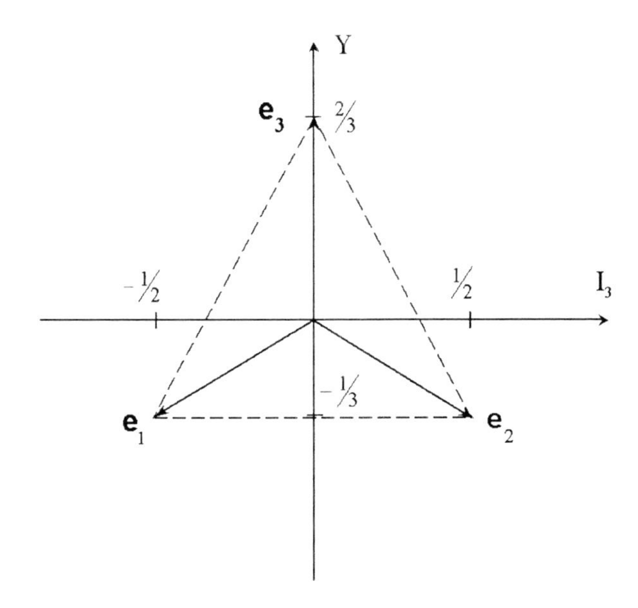

Figure 17. The basic vectors in the $E_{\overline{1}}$ space.

In case of the isotopic spin group two doublets, transforming according to the rules

$$\kappa' = U\kappa, \qquad \overline{\kappa}' = U^*\overline{\kappa}, \tag{4.175}$$

are equivalent, since U and U^* or, which is equivalent, σ_i and $-\sigma_i^*$ are connected through the similarity transformation. This statement breaks down in case of the matrices λ_a and $-\lambda_a^*$. For this reason the representations of the $SU(3)$-group are characterized by a number of the contravariant p and covariant q indices $D(p,q)$.

So, we got acquainted with the most simple representations: $D(0,0)$, $D(1,0)$ and $D(0,1)$. More complicated representations of the $SU(3)$-group are no longer irreducible. Consequently, the recipe is needed to divide these representations into direct sums of irreducible representations.

Let us first study some concrete examples, and then we shall try to summarize the results obtained. We start with the representation $U\otimes U$ in the space E_2 with the basis $\mathbf{e}_{kl}$. Arbitrary contravariant spinor of the second rank with components

$$\Phi^{kl} = \begin{pmatrix} \Phi^{11} & \Phi^{21} & \Phi^{31} \\ \Phi^{12} & \Phi^{22} & \Phi^{32} \\ \Phi^{13} & \Phi^{23} & \Phi^{33} \end{pmatrix}$$

can be expanded into symmetric part with six independent components

$$\Phi^{(s)kl} = \frac{1}{2}\left(\Phi^{kl} + \Phi^{lk}\right) \tag{4.176}$$

and into asymmetric part with three independent components

$$\Phi^{(a)kl} = \frac{1}{2}\left(\Phi^{kl} - \Phi^{lk}\right). \tag{4.177}$$

Notice, that $\mathrm{Sp}(\Phi^{kl})$ is not the $SU(3)$-scalar. Scalar can be obtained by convolutions of covariant and contravariant indices. According to this, the space E_2 expands into the two subspaces $E_2^{(6)}$ and $E_2^{(3)}$, bases of which are formed from six vectors

$$\mathbf{e}_{\{kl\}} = \mathbf{e}_{kl}^{(6)} = \begin{cases} \frac{1}{\sqrt{2}}(\mathbf{e}_{kl} + \mathbf{e}_{lk}), & k \neq l, \\ \mathbf{e}_{kl}, & k = l, \end{cases} \tag{4.178}$$

and three vectors

$$\mathbf{e}_{[kl]} = \mathbf{e}_{kl}^{(3)} = \frac{1}{\sqrt{2}}(\mathbf{e}_{kl} - \mathbf{e}_{lk}) \tag{4.179}$$

respectively. Using (4.178) and (4.179) we can write down any vector in the subspaces $E_2^{(6)}$ and $E_2^{(3)}$. For example, an arbitrary vector in $E_2^{(6)}$ has the form

$$\Phi^{(6)} = \sum \Phi^{kl}\mathbf{e}_{kl}^{(6)} = \frac{1}{\sqrt{2}}\sum_{k>l}\Phi^{kl}(\mathbf{e}_{kl} + \mathbf{e}_{lk}) + \sum_{k}\Phi^{kk}\mathbf{e}_{kk}. \tag{4.180}$$

The vector Φ, belonging to the space E_2, can be viewed as direct product of two contravariant spinors of the first rank, or in other words, the components of the Φ^{kl}-spinor is direct product of the components of the Φ^k- and Φ^l-spinors. The operation of antisymmetrization can be reduced to the multiplying the starting spinor either on the quantities ε^{kmn}, ε_{kmn} or on their production. Thus, for example, since the relations are fulfilled

$$V^p = \varepsilon^{pmn}\Phi_{mn}^{(a)} \qquad \Phi_{mn}^{(a)} = \frac{1}{2}\varepsilon_{mnp}V^p \tag{4.181}$$

the symmetric spinor of the second rank having the components $\Phi_{mn}^{(a)}$ is equivalent to the three-dimensional vector V_p. Consequently, we can from the outset deal not with $\Phi_{mn}^{(a)}$ but with the quantity $\varepsilon^{pmn}\Phi_{mn}$. Notice, that symmetrization of spinors of the highest ranks does not cause a transition from covariant to contravariant indices and vice versa. It is not true, however, in case of spinors antisymmetrization (see, for example (4.181). Thus we demonstrated that the product Φ^k and Φ^l is reducible and breaks down into the two irreducible representations, that is

$$\mathbf{3} \bigotimes \mathbf{3} = \mathbf{6} \bigoplus \overline{\mathbf{3}}.$$

Let us consider now the representation $U \otimes U^*$ in the space $E_{1\bar{1}}$ with the basis $\mathbf{e}_l^k$. In this space vectors represent mixed spinors of the second rank with components Φ_l^k

$$\Phi_l^k = \begin{pmatrix} \Phi_1^1 & \Phi_1^2 & \Phi_1^3 \\ \Phi_2^1 & \Phi_2^2 & \Phi_2^3 \\ \Phi_3^1 & \Phi_3^2 & \Phi_3^3 \end{pmatrix}. \tag{4.182}$$

The track of this spinor $\mathrm{Sp}(\Phi_l^k) = \Phi_1^1 + \Phi_2^2 + \Phi_3^3$ is an unitary scalar, which is not changed under the $SU(3)$-transformations. Subtracting the quantity $\delta_l^k\Phi_n^n/3$ from (4.182) we obtain the eight-component spinor

$$\Phi_l^k - \frac{1}{3}\delta_l^k\Phi_n^n =$$

$$= \begin{pmatrix} \frac{2}{3}\Phi^1_1 - \frac{1}{3}(\Phi^2_2+\Phi^3_3) & \Phi^2_1 & \Phi^3_1 \\ \Phi^1_2 & \frac{2}{3}\Phi^2_2 - \frac{1}{3}(\Phi^1_1+\Phi^3_3) & \Phi^3_2 \\ \Phi^1_3 & \Phi^2_3 & \frac{2}{3}\Phi^3_3 - \frac{1}{3}(\Phi^1_1+\Phi^2_2) \end{pmatrix}. \qquad (4.183)$$

which already is irreducible. According to this, the space $E_{1\bar{1}}$ is decomposed into the two irreducible subspaces $E^{(8)}_{1\bar{1}}$ and $E^{(1)}_{1\bar{1}}$. The basis vector of the one-dimensional subspace $E^{(1)}_{1\bar{1}}$ has the form

$$\mathbf{e}^{(1)k}_l = \frac{1}{\sqrt{3}}\left(\mathbf{e}^1_1+\mathbf{e}^2_2+\mathbf{e}^3_3\right). \qquad (4.184)$$

The basis vectors in the second subspace must be orthogonal to the vector (4.184). Obviously, six quantities $\mathbf{e}^k_l$ with $k \neq l$ meet this demand. The remaining two basis vectors have the form of linear combinations of vectors $\mathbf{e}^1_1$, $\mathbf{e}^2_2$ and $\mathbf{e}^3_3$. They will be orthonormalized and orthogonal to the basis vector of the subspace $E^{(1)}_{1\bar{1}}$, provided the following choice is made

$$\frac{1}{\sqrt{2}}\left(\mathbf{e}^1_1-\mathbf{e}^2_2\right), \qquad \frac{1}{\sqrt{6}}\left(\mathbf{e}^1_1+\mathbf{e}^2_2-2\mathbf{e}^3_3\right).$$

Thus the basis vectors of the space $E^{(8)}_{1\bar{1}}$ have the form

$$\mathbf{e}^1_2,\ \frac{1}{\sqrt{2}}\left(\mathbf{e}^1_1-\mathbf{e}^2_2\right),\ \mathbf{e}^2_1,\ \mathbf{e}^1_3,\ \mathbf{e}^2_3,\ \mathbf{e}^3_1,\ \mathbf{e}^3_2,\ \frac{1}{\sqrt{6}}\left(\mathbf{e}^1_1+\mathbf{e}^2_2-2\mathbf{e}^3_3\right). \qquad (4.185)$$

Since the vector Φ, belonging to the space $\mathcal{E}_{1\bar{1}}$, is a direct product of the contravariant and covariant spinors of the first rank, then corresponding to it representation breaks down into two irreducible ones

$$\mathbf{3}\bigotimes\bar{\mathbf{3}} = \mathbf{1}\bigoplus\mathbf{8}.$$

Thus from the very beginning we managed to find among irreducible representations of the $SU(3)$-group the eight-dimensional representation, which can be used to describe particles octets.

Let us determine isospin content of the octet obtained, considering, that a mixed spinor with components Φ^k_l is composed of a product of the fundamental triplet and the fundamental antitriplet. If we denote an isospin multiplet by (I,Y) then a triplet and antitriplet are presented as $(1/2,1/3)+(0,-2/3)$ and $(1/2,-1/3)+(0,2/3)$, respectively. The product of the two isospin multiplets (I_1,Y_1) and (I_2,Y_2) contains multiplets with the hypercharge $Y = Y_1+Y_2$ and the isospins, which according to the moment summation rule, take the values

$$I = \mid I_1-I_2 \mid, \mid I_1-I_2 \mid +1, .., I_1+I_2.$$

Thus, the isomultiplet multiplication rule has the form

$$(I_1,Y_1)(I_2,Y_2) = (\mid I_1-I_2 \mid, Y_1+Y_2)+(\mid I_1-I_2 \mid +1, Y_1+Y_2)+.....+$$

$$+(I_1+I_2,Y_1+Y_2). \qquad (4.186)$$

Using (4.186) we arrive at the relation

$$\mathbf{3}\bigotimes\bar{\mathbf{3}} = \mathbf{1}\bigoplus\mathbf{8} = (1/2,1)\bigoplus 2(0,0)\bigoplus(1,0)\bigoplus(1/2,-1). \qquad (4.187)$$

With the help of Eqs. (4.186) and (4.187) we finally obtain

$$\mathbf{8} = (1/2,1)\bigoplus(0,0)\bigoplus(1,0)\bigoplus(1/2,-1), \tag{4.188}$$

that is, the octet consists of the two isodoublets with $Y = 1$ and $Y = -1$, the isotriplet with $Y = 0$, and the isosinglet with $Y = 0$.

It could be shown that the number of the covariant and contravariant indices in irreducible representation is connected with the number of particles n in multiplet by the relation

$$n(p,q) = n(p,0)n(0,q) - n(q-1,0)n(0,p-1) = \frac{1}{2}(p+1)(q+1)(p+$$
$$+q+2). \tag{4.189}$$

The next step is to determine the explicit form of wave functions of unitary multiplets. Let us consider a space transforming under representation which is the product of the fundamental representations

$$\underbrace{U\bigotimes\cdots\bigotimes U}_{p\ \text{times}}\underbrace{\bigotimes U^*\cdots\bigotimes U^*}_{q\ \text{times}}. \tag{4.190}$$

We introduce the orthonormalized basis $\mathbf{e}_{k_1\cdots k_p}^{l_1\cdots l_q}$. Then an arbitrary vector in this space is defined by the formula

$$\Psi = \Psi_{l_1\cdots l_q}^{k_1\cdots k_p}\mathbf{e}_{k_1\cdots k_p}^{l_1\cdots l_q}. \tag{4.191}$$

The components of the vector Ψ are transformed according to the law

$$\Psi_{l_1\cdots l_q}^{\prime k_1\cdots k_p} = U_{k_1m_1}\cdots U_{k_pm_p}U_{l_1n_1}^*\cdots U_{l_qn_q}^*\Psi_{n_1\cdots n_q}^{m_1\cdots m_p}. \tag{4.192}$$

Acting the representation generators on the basis vectors is determined by the relations

$$S_a^{(un)}\mathbf{e}_{k_1\cdots k_p}^{l_1\cdots l_q} = \sum_{r=1}^{q}(\lambda_a)_{l_rl_r'}\mathbf{e}_{k_1\cdots k_p}^{l_1\cdots l_{r-1}l_r'l_{r+1}\cdots l_q} -$$

$$-\sum_{r=1}^{p}(\lambda_a^T)_{k_rk_r'}\mathbf{e}_{k_1\cdots k_{r-1}k_r'k_{r+1}\cdots k_p}^{l_1\cdots l_q}, \tag{4.193}$$

where we have taken into consideration the explicit form of the generators for the contravariant and covariant spinors of the first rank. Let us denote the numbers of the covariant (contravariant) indices, equal to one, two and three by q_1, q_2 and q_3 (p_1, p_2 and p_3), respectively. Then from Eq. (4.193) and the particular form of the λ_a-matrices one can see, that the basis vector with the given numbers of indices, equal to one, two, three correspond to the following eigenvalues of the $SU(3)$-group invariants

$$S_3^{(is)}\mathbf{e}_{k_1\cdots k_p}^{l_1\cdots l_q} = \{\frac{1}{2}[q_1-p_1] - \frac{1}{2}[q_2-p_2]\}\mathbf{e}_{k_1\cdots k_p}^{l_1\cdots l_q}, \tag{4.194}$$

$$Y\mathbf{e}_{k_1\cdots k_p}^{l_1\cdots l_q} = \{\frac{1}{2\sqrt{3}}[q_1-p_1] + \frac{1}{2\sqrt{3}}[q_2-p_2] - \frac{1}{\sqrt{3}}[q_3-p_3]\}\mathbf{e}_{k_1\cdots k_p}^{l_1\cdots l_q}. \tag{4.195}$$

As before, we shall consider the eigenvalues of the operators $S_3^{(is)}$ and Y to be components of two-dimensional vectors on the unitary spin plane, that is, the end of the vector $\mathbf{e}_{k_1\cdots k_p}^{l_1\cdots l_q}$ corresponds to the definite state of the $SU(3)$-multiplet.

As an example we examine the irreducible representation $D(1,1)$, according to which the mixed spinor of the second rank ψ_l^k with zero-trace is transformed. As we already know, the vectors

$$\mathbf{f}^a=(\psi_l^k)^a\mathbf{e}_k^l, \qquad a=1,2,...8, \tag{4.196}$$

making up the basis in the space $D(1,1)$ can be chosen in the form

$$\left.\begin{array}{llll}\mathbf{f}^1=\mathbf{e}_2^1, & \mathbf{f}^2=\frac{1}{\sqrt{2}}\left(\mathbf{e}_1^1-\mathbf{e}_2^2\right), & \mathbf{f}^3=\mathbf{e}_1^2, & \mathbf{f}^4=\mathbf{e}_3^1,\\ \mathbf{f}^5=\mathbf{e}_3^2, & \mathbf{f}^6=\mathbf{e}_1^3, & \mathbf{f}^7=\mathbf{e}_2^3, & \mathbf{f}^8=\frac{1}{\sqrt{6}}\left(\mathbf{e}_1^1+\mathbf{e}_2^2-2\mathbf{e}_3^3\right),\end{array}\right\} \tag{4.197}$$

Then the non-zero components of the vectors $(\psi_l^k)^a$ of basis $\mathbf{f}^a$ have the following values

$$\left.\begin{array}{llll}\left(\psi_1^2\right)^1=1; & \left(\psi_1^1\right)^2=\frac{1}{\sqrt{2}}, \ \left(\psi_2^2\right)^2=-\frac{1}{\sqrt{2}}; & & \\ \left(\psi_2^1\right)^3=1; & \left(\psi_1^3\right)^4=1; & \left(\psi_2^3\right)^5=1; & \left(\psi_3^1\right)^6=1;\\ \left(\psi_3^2\right)^7=1; & \left(\psi_1^1\right)^8=\left(\psi_2^2\right)^8=\frac{1}{\sqrt{6}}, \ \left(\psi_3^3\right)^8=-\frac{2}{\sqrt{6}}. & & \end{array}\right\} \tag{4.198}$$

Using Eqs. (4.194) and (4.195) it is easy to demonstrate that the vectors $\mathbf{f}^a$ describe the following isomultiplets of the baryon $SU(3)$-octet

$$\left.\begin{array}{lll}I=1,Y=0:\Sigma^+=\mathbf{f}^1, & \Sigma^0=\mathbf{f}^2, & \Sigma^-=\mathbf{f}^3;\\ I=\frac{1}{2},Y=1:p=\mathbf{f}^4, & n=\mathbf{f}^5; & \\ I=\frac{1}{2},Y=-1:\Xi^-=\mathbf{f}^6, & \Xi^0=\mathbf{f}^7; & \\ I=0,Y=0:\Lambda^0=\mathbf{f}^8. & & \end{array}\right\} \tag{4.199}$$

Now we introduce the quantity

$$\mathbf{B}_l^k=\left(\psi_l^k\right)^a\mathbf{f}^a,$$

which matrix components represent wave function of the baryon octet

$$(B_l^k)=\begin{pmatrix}\frac{1}{\sqrt{2}}\Sigma^0+\frac{1}{\sqrt{6}}\Lambda^0 & \Sigma^+ & p\\ \Sigma^- & -\frac{1}{\sqrt{2}}\Sigma^0+\frac{1}{\sqrt{6}}\Lambda^0 & n\\ \Xi^- & \Xi^0 & -\frac{2}{\sqrt{6}}\Lambda^0\end{pmatrix}. \tag{4.200}$$

To obtain a normalized wave function, corresponding to the definite octet particle, in Eq. (4.200) its symbol should be changed by 1 and all the unrelated to it elements of the matrix B_l^k must be set equal to zero.

Meson octets have a similar form. Thus, for example, the wave functions of the pseudoscalar and vector meson octets are defined by the expressions

$$(P_l^k)=\begin{pmatrix}\frac{1}{\sqrt{2}}\pi^0+\frac{1}{\sqrt{6}}\eta^0 & \pi^+ & K^+\\ \pi^- & -\frac{1}{\sqrt{2}}\pi^0+\frac{1}{\sqrt{6}}\eta^0 & K^0\\ K^- & \overline{K^0} & -\frac{2}{\sqrt{6}}\eta^0\end{pmatrix} \tag{4.201}$$

$$(V_l^k) = \begin{pmatrix} \frac{1}{\sqrt{2}}\rho^0 + \frac{1}{\sqrt{6}}\omega^0 & \rho^+ & K^{*+} \\ \rho^- & -\frac{1}{\sqrt{2}}\rho^0 + \frac{1}{\sqrt{6}}\omega^0 & K^{*0} \\ K^{*-} & \overline{K^{*0}} & -\frac{2}{\sqrt{6}}\omega^0 \end{pmatrix}, \qquad (4.202)$$

respectively.

With expressions for unitary wave functions near at hand, one can use Lagrangian formalism to describe behavior of baryon and meson superfamilies. However, there is one "but", connected with particles masses in a unitary multiplet. Thus, the free Lagrangian of the baryon octet has the form

$$L = \frac{i}{2}\left[\overline{B}_l^k(x)\gamma_\mu\partial^\mu B_k^l(x) - \partial^\mu\overline{B}_l^k(x)\gamma_\mu B_k^l(x)\right] - m_0\overline{B}_l^k(x)B_l^k(x), \qquad (4.203)$$

where we have assumed, that all the baryons possess the same mass m_0. However, in reality the $SU(3)$-symmetry has been violated and the particles masses in the multiplet differ from each other. Consequently, this factor should be necessarily taken into consideration under accomplishing the exact calculations.

The world is made in such a way, that the weaker is the interaction, the less symmetric it is. The more strong interaction behaves as though it not observe slight violations of definite conservation laws. As a result, such interaction conserves the given physical quantity and consequently, it is more symmetric. Total interaction between hadrons can be presented as a sum of a hypothetical super strong interaction (with the $SU(3)$-symmetry group), strong interaction (which violates the unitary symmetry, but conserves the isotopic one), and lastly electromagnetic and weak interactions (which both violate the isotopic invariance). The division of the strong interaction into super strong and normal strong interaction, is, of course, rather conditional. In fact, there is no super strong interaction at all. There are only high energy regions, where masses differences of particles in multiplets are insignificant. When the strong, electromagnetic and weak interactions are switched off, the exact the $SU(3)$-symmetry takes place and all the particles in the unitary multiplet are degenerated in mass (m_0 is the degeneracy mass). Switching on strong interaction makes the mass operator in free Lagrangians dependent on the isospin and hypercharge

$$(\Psi_\alpha^{(n)}, \mathcal{M}\Psi_\alpha^{(n)}) = (\Psi_\alpha^{(n)}, \mathcal{M}_0\Psi_\alpha^{(n)}) +$$

$$+(\Psi_\alpha^{(n)}, \Delta\mathcal{M}\Psi_\alpha^{(n)}) \equiv M = m_0 \cdot 1^{(n)} + \Delta m^{(n)}(I,Y), \qquad (4.204)$$

where $\Psi_\alpha^{(n)}$ denotes the unitary wave function written in the form of a column matrix, the indices n and α characterize a representation and a particle state in a multiplet, respectively. Switching on electromagnetic interaction initiates further mass splitting in a unitary multiplet and one more term is introduced in Eq. (4.204).

Everything, we have yet known about the $SU(3)$-symmetry and about symmetry in general, concerns only with kinematic aspects of symmetry. However the real power of symmetry reveals itself under investigation of physical systems evolution. For example, symmetry allows us to obtain relationships between the cross sections of different reactions without using the motion equations.

A wave function of the unitary multiplets are eigenfunctions of the Casimir operators $S_a^{(un)2}$ and $S_a^{(un)}D_a$. Thus, every irreducible representation can be marked with the eigenvalues of these operators. Labeling of $SU(3)$-multiplet only with the eigenvalues of the operator of the unitary spin square is generally accepted. From Eq. (4.193) it is easy to obtain

$$S_a^{(un)2}D(p,q) = gD(p,q) = [p+q+\frac{1}{3}(p^2+pq+q^2)]D(p,q). \tag{4.205}$$

Then, for the unitary multiplets of the lowest dimensions we have

$$g = \begin{cases} 0 & \text{for} \quad \mathbf{1}, \\ \frac{4}{3} & \text{for} \quad \mathbf{3}, \overline{\mathbf{3}}, \\ 3 & \text{for} \quad \mathbf{8}, \\ 6 & \text{for} \quad \mathbf{10}. \end{cases} \tag{4.206}$$

The classification of hadrons according to unitary multiplets resembles one of the chemical elements in Mendeleev periodic table. Like Mendeleev table, the hadrons classification unostentatiously points up (but again for the experienced mind) a composite structure of hadrons. In this sense the $SU(3)$-group of the isospin and the hypercharge has served its historical mission. It has prepared all the conditions for the next step up the Quantum Stairway, going out on the quark-lepton level of matter structure. However unlike Moor, who had to disappear after his task was realized, the $SU(3)$-symmetry, as we shall see later, settles down firmly in physics of strong interaction. True enough, it must slightly change its role.

Chapter 5

Quark-Lepton Level of Matter Structure

5.1. Quark "Atoms"

In 1964 M. Gell-Mann and G. Zweig independently from each other hypothesized that all the hadrons are built of three particles of an unitary triplet. Wishing to emphasize the unusual properties of new blocks of matter, Gell-Mann called them "quarks". The term was borrowed from "Finnegans Wake" by J. Joyce. If one compares this novel with "Peace and War" by L. Tolstoy, the first thing, that comes to mind, is that "Finnegans Wake" was written in at least 26-dimensional space-time, which never experienced the joy of compactification. During the act the protagonist Hemphree Chaampden Ervicher is constantly changing his appearance. He is reincarnated at one moment into Mark, the king of Cornwell, at another into his sons, Sham and Shaun, and so on. Erwicher's children (he has a daughter as well) are from being not simple and they can be also transformed into their father. There is an episode in the novel, when the protagonist being reincarnated into king Mark, sends his nephew, knight Tristan, by wedding boat to bring the king bride Isolda. As one should be expected, during the travel Tristan and Isolda happen to have been struck down by Cupid arrows practically on the spot. The seagulls circling above the ship are completely informed about the events to be taking place at the ship, as demonstrated by their song started with the words: "Three quarks for mister Mark". If one distracts from the remaining part of the song, then the above mentioned phrase can be unambiguously viewed as a prediction of the fundamental triplet. Setting imagination free, one can assume, that mister Erwicher with his transformations chain reproduces the hadrons spectrum and his children are nothing else but three quarks. However, the whole content of the bird's opus suggests that the phrase "three quarks" may be treated as evidence that the old king was deceived triply. Time will show whether that is the excited indication on the analogous fate of the quark hypothesis, developing of which physicists have been devoting already for half a century.

Earlier we demonstrated, how to build singlets, octets and decuplets from unitary triplets with precisely the same isotopic structure as the experimentally observed hadrons unitary multiplets. It appeared, that fundamental triplet hypothesis was obvious even for philosophers who are constantly loitering around the building of Modern Physics and whose basic activities consist in abusing their own specially invented terminology. However, such models were constantly neglected due to inertia of thinking because they demanded fragmentation of electric and baryon charges of particles entering into the fundamental triplet.

Originally a quark family included only three particles and three corresponding antiparticles. Two fundamental triplets, which we denote with 3 and $\overline{3}$, had the following form

$$q = \begin{pmatrix} q_1 \\ q_2 \\ q_3 \end{pmatrix} = \begin{pmatrix} u \\ d \\ s \end{pmatrix}, \qquad \overline{q} = (\overline{q}^1, \overline{q}^2, \overline{q}^3) = (\overline{u}, \overline{d}, \overline{s}), \tag{5.1}$$

where symbols u, d, and s ("up", "down" and "strange") are used for components of quark triplet q_α ($\alpha = 1,2,3$). It is also possible to say, that a quark dwells in three aroma (flavor) states and attribute the flavor meaning to the index α. u- and d-quark form the isotopic doublet ($I_3^u = 1/2$ and $I_3^d = -1/2$), while s-quark does the isotopic singlet. Only s-quark has a non-zero strangeness ($s = -1$). Since we are going to construct particles with arbitrary spin values, then the quarks must have spin $1/2$. We also ascribe to all the three quarks baryon charge $1/3$. Then it becomes obvious, that in the unitary spin plane the quark and antiquark triplets are denoted by the same triangles as the fundamental triplet and antitriplet of the $SU(3)$ group, respectively (see, Fig.18).

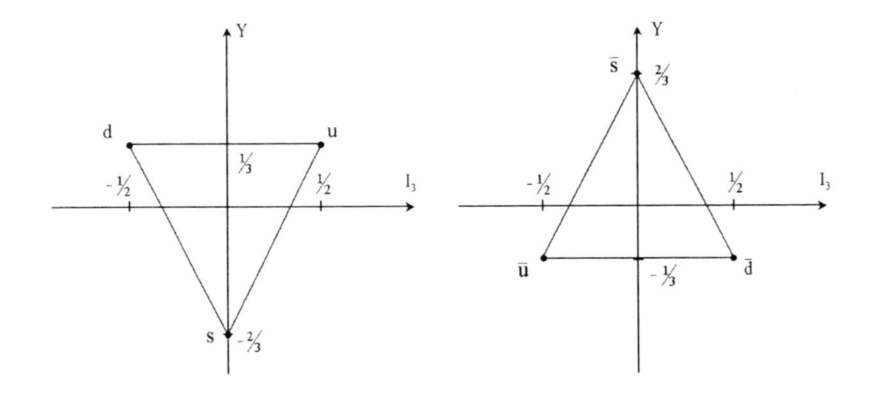

Figure 18. The quarks and the antiquarks triplets.

In its turn, from Gell— Mann — Nishijima formula it follows, that the quarks charges in units $|e|$ are fractional quantities

$$Q_u = 2/3, \qquad Q_d = Q_s = -1/3.$$

Further we should define quark contents of hadrons. It is evident, that mesons must contain even number of quarks, while baryons must contain odd number of quarks. Let us assume, that all the mesons are built from quark-antiquark pairs

$$M_k^i = \overline{q}^i q_k,$$

and all the baryons are built from three quarks

$$B_{ikl} = q_i q_k q_l.$$

Let us see, how to build some hadrons with the help of this simple scheme. We represent u-, d-, s-quarks and corresponding to them antiquarks by symbols depicted in Fig. 19, where the arrows show spin directions.

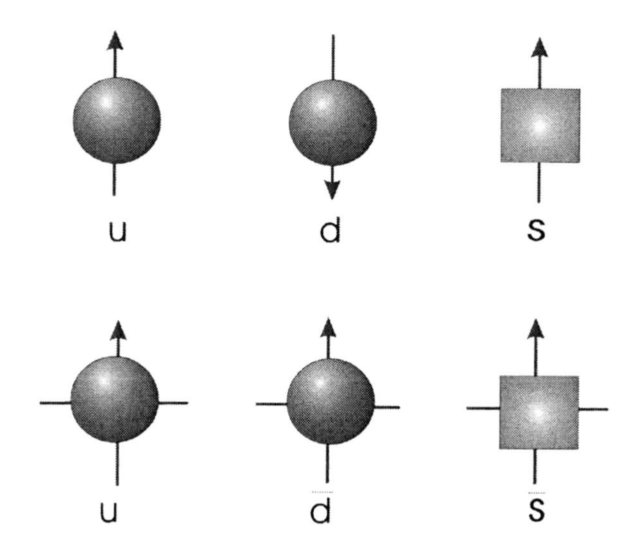

Figure 19. The u-, d-, and s-quarks.

In Fig. 20 we display quark filling of the proton, Λ^0-hyperon and one of the hero of Yukawa's Odyssey π^+-meson.

Notice, that we would ensure zero-spin of the π^+-meson by setting spins of the $(u\overline{d})$-pair in the same direction and accepting orbital angular moment L being equal to 1. However, since parity P_π is defined by the expression

$$P_\pi = (-1)^{L+1},$$

where we took into account that q and $\overline{q}$ have different parities, then in this case π^+-meson would be a scalar particle, but not a pseudoscalar one.

Further on, hadrons made of the quarks should be placed into the corresponding unitary multiplets. In our forthcoming design activity we are going to use the law of composition of the unitary spins as our basic instrument. This law must be a generalization of the composition rule for ordinary spins (4.40). Let us formulate this law in such a way, that it become applicable to any spin, whether it is the ordinary, the isotopic or the unitary one, etc. So, to obtain all the possible spin states of a coupled two-particle system, it is necessary: 1) to superpose the center of a spin diagram of the first particle on every state of a spin diagram of the second particle; 2) to mark the states obtained; 3) to single out diagrams with equal spin value out of all states.

Let us begin hadron construction with meson sector. Making center of diagram $\overline{3}$ coincident with every out of three states of diagram 3, we obtain nine possible states of the $\overline{3}3$-system (Fig. 21).

These states are separated into the $SU(3)$ octet and $SU(3)$ singlet. Three of them, namely, *I*,*II*,*III* being linear combinations $u\overline{u}$, $d\overline{d}$ and $s\overline{s}$ hit one and the same point of the

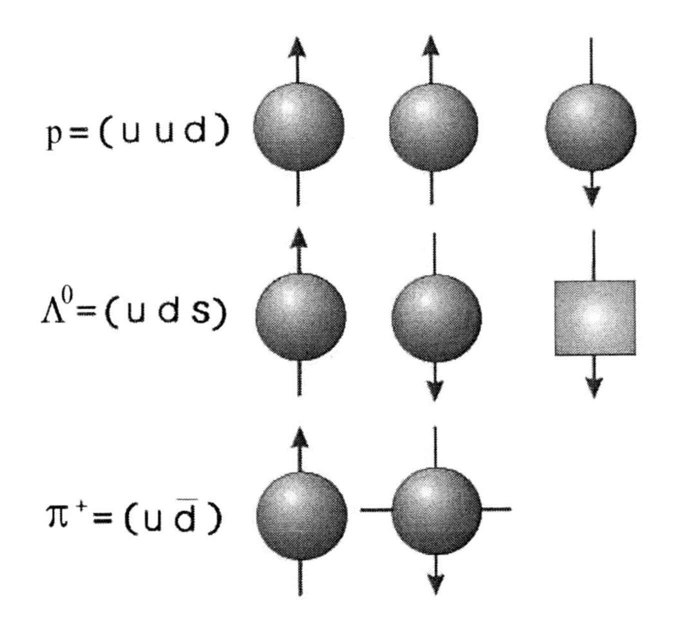

Figure 20. The quarks filling of hadrons.

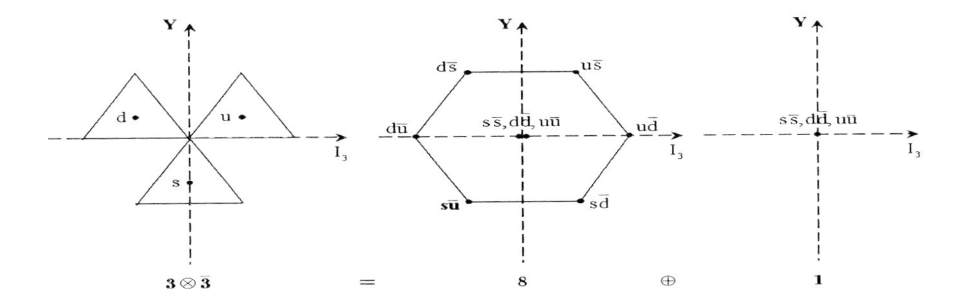

Figure 21. The meson multiplets.

diagram with $I_3 = Y = 0$. Consequently, they demand our great attention. The unitary singlet I is nothing else but

$$\text{const}\left(\overline{u},\overline{d},\overline{s}\right)\begin{pmatrix} u \\ d \\ s \end{pmatrix},$$

where const is defined from the normalization condition. It gives

$$I = \frac{1}{\sqrt{3}}\left(u\overline{u} + d\overline{d} + s\overline{s}\right). \tag{5.2}$$

Two remaining states belong to the octet. The state II we assign to the isospin triplet, whose contents we can reconstruct by the replacement $p \to u$ and $n \to d$ in formulas (4.76). So,

$$(d\overline{u}, II, -u\overline{d}), \tag{5.3}$$

where

$$II = \frac{1}{\sqrt{2}}\left(u\bar{u} - d\bar{d}\right).$$

On the strength of the demand of the orthogonality with I and II the singlet in isospin state III has the form

$$III = \frac{1}{\sqrt{6}}\left(u\bar{u} + d\bar{d} - 2s\bar{s}\right). \tag{5.4}$$

Since the spin of the $(q\bar{q})$ system can be equal either to 1 or to 0, the quark model predicts the existence of the following meson octets and meson singlets, which with the use of spectroscopic symbolics $^{2S+1}L_j$ can be presented as follows

$$L = 0, \qquad J = 0,1 \rightarrow \begin{array}{cc} 0^- & {}^1S_0 \\ 1^- & {}^3S_1 \end{array} \Bigg\},$$

$$L = 1, \qquad J = 0,1,2 \rightarrow \begin{array}{cc} 0^+ & {}^3P_0 \\ 1^+ & {}^3P_1 \\ 1^+ & {}^1P_1 \\ 2^+ & {}^3P_2 \end{array} \Bigg\},$$

and so on with higher values of L. The states with $L = 0$ can be viewed as the basic ones, while the states with $L > 0$ can be viewed as orbital excitations. It is quite natural to expect recurrence of the octet and singlet $L = 0$ with bigger mass values. So, all the discovered mesons can be placed into the $SU(3)$ multiplets $q\bar{q}$.. Up to 1971, when the first reliable data, confirming the existence of quarks into hadrons, were obtained, such tests have been the main argument in favor of the quark hypothesis.

We are coming now to investigation of quark structure of baryons. First we combine two quarks. We plot the already familiar to us result

$$3 \bigotimes 3 = 6 \bigoplus \bar{3}$$

in Fig. 22.

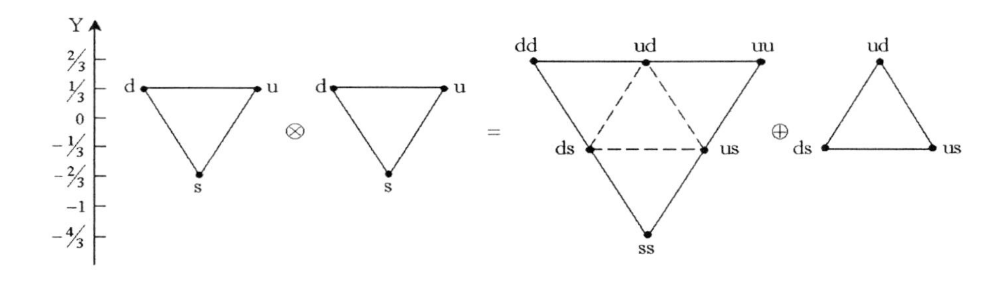

Figure 22. The graphical interpretation of the formula $3 \bigotimes 3 = 6 \bigoplus \bar{3}$.

We remind, that the sextet 6 is symmetric with respect to transposition of two quarks while the triplet $\bar{3}$ is antisymmetric. In Fig. 22 we specify only the quark filling in every point of the diagram. For the wave function of particles to be exactly defined we must take into account the symmetry properties of multiplets. So, the state ud belonging to the sextet is described

$$\psi_s(ud) = \frac{1}{\sqrt{2}}(ud+du),$$

whereas the wave function of the analogous state in the triplet has the form

$$\psi_t(ud) = \frac{1}{\sqrt{2}}(ud-du).$$

Let us add one more quark. The final result of decomposition

$$3\bigotimes 3\bigotimes 3 = (6\bigoplus\overline{3})\bigotimes 3 = (6\bigotimes 3)\bigoplus(\overline{3}\bigotimes 3) = 10\bigoplus 8\bigoplus 8\bigoplus 1, \qquad (5.5)$$

is displayed in Fig. 23, where the octet following decuplet appears under summarizing the unitary spins of the sextet and the triplet, while the second octet comes from $\overline{3}\bigotimes 3$.

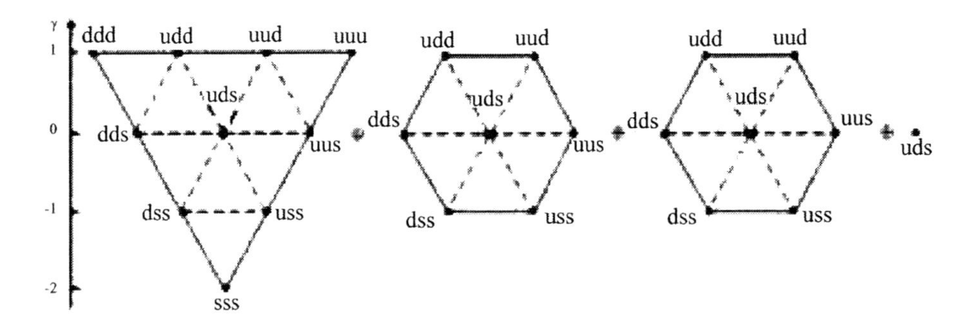

Figure 23. The graphical interpretation of the formula $3\bigotimes 3\bigotimes 3 = 10\bigoplus 8\bigoplus 8\bigoplus 1$.

We denote the relative momentum moment of two quarks by $\mathbf{L}_1$ and the momentum moment of the third quark as related to the masses center of the first two quarks by $\mathbf{L}_2$. Then the total momentum moment of the three quarks is determined by the expression

$$\mathbf{L} = \mathbf{L}_1 + \mathbf{L}_2.$$

Assuming the quarks parities to be positive we find that for the low-laying baryons states ($\mathbf{L} = \mathbf{L}_1 = \mathbf{L}_2 = 0$) the parity is equal to $(-1)^L = +1$. According to the vector summation rules the resulting spin of the three quarks may equal either $3/2$ or $1/2$. Thus the low-laying baryons multiplets are characterized by the values $3/2^+$ and $1/2^+$. Baryons having the lowest masses values are precisely placed into the decuplet $3/2^+$ and the octet $1/2^+$. More heavier baryons must have $L = 1$, that is, either $L_1 = 1$ and $L_2 = 0$ or $L_1 = 0$ and $L_2 = 1$ (in both cases their parity is negative $(-1)^L = -1$). And really, the existing baryon resonances are finely placed into the following multiplets: singlets and decuplets with $\frac{1}{2}^-$ and $\frac{3}{2}^-$, octets with $\frac{1}{2}^-$, $\frac{3}{2}^-$ and $\frac{5}{2}^-$.

5.2. Elastic Formfactors of Nucleons

Today building the quark model of hadrons could be viewed as some kind of analog of establishing the nucleon structure of atom nuclei. However, initially the quark model is considered as only the formal scheme which was very convenient to systematize hadrons. In the end of 60th of XX century physicists have obtained at their disposal new possibilities to investigate hadrons structure. The created sources of high-energy electrons allow to probe the distances up to 10^{-15} cm, that is, on two order smaller then the hadron size. To use the electrons is convenient on two reasons. Firstly, they are structureless particles and, secondly, they do not participate in strong interaction. Since electromagnetic interaction of point particles has studied thoroughly the theoretical analysis of the experimental results is greatly facilitated. At that time it was already known that hadrons have specific structure which is described by electromagnetic formfactors. By formfactors we agreed to understand the function characterizing the space distribution of the electric charge and multipole moments inside hadrons (further, for the sake of simplicity, we shall talk about the magnetic dipole moment only). Thus, carrying out "Roentgen" of proton with the help of scattered electrons one may investigate the proton structure more carefully and, by doing so, one gets down to the experimental checkout of the quark hypothesis.

The first stage in similar kind of investigations consists of the analysis of the elastic scattering. Consequently, we should start with the reaction

$$e^- + p \to \gamma^*, Z^* \to e^- + p, \tag{5.6}$$

where the intermediate step in (5.6) means that interaction is performed by exchanges of the virtual photon and Z-boson. For the sake of simplicity, we shall take into account the photon exchange only and shall be constrained by the second order of perturbation theory. The corresponding Feynman diagram is given in Fig. 24.

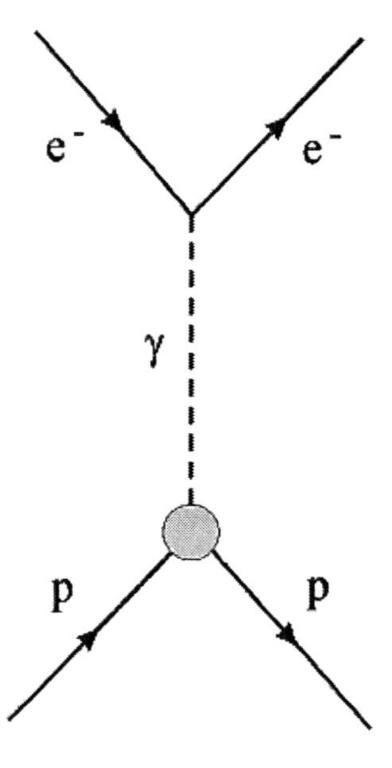

Figure 24. The Feynman diagram for the process $e^- + p \to \gamma^*, Z^* \to e^- + p$.

The circle in the vertex, describing interaction of the virtual photon with the electron, stresses the circumstance that the proton, unlike the electron, has the internal structure. When the proton were the point particle with the charge $|e|$ and the magnetic moment

$e\hbar/2m_pc$ predicted by Dirac theory, then the cross section of the elastic electron-proton scattering would follow from the cross section of the process

$$e^- + \mu^+ \to e^- + \mu^+, \tag{5.7}$$

whose Feynman diagram is depicted in Fig. 25, under changing the muon mass by the proton mass.

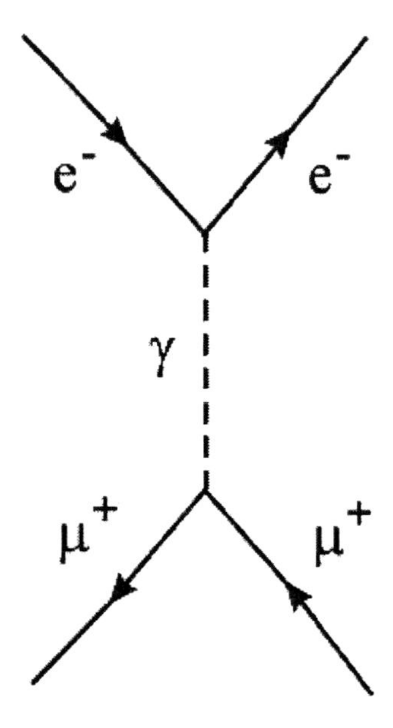

Figure 25. The Feynman diagram for the process $e^- + \mu^+ \to e^- + \mu^+$.

Passing to the laboratory system (the initial muon is in rest) we obtain the following expression for the differential cross section of the reaction (5.7)

$$\frac{d\sigma}{d\Omega} = \left(\frac{\alpha}{2E_i^{(e)}\sin^2(\theta/2)}\right)^2 \frac{E_f^{(e)}}{E_i^{(e)}} \left[\cos^2\frac{\theta}{2} - \frac{q^2}{2m_\mu^2}\sin^2\frac{\theta}{2}\right], \tag{5.8}$$

where $E_i^{(e)}$ ($E_f^{(e)}$) is the initial (final) energy of the electron, $q_\nu = (p_i^{(e)} - p_f^{(e)})_\nu = (p_f^{(\mu)} - p_i^{(\mu)})_\nu$, θ is the angle between momentum of the incoming and outgoing electron.

Before proceeding to researching the expression (5.8), let us agree about the choice of the unit of measurement. The reasonably chosen units comprise convenient tool under description of definite phenomena region. The fundamental constants of the quantum field theory (QFT) $\hbar$ and c enter into majority of theory equations. One may "unload" the QFT formulae when one set these constants to be equal 1, that is, one chooses the action quantum $\hbar$ as a action unit and does the light velocity c as a velocity unit. The system of units with $\hbar = c = 1$ derives the name of the natural system of units (NSU). The description of the NSU is given in Appendix. Further, unless otherwise specified, we shall use the NSU over the course of all the book. It is self-evident that, under comparing between theory and experiment we should pass to one of the ordinary system of units in obtained formulae.

The deduction of the expression (5.8) (it is named by Mott formula) needs knowledge of the Feynman diagram formalism the consecutive presentation of which is beyond the framework of the given book. However, in order to understand the basic details of obtaining Eq. (5.8) the quantum mechanics bases would be ample.

Recall the time dependent perturbation theory of the nonrelativistic quantum mechanics (NQM). In the first order in interaction $H_{int} \equiv V(\mathbf{r},t)$ the amplitude of the transition from the initial state with the wave function Φ_i into the final state with the wave function Φ_f is given by the expression

$$\mathcal{A}^{(1)}_{i\to f} = -i\int V_{if}(t)e^{i(E_f-E_i)t}dt, \tag{5.9}$$

where

$$V_{if}(t) = \int d^3x\Phi^*_f(\mathbf{r})V(\mathbf{r},t)\Phi_i(\mathbf{r}).$$

The expression (5.9) may be also represented in the covariant form

$$\mathcal{A}^{(1)}_{i\to f} = -i\int d^4x\Phi^*_f(x)V(x)\Phi_i(x),$$

where we have used the designations

$$\Phi(\mathbf{r})e^{-iEt} = \Phi(\mathbf{r},t) = \Phi(x), \qquad V(\mathbf{r},t) = V(x).$$

In the second order of perturbation theory $\mathcal{A}^{(2)}_{i\to f}$ takes the form

$$\mathcal{A}^{(2)}_{i\to f} = \mathcal{A}^{(1)}_{i\to f} + (-i)^2\sum_{n\neq i}\int_{-\infty}^{\infty} dtV_{fn}(t)e^{i(E_f-E_n)t}\int_{-\infty}^{t} dt'V_{ni}(t')e^{i(E_n-E_i)t'}.$$

Note, that $\mathcal{A}^{(2)}_{i\to f}$ may be rewritten in the relativistic invariant form too. To take the integral over dt' one needs to make it finite. That is achieved by including the small positive quantity ε into the exponent. After integration ε must be approached to zero.

If, for the sake of simplicity, we assume, that $V(\mathbf{r})$ does not depend on time and then, taking into account the high orders of perturbation theory, we obtain the following expression for the transition amplitude

$$\mathcal{A}_{i\to f} = \mathcal{A}^{(1)}_{i\to f} + (-i)^2 2\pi i\delta(E_f - E_n)\sum_{n\neq i} V_{fn}\frac{1}{E_i - E_n}V_{ni} + \ldots. \tag{5.10}$$

From (5.10) follows, the factor of the kind V_{ni} corresponds to every interaction vertex and the factor of the kind $1/(E_i - E_n)$ corresponds to every intermediate state. Intermediate states are "virtual" in the sense that energy is not conserved in these states ($E_n \neq E_i$). The energy conservation law is fulfilled only for the initial and final states $E_i = E_f$ as indicated by delta function. Of course, in the limit of low energies we can study the reaction (5.7) in the framework of the NQM. In doing so, we should suppose that the electron scattering occurs on the electromagnetic potential produced by the positive charged muon.

In the QED wave functions of fields are operators which describe destruction and creation of particles. For the electron-positron field $\psi^{(e)}(x)$ is represented by the sum of two operators

$$\psi^{(e)}(x) = \psi^{(e^-)}(x) + \psi^{(e^+)}(x),$$

where the former is the destruction operator of the electron and the latter is the creation operator of the positron. Then the quantity $\overline{\psi}^{(e)}(x) = \psi^{(e)\dagger}(x)\gamma_0$ will contain the creation operator of the electron $\overline{\psi}^{(e^-)}(x)$ and the destruction operator of the positron $\overline{\psi}^{(e^+)}(x)$. Since the photon field is neutral ($A_\tau^*(x) = A_\tau(x)$), then $A_\tau(x)$ is the sum of the operators of the creation and destruction of the photon

$$A_\tau(x) = A_\tau^{(+)}(x) + A_\tau^{(-)}(x).$$

The amplitude structure of the reaction (5.7) is defined by the form of Hamiltonians describing interactions of electromagnetic field with electron-positron and muon-antimuon fields. These Hamiltonians shall look like

$$H_{int}^{(e)}(x) = e\overline{\psi}^{(e)}(x)\gamma_\nu\psi^{(e)}(x)A^\nu(x), \qquad H_{int}^{(\mu)}(y) = e\overline{\psi}^{(\mu)}(y)\gamma_\tau\psi^{(\mu)}(y)A^\tau(y).$$

According to the correspondence principle the amplitude of the process (5.7) will have the same form as that in the NQM, that is, in the second order of perturbation theory it will be proportional to the product $H_{int}^{(e)}(x)H_{int}^{(\mu)}(y)$. However, in this case, there is one "but". In the NQM interaction is carried out through intermediate (virtual) states. The factors $1/(E_i - E_n)$, whose procedure of appearance is noncovariant, correspond to these states. In the QED the interaction carriers are virtual particles and their description must be relativistic covariant. Let us use the symbol V for the operation, which leads to creation and subsequent destruction of virtual particles. In the result of such a operation one or more the Green functions appear. As this takes place, every Green function multiplied by i corresponds disappearance of operators pair of one and the same field from under the symbol of the V-operator. Then, in the second order of perturbation theory the amplitude of the reaction (5.7) is determined by the expression

$$f_{i\to f}^{(2)} = (-i)^2 \int d^4x \int d^4yV \left[H_{int}^{(e)}(x)H_{int}^{(\mu)}(y)\right], \tag{5.11}$$

where we have taken into account that the reaction (5.7) does not go in the first order of perturbation theory. In the case of the reaction (5.7) interaction between leptons is caused by the virtual photon, consequently, the symbol V leads to appearance of the photon Green function on the place of two electromagnetic field operators. Note, in this case we are interested in the operators of the electron and the positive muon fields only.

Special attention must be given to the direction of arrows in the muon part of the diagram. In the Feynman diagram formalism antiparticles are moving backwards in time. The easiest way to understand that is to address to the hole theory which was proposed by Dirac to overcome difficulties connected with appearance of the negative energies. In this theory the states with the positive energy, i. e. the states, whose dependence on time has the form $\sim \exp(-iEt)$, are identified with electrons while the states ($\psi \sim \exp(-iEt)$) having the negative energies are identified with positrons. It is obvious that changing the time direction one can achieve changing the sign in the exponent $\exp(iEt)$, that is, one passes to solutions with positive energy.

Let us point the way to correspondence between the elements of the diagram displayed in Fig. 25 and the operators entering into $H_{int}^{(e)}(x)H_{int}^{(\mu)}(y)$. So, the diagram of Fig. 25 states the following. In the point x the initial electron is destroyed (the factor $\psi^{(e^-)}(x)$) while the final electron (factor $\overline{\psi}^{(e^-)}(x)$) and the virtual photon are created. Two operators $A^{\nu}(x)$ and $A^{\tau}(y)$ from $H_{int}^{(e)}(x)H_{int}^{(\mu)}(y)$ proves to be involved in description of the virtual photon moving from the point x to the point y. They lead to appearance of the photon Green function $G_{\mu\nu}^{(\gamma)}(x-y)$ which is called the photon propagator. In the point y the virtual photon annihilates with the initial antimuon (the factor $\overline{\psi}^{(\mu^+)}(y)$) and in so doing the final antimuon is created (the factor $\psi^{(\mu^+)}(y)$). The factors $-ie\gamma_{\nu}$ and $-ie\gamma_{\tau}$ correspond to two vertices in which electromagnetic interaction of the point-like leptons occur. To pass to the momentum space one should use the expansions in the Fourier integrals

$$\psi^{(e^-)}(x) = \sqrt{\frac{m_e}{E_e}}\int d^4p\psi^{(e)}(p)e^{-ipx}, \qquad \psi^{(\mu^+)}(x) = \sqrt{\frac{m_\mu}{E_\mu}}\int d^4p\psi^{(\mu)}(p)e^{ipx},$$

$$G_{\mu\nu}^{(\gamma)}(x-y) = \frac{1}{(2\pi)^4}\int d^4kG_{\mu\nu}^{(\gamma)}(k)e^{-ik(x-y)},$$

and one should also take into account the integral representation of the delta function

$$\delta^{(4)}(x) = \frac{1}{(2\pi)^4}\int d^4ke^{-ikx}.$$

Proceeding such a way we obtain the following expression for the amplitude of the reaction (5.7)

$$f_{i\to f}^{(2)} = (2\pi)^4\mathcal{M}_{i\to f}\delta^{(4)}\left(\sum p_i - \sum p_f\right), \tag{5.12}$$

where the delta function expresses the four-momentum conservation law

$$M_{i\to f}^{(2)} = \left(\overline{\psi}(p_f^{(e)})\sqrt{\frac{m_e}{E_f^{(e)}}}\right)(-ie\gamma^{\nu})\left(\psi(p_i^{(e)})\sqrt{\frac{m_e}{E_i^{(e)}}}\right)\left[\frac{-ig_{\nu\tau}}{(p_f^{(e)}-p_i^{(e)})^2}\right]\times$$

$$\times\left(\overline{\psi}(p_i^{(\mu)})\sqrt{\frac{m_\mu}{E_i^{(\mu)}}}\right)(-ie\gamma^{\tau})\left(\psi(p_f^{(\mu)})\sqrt{\frac{m_\mu}{E_f^{(\mu)}}}\right), \tag{5.13}$$

and we have accepted the designations $\psi^{(l)}(p^{(l)}) \equiv \psi(p^{(l)})$. The order of writing the spinor matrices in (5.13) is determined by the direction of the diagram detour which is opposite to the fermion lines direction, that is, the diagram detour is performed from the final fermion state to the initial one (for antifermions, on the contrary). The cofactor standing in the square bracket comprises the Fourier transform of the photon Green function and describes the propagation of the virtual photon. To establish its explicit form one sufficiently writes the equation for the Green function of the photon field

$$\Box G_{\mu\nu}^{(\gamma)}(x) = g_{\mu\nu}\delta(x).$$

Substituting the expansions in the Fourier integral of $G_{\mu\nu}^{(\gamma)}(x)$ and the delta function in this equation we arrive at the result

$$G_{\mu\nu}^{(\gamma)}(k) = -\frac{g_{\mu\nu}}{k^2}.$$

Since the probability of the process $P_{i\to f}$ is connected with its amplitude by the expression

$$P_{i\to f} = | f_{i\to f} |^2,$$

then we have in our disposal two delta functions. One of them is needed to provide the four-momentum conservation while the destiny of the second one seems not to be quite obvious. Let us change it by the integral

$$\delta^{(4)}(k) = \frac{1}{(2\pi)^4}\int d^4x \exp(ikx), \qquad k = \sum p_i - \sum p_f$$

and shall consider the integration region to be finite. This immediately gives

$$\delta^{(4)}(k) = \frac{VT}{(2\pi)^4},$$

where V is the integration space volume and T is the time interval of the integration. Introduce the probability transition in the time unit and in the volume unit

$$W_{i\to f} = \frac{P_{i\to f}}{VT}.$$

Connection of $W_{i\to f}$ with the cross section which directly measured in experiment is defined by the expression

$$d\sigma = \frac{W_{i\to f}}{j_0} dN_f,$$

where j_0 is the initial density of the flux of particles participating in the reaction and dN_f is the number of the final states. To find dN_f one should recall the energy quantization rule in quasiclassical approximation

$$\oint p(x)dx = 2\pi(n + \frac{1}{2}). \qquad (5.14)$$

The integral standing in the left hand side of Eq. (5.14) represents the area, covered by the closed classical phase trajectory of particle with the energy E, on the phase plane. According to Eq. (5.14) at $n \gg 1$ this area is equal to $2\pi n$. Thus, the area being equal to $2\pi n$ corresponds to every quantum state in the phase space. To put it otherwise, the number of states conforming to the area $\Delta x \Delta p$ will be

$$\frac{\Delta x \Delta p}{(2\pi)}.$$

Generalizing this relation on the three-dimensional case, we obtain that for the particle being in volume V, the maximum number of states with momenta confined in the element d^3p is equal to

$$\frac{Vd^3p}{(2\pi)^3}.$$

Further, for the sake of simplicity, we shall believe that the volume V, on which the wave functions are normalized, is equal to 1. Then, in the case of the reaction (5.7) we have

$$dN_f = \frac{d^3p_f^{(e)} d^3p_f^{(\mu)}}{(2\pi)^6}.$$

Since in the laboratory system (the initial muon rests) the number of particles transiting the area unit in the time unit is equal to $|\mathbf{v}^{(\mathbf{e})}|$ and the number of targets in the volume unit is equal to 1, then

$$j_0 = |\mathbf{v}^{(\mathbf{e})}|.$$

Gathering the results obtained we get the following expression for the differential cross section of the process (5.7)

$$d\sigma = |\mathcal{M}_{if}|^2 \frac{d^3p_f^{(e)}d^3p_f^{(\mu)}}{(2\pi)^2 |\mathbf{v}^{(\mathbf{e})}|}\delta^{(4)}(p_i^{(\mu)} + p_i^{(e)} - p_f^{(\mu)} - p_f^{(e)}). \tag{5.15}$$

When we are not interested by the particles polarization in the initial and final states then in $|\mathcal{M}_{if}|^2$ one should fulfill averaging in the initial polarizations and summing in the final ones. It means that we must pass to the quantity

$$\frac{1}{4}\sum_{i,f}|M_{i\to f}|^2 = g_0\sum_{i,f} |[\overline{\psi}_\alpha(p_f^{(e)})(\gamma^\nu)_{\alpha\beta}\psi_\beta(p_i^{(e)})][\overline{\psi}_m(p_f^{(\mu)})(\gamma_\nu)_{mn}\psi_n(p_i^{(\mu)})]|^2, \tag{5.16}$$

where

$$g_0 = \frac{e^4 m_e^2 m_\mu^2}{4q^4 E_i^{(\mu)}E_i^{(e)}E_f^{(\mu)}E_f^{(e)}}.$$

and the factor 1/4 appears at the cost of averaging in polarizations of the initial electron and muon. Since in Eq. (5.16) the matrix indices, which define the multiplication order, are represented in explicit form then the factors, entering into this expression, may be interchanged arbitrarily. Gathering the electron and muon parties separately, we obtain

$$\frac{1}{4}\sum_{i,f}|\mathcal{M}_{if}|^2 = g_0\sum_{i,f}[\overline{\psi}_\alpha(p_f^{(e)})(\gamma^\nu)_{\alpha\beta}\psi_\beta(p_i^{(e)})\overline{\psi}_{\alpha'}(p_i^{(e)})(\gamma^\tau)_{\alpha'\beta'}\psi_{\beta'}(p_f^{(e)})]\times$$

$$\times[\overline{\psi}_m(p_f^{(\mu)})(\gamma_\nu)_{mn}\psi_n(p_i^{(\mu)})\overline{\psi}_{m'}(p_i^{(\mu)})(\gamma_\tau)_{m'n'}\psi_{n'}(p_f^{(\mu)})] =$$

$$= g_0\sum_{i,f}[\psi_{\beta'}(p_f^{(e)})\overline{\psi}_\alpha(p_f^{(e)})(\gamma^\nu)_{\alpha\beta}\psi_\beta(p_i^{(e)})\overline{\psi}_{\alpha'}(p_i^{(e)})(\gamma^\tau)_{\alpha'\beta'}]\times$$

$$\times[\psi_{n'}(p_f^{(\mu)})\overline{\psi}_m(p_f^{(\mu)})(\gamma_\nu)_{mn}\psi_n(p_i^{(\mu)})\overline{\psi}_{m'}(p_i^{(\mu)})(\gamma_\tau)_{m'n'}]. \tag{5.17}$$

To summarize over the spin leptons states (the factor 1/2 converts such summarizing into averaging) is fulfilled with the help of the relations

$$\sum\psi_\alpha^\varepsilon(p)\overline{\psi}_\beta^\varepsilon(p) = \left(\frac{\varepsilon\hat{p}+m}{2m}\right)_{\alpha\beta}, \tag{5.18}$$

where $\hat{p} = \gamma_\nu p^\nu$, $\varepsilon = 1$ for particles and $\varepsilon = -1$ for antiparticles. To deduce the relation (5.18) we address to Dirac equation

$$(i\gamma_\mu\partial^\mu - m)\psi(x) = 0, \tag{5.19}$$

where

$$\{\gamma_\mu, \gamma_\nu\} = 2g_{\mu\nu}. \tag{5.20}$$

Inasmuch as any free particle must be described by de Broglie wave then the Dirac equation solution should be sought in the form

$$\psi^{(+)}(x) = \psi^{+}(p)\exp(-ipx) \qquad \text{for positive energy,} \tag{5.21}$$

$$\psi^{(-)}(x) = \psi^{-}(p)\exp(ipx) \qquad \text{for negative energy.} \tag{5.22}$$

To substitute Eqs. (5.21) and (5.22) into the Dirac equation gives the equations determining the bispinors $\psi^{+}(p)$ and $\psi^{-}(p)$

$$(\hat{p}-m)\psi^{+}(p) = 0, \tag{5.23}$$

$$(\hat{p}+m)\psi^{-}(p) = 0. \tag{5.24}$$

Now it is necessary to establish the procedure which allows to distinguish between the particles states and the antiparticles ones. Eqs. (5.23) and (5.24) suggest that the operator of projecting on the particles states may be presented in the form

$$P_{+} = \frac{\hat{p}+m}{2m}, \tag{5.25}$$

as thanks to Eqs. (5.23) and (5.24) it has the properties

$$P_{+}\psi^{+}(p) = \psi^{+}(p), \qquad P_{+}\psi^{-}(p) = 0. \tag{5.26}$$

When we are interested in the negative energy states, that is, the antiparticles states, then, by analogy, we may define the operator

$$P_{-} = \frac{-\hat{p}+m}{2m} \tag{5.27}$$

with properties

$$P_{-}\psi^{-}(p) = \psi^{-}(p), \qquad P_{-}\psi^{+}(p) = 0. \tag{5.28}$$

Since the standard relations for the projection operators are fulfilled

$$P_{\varepsilon}P_{\varepsilon'} = \delta_{\varepsilon\varepsilon'}, \qquad P_{+}+P_{-} = I, \tag{5.29}$$

where

$$P_{\varepsilon} = \frac{\varepsilon\hat{p}+m}{2m} \tag{5.30}$$

then P_{+} and P_{-} represent the desired projection operators.

Further, by virtue of the orthonormalization condition, we have

$$\sum_{all}\psi_{\alpha}(p)I_{\alpha\gamma}\overline{\psi}_{\beta}(p) = I_{\gamma\beta}, \tag{5.31}$$

where $I_{\alpha\gamma}$ is the identity matrix and the sum is already taken over all the possible states (two electron states with the opposite spin directions and two positron states, the spins of which are also antiparallel). Now the action of the operators P_{ε} on the both sides of Eq. (5.31) gives the relation (5.18).

Let us rewrite Eq. (5.17) in the following form

$$\frac{1}{4}\sum_{i,f}|M_{i\to f}|^2 = g_0 L^{\nu\tau} M_{\nu\tau}, \tag{5.32}$$

where we shall label $L^{\nu\tau}$ and $M_{\nu\tau}$ as the electron and muon tensor respectively. For the electron tensor we have

$$L^{\nu\tau} = \sum_{i,f}\psi_{\beta'}(p_f^{(e)})\overline{\psi}_{\alpha}(p_f^{(e)})(\gamma^{\nu})_{\alpha\beta}\psi_{\beta}(p_i^{(e)})\overline{\psi}_{\alpha'}(p_i^{(e)})(\gamma^{\tau})_{\alpha'\beta'} =$$

$$= \left(\frac{\hat{p}_f^{(e)}+m_e}{2m_e}\right)_{\beta'\alpha}(\gamma^{\nu})_{\alpha\beta}\left(\frac{\hat{p}_i^{(e)}+m_e}{2m_e}\right)_{\beta\alpha'}(\gamma^{\tau})_{\alpha'\beta'} =$$

$$= \frac{1}{4m_e^2}\mathrm{Sp}[(\hat{p}_f^{(e)}+m_e)\gamma^{\nu}(\hat{p}_i^{(e)}+m_e)\gamma^{\tau}] =$$

$$= \frac{1}{4m_e^2}\mathrm{Sp}[\gamma^{\nu}(\hat{p}_i^{(e)}+m_e)\gamma^{\tau}(\hat{p}_f^{(e)}+m_e)], \tag{5.33}$$

where the symbol "Sp" means the diagonal elements sum of the matrices product and the cyclic property

$$\mathrm{Sp}[ABCD] = \mathrm{Sp}[DABC] = \mathrm{Sp}[CDAB] = \mathrm{Sp}[BCDA]$$

has been taken into consideration. As far as the muon tensor is concerned, the analogous operations give

$$M_{\nu\tau} \to L_{\nu\tau}(p^{(e)} \to -p^{(\mu)}, m_e \to m_\mu). \tag{5.34}$$

The spur in the expressions (5.33) and (5.34) could be found by application of the formulae

$$\mathrm{Sp}(\gamma^{\mu}\gamma^{\nu}) = 4g^{\mu\nu}, \qquad \mathrm{Sp}(\gamma^{\mu}\gamma^{\nu}\gamma^{\lambda}\gamma^{\sigma}) = 4(g^{\mu\nu}g^{\lambda\sigma}+g^{\mu\sigma}g^{\nu\lambda}-g^{\mu\lambda}g^{\nu\sigma}) \tag{5.35}$$

and by considering the fact, that the spur of the odd number of γ-matrices is equal to zero. The calculations give

$$L^{\nu\tau} = [g^{\nu\tau}(m_e^2 - p_i^{(e)}p_f^{(e)}) + p_i^{(e)\nu}p_f^{(e)\tau} + p_i^{(e)\tau}p_f^{(e)\nu}]. \tag{5.36}$$

Multiplying the lepton tensors and neglecting the electron mass, we obtain the following expression for the (5.17) in the laboratory system

$$\frac{1}{4}\sum_{i,f}|M_{i\to f}|^2 = \frac{e^4}{2q^4E_i^{(\mu)}E_i^{(e)}E_f^{(\mu)}E_f^{(e)}}\left[m_\mu^2(q^2+4E_i^{(e)}E_f^{(e)})-\right.$$

$$\left.-q^2m_\mu(E_i^{(e)}-E_f^{(e)})\right] = \frac{2e^4m_\mu^2}{q^4E_i^{(\mu)}E_f^{(\mu)}}\left(\cos^2\frac{\theta}{2}-\frac{q^2}{2m_\mu^2}\sin^2\frac{\theta}{2}\right). \tag{5.37}$$

Under deduction of (5.37) we have made use of the relations

$$q^2 = (p_i^{(e)} - p_f^{(e)})^2 \approx -2(p_f^{(e)} p_i^{(e)}) \approx$$

$$\approx -2E_f^{(e)} E_i^{(e)}(1-\cos\theta) = -4E_f^{(e)} E_i^{(e)} \sin^2\frac{\theta}{2} = -\frac{4E_i^{(e)2}\sin^2\frac{\theta}{2}}{1+\frac{2E_i^{(e)}}{m_\mu}\sin^2\frac{\theta}{2}}, \tag{5.38}$$

$$q^2 = -2(qp_i^{(\mu)}) = -2(E_i^{(e)} - E_f^{(e)})m_\mu. \tag{5.39}$$

The latter is the consequence of the operation of taking the square of the identity $q + p_i^{(\mu)} = p_f^{(\mu)}$.

Let us find the differential cross section using Eq. (5.15). The presence of the delta functions in Eq. (5.15) gives an opportunity to make integration over the three-dimensional momenta of final particles. Taking into consideration the delta function property

$$\delta(x^2 - a^2) = \frac{1}{2a}[\delta(x-a) + \delta(x+a)], \tag{5.40}$$

and accounting that the muon is on the mass shell and its energy is positive, we can rewrite the integral over $\mathbf{p}_f^{(\mu)}$ for the cross section part, depending on $\mathbf{p}_f^{(\mu)}$, in the form

$$I_{p_f^{(\mu)}} = \int \frac{d^3 p_f^{(\mu)}}{2E_f^{(\mu)}} \delta^{(4)}(p_i^{(\mu)} + q - p_f^{(\mu)}) =$$

$$= \int d^3 p_f^{(\mu)} dE_f^{(\mu)} \delta^{(4)}(p_i^{(\mu)} + q - p_f^{(\mu)})\theta(E_f^{(\mu)})\delta(p_f^{(\mu)2} - m_\mu^2),$$

where

$$\theta(x) = \begin{cases} 1, & x > 0, \\ 0, & x < 0. \end{cases}$$

Further calculations of $I_{p_f^{(\mu)}}$ are trivial and the final result is given by the expression

$$I_{p_f^{(\mu)}} = \delta\left((p_i^{(\mu)} + q)^2 - m_\mu^2\right) = \delta\left(2p_i^{(\mu)} q + q^2\right) = \frac{1}{2m_\mu}\delta\left(E_i^{(e)} - E_f^{(e)} + \frac{q^2}{2m_\mu}\right) =$$

$$= \frac{1}{2m_\mu}\delta\left(E_i^{(e)} - E_f^{(e)} - \frac{2E_i^{(e)}E_f^{(e)}}{m_\mu}\sin^2\frac{\theta}{2}\right) = \frac{1}{2bm_\mu}\delta\left(E_f^{(e)} - \frac{E_i^{(e)}}{b}\right), \tag{5.41}$$

where

$$b = 1 + \frac{2E_i^{(e)}}{m_\mu}\sin^2\frac{\theta}{2}. \tag{5.42}$$

So, our cross section acquires the form

$$\frac{d\sigma}{d^3 p_f^{(e)}} = \frac{e^4}{q^4(2\pi)^2 b}\left(\cos^2\frac{\theta}{2} - \frac{q^2}{2m_\mu^2}\sin^2\frac{\theta}{2}\right)\delta\left(E_f^{(e)} - \frac{E_i^{(e)}}{b}\right). \tag{5.43}$$

In what follows to pass into the spherical coordinate system with the vector $\mathbf{p}_i^{(e)}$ directing along the axis z allows to present $d^3p_f^{(e)}$ in the form

$$d^3p_f^{(e)} = |\,\mathbf{p}_f^{(e)}\,|^2\, d\,|\,\mathbf{p}_f^{(e)}\,|^2\, d\varphi d(\cos\theta^{(e)}) = (E_f^{(e)})^2 dE_f^{(e)} d\Omega.$$

Using the expression obtained one may carry out integrating the expression (5.43) without any trouble and obtains the Mott formula.

Notice, when the electron scattering were on a spinless point particle then in the relation (5.33) one would make the replacement

$$M_{\nu\tau} \to (p_i + p_f)_\nu (p_i + p_f)_\tau,$$

which would lead to the following value of the cross section

$$\left(\frac{d\sigma}{d\Omega}\right)_0 = \left(\frac{\alpha}{2E_i^{(e)}\sin^2(\theta/2)}\right)^2 \frac{E_f^{(e)}}{E_i^{(e)}} \cos^2\frac{\theta}{2}. \tag{5.44}$$

Thus the factor $-ie\gamma_\mu$ corresponds to the Feynman diagram vertex describing the electromagnetic interaction of the point particles with the spin $1/2$. For particles having the composite structure, such as the proton, form of the vertex has more complicated view and this form is not known on the whole. However, one may define the form of the proton electromagnetic vertex (Fig. 26) from conditions which are enough natural for the quantum theory. We mean the relativistic covariance and the four-dimensional current conservation law or, what is the same, gradient invariance.

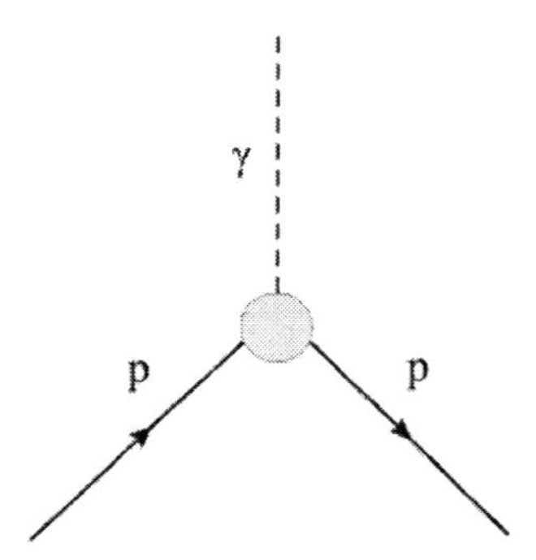

Figure 26. The electromagnetic proton vertex.

The number of elements from which one may construct the Lorentz covariant vector operator of vertex is not quite large. They are:
four-vectors

$$p_{-\nu} = (p_f - p_i)_\nu, \qquad p_{+\nu} = (p_f + p_i)_\nu,$$

four-tensors of the second rank

$$\varepsilon_{\mu\nu\lambda\sigma}, \qquad g^{\mu\nu}$$

and the Dirac matrices

vectors	pseudovectors
γ_μ	$\gamma_5\gamma_\mu$
tensors	pseudotensors
$\sigma_{\mu\nu}=\frac{i}{2}[\gamma_\mu,\gamma_\nu]$	$\gamma_5\sigma_{\mu\nu}$
$\gamma_\mu\gamma_\nu\gamma_\lambda$	$\gamma_5\gamma_\mu\gamma_\nu\gamma_\lambda$

where $\gamma_5=i\gamma^0\gamma^1\gamma^2\gamma^3$, (all the other Dirac matrices products may be reduced to the one of the listed above fourteen combinations by application of the anticommutation relations (5.20)). From these quantities one can build twelve independent four-vectors $v_{i\nu}$:

1. $p_{+\nu}$	**2**. $p_{-\nu}$	**3**. γ_ν
4. $\sigma_{\nu\mu}p^{+\mu}$	**5**. $\sigma_{\nu\mu}p^{-\mu}$	**6**. $\hat{p}_+p_{-\nu}$
7. $\hat{p}_-p_{+\nu}$	**8**. $\hat{p}_+p_{+\nu}$	**9**. $\hat{p}_-p_{-\nu}$
10. $\gamma_5\varepsilon^{\nu\mu\lambda\tau}\gamma_\mu p_{+\lambda}p_{-\tau}$	**11**. $\hat{p}_+\hat{p}_-p_{+\nu}$	**12**. $\hat{p}_+\hat{p}_-p_{-\nu}$

(multiplying every vector on γ_5 one may also introduce twelve pseudovectors which would prove to be claimed when the parity violation effects are being taken into account, that is, when weak interaction is switched on). We insert the same amount of the analytical functions F_i of invariant variable $q^2=p_-^2$ and represent the proton electromagnetic vertex in the form:

$$ie\Lambda_\nu=ie\sum_{i=1}^{12}F_i(q^2)v_{i\nu}. \tag{5.45}$$

From Fig. 26 it is obvious that the vertex operator, entering into the reaction amplitude, is confined between the wave functions of the free proton. Consequently, we are interested in the quantity

$$J_\nu=e\overline{\psi}(p_f)\Lambda_\nu\psi(p_i), \tag{5.46}$$

which is nothing more nor less than the four-vector of the proton electromagnetic current under $p_f\to p_i$. We call this quantity by the transition current. Then, using the free Dirac equation for the spinors $\psi(p)$ and $\overline{\psi}(p)$

$$(\hat{p}-m_p)\psi(p)=0,\qquad \overline{\psi}(p)(\hat{p}-m_p)=0, \tag{5.47}$$

one may reduce the number of the four-vectors $v_{i\nu}$ at the cost of redefining arbitrary functions $F_i(q^2)$.

Really, to apply Eqs. (5.47) reduces **6,8** to **1,2** and converts **7,9** into zero. Further, in consequence of the relations

$$\overline{\psi}(p_f)\hat{p}_+\hat{p}_-\psi(p_i)=\overline{\psi}(p_f)[-2m_p^2+2(p_ip_f)]\psi(p_i),$$

11,12 and **4,5** are also transformed to the kind of **1,2**. The inclusion of the identity

$$\varepsilon^{\nu\mu\lambda\tau}\gamma_5\gamma_\mu=\frac{1}{3}(\sigma^{\lambda\tau}\gamma^\nu+\sigma^{\tau\nu}\gamma^\lambda+\sigma^{\nu\lambda}\gamma^\tau) \tag{5.48}$$

and the Dirac equation reduces **10** to **1** and **2**. All this allows to exhibit the transition current in the form

$$e\overline{\psi}(p_f)\Lambda_\nu\psi(p_i)=e\overline{\psi}(p_f)[F_1(q^2)p_{+\nu}+F_2(q^2)p_{-\nu}+F_3(q^2)\gamma_\nu]\psi(p_i). \tag{5.49}$$

Since the current J_ν is conserved then multiplying the expression (5.49) on $p_{-\nu}$ gives

$$\overline{\psi}(p_f)F_2(q^2)q^2\psi(p_i) = 0.$$

This relation, in its turn, means that $F_2(q^2)$ is equal to zero for all real values of $q^2 \neq 0$. Then, from analyticity condition of formfactors it follows

$$F_2(q^2) = 0$$

for any q^2.

Carrying out the Hermitian conjugation of the space components of the expression (5.49) we obtain

$$\{\overline{\psi}(p_f)[F_1(q^2)\mathbf{p}_+ + F_3(q^2)\gamma]\psi(p_i)\}^\dagger = \{\psi^\dagger(p_f)\gamma_0[F_1(q^2)\mathbf{p}_+ + F_3(q^2)\gamma]\psi(p_i)\}^\dagger =$$
$$= \psi^\dagger(p_i)[F_1^*(q^2)\mathbf{p}_+ - F_3^*(q^2)\gamma]\gamma_0\psi(p_f) =$$
$$= \overline{\psi}(p_i)[F_1^*(q^2)\mathbf{p}_+ + F_3^*(q^2)\gamma]\psi(p_f), \tag{5.50}$$

where we have chosen the γ-matrices representation with anti-Hermitian γ-matrices and taken into consideration that the matrices γ and γ_0 anticommute with each other. When $p_i = p_f$ the transition current is nothing more nor less than the electromagnetic current J_ν. Since the three-dimensional electromagnetic current is Hermitian ($\mathbf{J}^\dagger = \mathbf{J}$), then comparing the left-hand and right-hand sides of Eq. (5.50) we have drawn the conclusion

$$F_1(q^2) = F_1^*(q^2), \qquad F_3(q^2) = F_3^*(q^2), \tag{5.51}$$

that is, F_1 and F_3 are the real quantities. Further, the relation

$$p_\nu\psi(p) = (i\sigma_{\nu\mu}p^\mu + m\gamma_\nu)\psi(p). \tag{5.52}$$

appears to be useful. To prove it one should multiply the equation

$$\gamma_\mu p^\mu\psi(p) = m\psi(p),$$

by γ_ν on the left and transform the left-hand side by the following manner

$$\gamma_\nu\gamma_\mu p^\mu\psi(p) = \frac{1}{2}([\gamma_\nu,\gamma_\mu] + \{\gamma_\nu,\gamma_\mu\})p^\mu\psi(p) = (-i\sigma_{\nu\mu} + g_{\nu\mu})p^\mu\psi(p).$$

It is convenient to represent the quantities $F_1(q^2)$ and $F_3(q^2)$ as follows

$$F_1(q^2) = -\frac{a^{(p)}\mathcal{F}_2(q^2)}{2m_p}, \qquad F_3(q^2) = \mathcal{F}_1(q^2) + a^{(p)}\mathcal{F}_2(q^2), \tag{5.53}$$

where $a^{(p)}$ is the value of the proton anomalous magnetic moment expressed in the nuclear magneton units. Then, based on Eq. (5.52) the expression for the transition current takes the form

$$J_\nu = e\overline{\psi}(p_f)[F_1(q^2)p_{+\nu} + F_3(q^2)\gamma_\nu]\psi(p_i) = e\overline{\psi}(p_f)\{[\mathcal{F}_1(q^2) + a^{(p)}\mathcal{F}_2(q^2)]\gamma_\nu -$$

$$-\frac{a^{(p)}\mathcal{F}_2(q^2)}{2m_p}p_{+\nu}\}\psi(p_i)=e\overline{\psi}(p_f)\left(\mathcal{F}_1(q^2)\gamma_\nu+i\frac{a^{(p)}}{2m_p}\mathcal{F}_2(q^2)\sigma_{\nu\mu}q^\mu\right)\psi(p_i)=$$

$$=e\overline{\psi}(p_f)\Lambda_\nu\psi(p_i), \tag{5.54}$$

where

$$\Lambda_\nu=\left(\mathcal{F}_1(q^2)\gamma_\nu+i\frac{a^{(p)}}{2m_p}\mathcal{F}_2(q^2)\sigma_{\nu\mu}q^\mu\right). \tag{5.55}$$

So, the proton electromagnetic vertex $ie\Lambda_\nu$ has been represented through two the formfactors $\mathcal{F}_{1,2}$ which hold all information concerning the proton structure. From Eq. (5.55) meaning of the transition $F_i(q^2)\to\mathcal{F}_i(q^2)$ has become clear as well. Now, the electromagnetic vertex operator consists of two terms where the former describes the Dirac type interaction and the latter does the Pauli type interaction. When $\mathcal{F}_i(q^2)\to 1$ the quantity Λ_ν transfers into the vertex operator of the point particle having the spin 1/2 and the magnetic moment $(1+a^{(p)})\mu_N$, as demands the correspondence principle.

When $q^2\to 0$, then gamma-raying of the proton is performed by long-wavelength photons which do not "see" internal proton structure. In this case we simply observe the point particle. By this reason the nucleons formfactors must be chosen such a way that the following conditions are fulfilled

$$\left.\begin{array}{lll}\mathcal{F}_1(0)=1, & \mathcal{F}_2(0)=1 & \text{for proton,}\\ \mathcal{F}_1(0)=0, & \mathcal{F}_2(0)=1 & \text{for neutron.}\end{array}\right\}.$$

When calculating the differential cross section of the elastic electron-proton scattering we shall make use the expression (5.55) as the proton electromagnetic vertex. Then, the proton tensor is defined by

$$Pr^{\lambda\sigma}=\frac{1}{4}\mathrm{Sp}[\Lambda^\lambda(\hat{p}_f+m_p)\Lambda^\sigma(\hat{p}_i+m_p)]. \tag{5.56}$$

Multiplying (5.33) on (5.56) we obtain the final result, namely, Rosenbluth formula

$$\frac{d\sigma}{d\Omega}=\left(\frac{d\sigma}{d\Omega}\right)_0\left\{\mathcal{F}_1^2(q^2)-\frac{a^{(p)2}q^2}{4m_p^2}\mathcal{F}_2^2(q^2)-\frac{q^2}{2m_p^2}\left[\mathcal{F}_1(q^2)+\right.\right.$$

$$\left.\left.+a^{(p)}\mathcal{F}_2(q^2)\right]^2\tan^2(\theta/2)\right\}. \tag{5.57}$$

The factor, standing in the braces of Eq. (5.57), describes the manner in which the scattering process is changed because of the proton structure. When the proton were the point particle, like the muon, then at any q^2 one would have

$$a^{(p)}=0 \qquad \text{and} \qquad \mathcal{F}_1(q^2)=1$$

and the expression (5.57) would coincide with Mott formula. As one should expect, deviations from Mott formula are most large in the region of small wavelength of the virtual photon, that is, in the region of the big values of q^2. For example, when the electric charge distribution inside the proton is described by the exponential law, then at $q^2 \to \infty$ we obtain

$$\frac{d\sigma}{d\Omega} = \left(\frac{d\sigma}{d\Omega}\right)_0 q^{-8}.$$

In order to measure the formfactors experimentally it is convenient to redefine them in such a way that the interference term $\mathcal{F}_1(q^2)\mathcal{F}_2(q^2)$ will be absent in the cross section (5.57). With this object in mind we introduce the electric and magnetic nucleon formfactors (they are called Sachs formfactors as well)

$$G_E^N(q^2) = \mathcal{F}_1^N + \frac{a^{(N)}q^2}{4m_p^2}\mathcal{F}_2^N, \qquad G_M^N(q^2) = \mathcal{F}_1^N + a^{(N)}\mathcal{F}_2^N, \tag{5.58}$$

where $N = n, p$. Then, the formula (5.57) takes the form

$$\frac{d\sigma}{d\Omega} = \left(\frac{d\sigma}{d\Omega}\right)_0 \left[\frac{(G_E^p)^2 - q^2(G_M^p)^2/(2m_p)^2}{1 - q^2/(2m_p)^2} - \frac{q^2}{2m_p^2}(G_M^p)^2\tan^2\frac{\theta}{2}\right]. \tag{5.59}$$

From definition of $G_E^N(q^2)$ and $G_M^N(q^2)$ follows

$$\left.\begin{array}{ll} G_E^p(0) = 1, & G_M^p(0) = 1 + a^{(p)} = 2.79\mu_N, \\ G_E^n(0) = 0, & G_M^p(0) = a^{(n)} = -1.91\mu_N. \end{array}\right\} \tag{5.60}$$

This suggests that $G_E^N(q^2)$ and $G_M^N(q^2)$ must be related with distributions of the charge and magnetic moment of nucleon.

The experimental method of determining the formfactors is simple in description at least. Let us fix q^2 and plot the quantity

$$f(\tan^2\frac{\theta}{2}) = \frac{d\sigma}{d\Omega}\left(\frac{d\sigma}{d\Omega}\right)_0^{-1}$$

along the ordinate whereas the values of $\tan^2(\theta/2)$ along the abscissa. Then the function f is represented as the straight line (Rosenbluth straight line) whose slope is $-q^2[G_M^p(q^2)]^2/(\sqrt{2}m_p)^2$ and whose ordinate in the point

$$\tan^2\frac{\theta}{2} = -\frac{1}{2[1 - q^2/(2m_p)^2]}$$

equals $(G_E^p)^2[1 - q^2/(2m_p)^2]^{-1}$. In other words, the straight line slope gives the value of $(G_M^p)^2$ while the ordinate does the value of $(G_E^p)^2$. Repeating this procedure under the different values of q^2 one may define the formfactors as the functions on q^2. True enough, since we are dealing with the formfactors squares then the confused ambiguity, concerning the formfactors signs, is being left. However, it will disappear if we take into account the formfactors values at $q^2 = 0$ (the sole exception is provided by $G_E^n(q^2)$ because $G_E^n(0) = 0$).

The neutron formfactors are found from the data of scattering off electrons on deuteron. For example, under the analysis of the reaction

$$e^- + d \to n + p + e^-$$

one should subtract the contribution coming from the electron-proton scattering and make the small correction on the nucleon coupling. The obtained data are in accordance with assumption

$$G_E^n(q^2) \approx 0, \qquad G_M^n(q^2) \approx a^{(n)} G_E^p(q^2).$$

In Fig. 27 we display the proton formfactor dependence on the square of the transferred momentum.

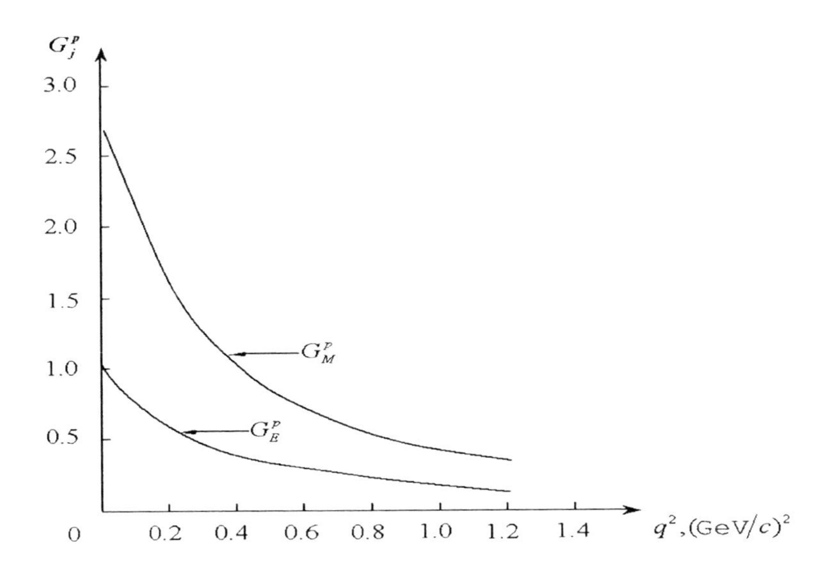

Figure 27. The transferred momentum square dependence of the proton formfactors.

For point particles the formfactor is the constant. The formfactor dependence on q^2 which is observed in experiments means that the nucleons have a specific structure. The experimental data are agreed with the so-called scale relations (the scaling law)

$$G_E^p(q^2) = \frac{G_M^p(q^2)}{2.79} = G(q^2), \tag{5.61}$$

where in the region $q^2 \leq 0.5$ (GeV)^2 the uniform formfactor $G(q^2)$ is well described by the empiric dipole formula

$$G(q^2) = \left(1 - q^2/m_0^2\right)^{-2}, \tag{5.62}$$

$m_0 = 0.71$ GeV. At $q^2 \geq 0.5$ GeV^2 deviations of $\leq 20\%$ from the dipole formula are observed. Notice, all the experimental data concerning the elastic ep-scattering may be described if one assumes that the scaling law is valid and the uniform formfactor takes the form of the two poles sum

$$G(q^2) = \frac{b}{1 - q^2/m_1^2} + \frac{1-b}{1 - q^2/m_2^2}, \tag{5.63}$$

where

$$b = -0.33, \qquad m_1 = 1.31 \text{ GeV}, \qquad m_2 = 0.64 \text{ GeV}.$$

Basing on the experimental data by measuring the formfactors it is easy to determine the proton space size and the charge distribution character of the proton. In nonrelativistic limit $\mathbf{q}^2 \ll m_p^2$ we may neglect the change of the proton energy under scattering. Then, in the rest system of the proton, the formfactor $G_E^p(q^2)$ may be interpreted as the Fourier transform of the static distribution of the proton electric charge

$$G_E^p(q^2) = \int \rho(\mathbf{r}) \exp(i\mathbf{q}\cdot\mathbf{r}) d\mathbf{r}, \tag{5.64}$$

(for the point particle we have $\rho(\mathbf{r}) = \delta(\mathbf{r})$ and the formfactor simply equals 1). Then, the exponent entering into (5.64), may be expanded in a series

$$G_E^p(q^2) \approx \int \rho(\mathbf{r}) \left(1 + i(\mathbf{q}\cdot\mathbf{r}) - \frac{(\mathbf{q}\cdot\mathbf{r})^2}{2} + ...\right) d\mathbf{r}.$$

Assuming the charge distribution to be spherically symmetric ($\rho(\mathbf{r}) = \rho(r)$) we obtain

$$G_E^p(q^2) \approx 1 - \frac{1}{2}\int \rho(r)(\mathbf{q}\cdot\mathbf{r})^2 d\mathbf{r} \approx 1 - \frac{\mathbf{q}^2}{6}\int \rho(r) r^2 (4\pi r^2 dr) \approx$$

$$\approx 1 - \frac{\mathbf{q}^2}{6} < r^2 >,$$

where $< r^2 >$ is the average value of the proton radius square. Measurements of the nucleons formfactors lead to the conclusion that the average radius of the proton and neutron has the order of 0.8 Fermi.

To establish the electromagnetic structure of the proton we should check what forms of the charge distribution lead to the dipole formula (5.62) which well works in the region of small values of q^2. It turns out that the positive result is provided by the exponential distribution

$$\rho(\mathbf{r}) = \rho(r) = \frac{m_0^3}{8\pi} \exp(-m_0 r), \tag{5.65}$$

where m_0 has meaning of mass of particle carrying interaction between nucleons. Really, substituting (5.65) into (5.64) and choosing the spherical coordinate system with the axis z along the vector $\mathbf{q}$, we arrive at the result (θ is the azimuthal angle!)

$$G_E^p(q^2) = \frac{m_0^3}{8\pi}\int_0^\infty dr \int_0^{2\pi} d\varphi \int_0^\pi d\theta r^2 \sin\theta \exp(i\,|\,\mathbf{q}\,|\,r\cos\theta - m_0 r) =$$

$$= \left(1 - q^2/m_0^2\right)^{-2}. \tag{5.66}$$

So, the distribution of the charge and the magnetic moment for the proton is described by the sufficiently simple function, the exponent. Since the quantity $\rho(r)$ tends to constant under $r \to 0$, then it is obvious, that a nucleon does not have any hard core, that is, there are no charges congestion in the center as it was in the case of atom. From it does not follow in any way that structure elements are generally absent inside a nucleon. On the contrary, distribution nonhomogeneity of the charge and the magnetic moment testifies to doubtless presence of such objects. In order to "see" and investigate the properties of blocks which constitute a nucleon one needs to increase the resolution of our devices. In the quantum language it means the following increasing de Broglier wavelength of the virtual photon, that is, the increase of q^2.

5.3. Bjorken Scaling

Without doubt that the energy growth of incoming electrons leads to changing the character of the collisions with nucleons. So, at definite values of q^2 the proton may be excited into one of the nucleons resonances ($\Delta(1238), N^*(1520), N^*(1688)$, and the like) and emits the π^0-meson under returning to the final state (quasi-elastic scattering). On further increasing the values of q^2 multiple hadrons generation, a proton break-up into numerous fragments (deep inelastic scattering), will take place. In such a situation, the identification of the final proton state is out of the question. In Fig. 28 we represent the Feynman diagram for this case.

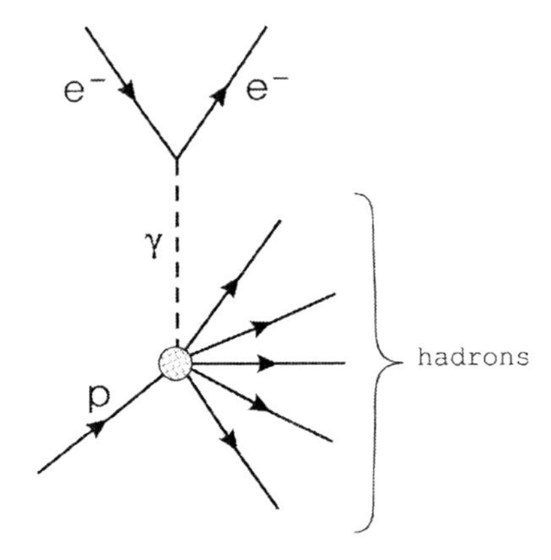

Figure 28. Deep inelastic collisions of electrons with hadrons.

Here, P is initial four-momentum of the proton and P_f is four-momentum of final hadron state. It appears, that now we already have two independent variables, as distinguished from the elastic scattering. To make sure in this we introduce the quantity

$$\nu = \frac{(q \cdot P)}{m_p}, \tag{5.67}$$

which simply is the electron energy loss in the proton rest system, that is,

$$\nu = E_i^{(e)} - E_f^{(e)}.$$

Let us see what a role plays the quantity ν in the electron-proton collisions. Mass of a final state is defined according to

$$M_f^2 = P_f^2 = (q+P)^2 = m_p^2 + q^2 + 2m_p\nu. \tag{5.68}$$

From Eq. (5.68) we find the values of q^2 for the elastic scattering

$$q^2 = -2m_p\nu, \tag{5.69}$$

for exciting a nucleon resonance with the mass m_N

$$q^2 = -2m_p\nu - m_p^2 + m_N^2, \tag{5.70}$$

and for deep inelastic scattering (the proton converts into continuous spectrum through the resonances region)

$$q^2 = -2m_p\nu - m_p^2 + M_f^2, \tag{5.71}$$

where M_f^2 forms continuous spectrum. We select the scattered electron in the final state and do not concretize hadrons system on which the proton is decayed. Such reactions are called inclusive ones, since they include all what is possible into hadrons system. Then the measured cross section is the cross sections sum including different final states of hadrons. From (5.71) follows, since M_f^2 may take any values then the quantities q^2 and ν are already independent variables in the case of deep inelastic scattering. Therefore, the additional kinematic degree of freedom appears under the inelastic electron-proton scattering. One is clearly expecting, thanks to this circumstance, inelastic scattering investigation gives more information than elastic processes investigation. In the case of inelastic scattering the electron-proton interaction dynamics will already define the formfactors $W_{1,2}(\nu, q^2)$ (since they define the proton structure they are called the structure functions). Their precise form can be established only within sequential theory of strong interaction, that, is, the QCD. However, in order to determine $W^{\lambda\sigma}$ we again call to relativistic and gauge invariance for help. Let us go in this way.

There are two vectors P_λ and q_λ in our disposal. As we are going to parameterize the cross section, which has been already summed up and averaged in spins, the matrices γ_λ are not included in consideration. The set of independent tensors which are built from the vectors P_λ and q_λ is as follows:

$$P^\lambda P^\sigma, \qquad q^\lambda q^\sigma, \qquad P^\lambda q^\sigma, \qquad q^\lambda P^\sigma. \tag{5.72}$$

Based on the fact that the metric tensor $g^{\lambda\sigma}$ can also participate in our constructions, the most general expression for the hadron tensor takes the form

$$W^{\lambda\sigma} = a_1 g^{\lambda\sigma} + a_2 P^\lambda P^\sigma + a_3 q^\lambda q^\sigma + a_4 P^\lambda q^\sigma + a_5 q^\lambda P^\sigma, \tag{5.73}$$

where a_i is the function of the scalars q^2 and ν. From the current conservation law follows

$$q_\lambda W^{\lambda\sigma} = q_\sigma W^{\lambda\sigma} = 0. \tag{5.74}$$

Then, multiplying (5.73) on q_λ (q_σ) and equaling the coefficients at q^σ and P^σ (q^λ and P^λ), we arrive at four equations:

$$a_1 + a_3 q^2 + a_4(Pq) = 0, \tag{5.75}$$

$$a_2(Pq) + a_5 q^2 = 0, \tag{5.76}$$

$$a_1 + a_3 q^2 + a_5(Pq) = 0, \tag{5.77}$$

$$a_2(Pq) + a_4 q^2 = 0. \tag{5.78}$$

Subtracting (5.76) from (5.78), we obtain $a_4 = a_5$. That, in its turn, means that Eqs. (5.75) and (5.77) coincide, i.e. there are only two equations to define four quantities. We choose a_1 and a_2 as the independent quantities. From Eqs. (5.77) and (5.78) we find:

$$a_4 = -\frac{a_2(Pq)}{q^2}, \qquad a_3 = -\frac{a_1}{q^2} + \left[\frac{(Pq)}{q^2}\right]^2 a_2. \tag{5.79}$$

Using the relations (5.79) and introducing the designations

$$a_1 = -W_1(\nu, q^2), \qquad a_2 = \frac{W_2(\nu, q^2)}{m_p^2},$$

we obtain the final expression for the hadron tensor $W^{\lambda\sigma}$:

$$W^{\lambda\sigma} = \left(-g^{\lambda\sigma} + \frac{q^\lambda q^\sigma}{q^2}\right) W_1(\nu, q^2) + \frac{1}{m_N^2}\left[P^\lambda - \frac{(Pq)q^\lambda}{q^2}\right] \times$$

$$\times \left[P^\sigma - \frac{(Pq)q^\sigma}{q^2}\right] W_2(\nu, q^2), \tag{5.80}$$

where $W_{1,2}(\nu, q^2)$ are the functions defining the nucleon structure. Multiplying (5.80) with the electron tensor $L_{\lambda\sigma}$ (Eg. (5.33)) and neglecting the electron mass we obtain the following expression for the doubly differential cross section of the inclusive ep-scattering in the laboratory system

$$\frac{d^2\sigma}{d\Omega dE_f^{(e)}} = \sigma_0 \left[2W_1(\nu, q^2)\tan^2\frac{\theta}{2} + W_2(\nu, q^2)\right], \tag{5.81}$$

where

$$\sigma_0 = \left(\frac{\alpha}{2E_i^{(e)}\sin^2(\theta/2)}\right)^2 \cos^2\frac{\theta}{2}. \tag{5.82}$$

From (5.81) it follows that the functions $W_{1,2}$ have the dimensionality of length. The comparison of (5.82) with (5.44) makes obvious the fact that σ_0 is nothing else but the cross section of the elastic scattering off electrons on the point spinless particle having infinitely large mass.

The structure functions $W_{1,2}(\nu, q^2)$ represent the inelastic analog of the formfactors of the elastic scattering. Technique of their experimental determination is also simple. When one fixes the variables q^2, ν and plots the values of

$$f(\tan^2(\theta/2)) = \frac{d^2\sigma}{d\Omega dE_f^{(e)}}\sigma_0^{-1}$$

along the ordinate whereas the values of $\tan^2(\theta/2)$ along the abscissa, then $f(\tan^2(\theta/2))$ is displayed as a straight line. Its slope is equal to $2W_1(\nu, q^2)$ and its ordinate is equal to $W_2(\nu, q^2)$ at the point $\tan^2(\theta/2) = 0$. In Fig. 29 we represent $W_2(\nu, q^2)$ as a function on ν for different values of q^2 at $\theta = 6^0$.

At small values of ν the peaks of $W_2(\nu, q^2)$ correspond to the elastic formfactor and the resonances excitation. At $\nu \geq 3$ (GeV)^2 the final states hit on the continuous spectrum region and the curve is flatten. When $\nu \geq 4$ (GeV)^2 all the experimental points of Fig. 27 lay on the same curve for any values of q^2. Let us try to understand all the following consequences.

We define the dimensionless variable

$$x = -\frac{q^2}{2m_p\nu}, \qquad 0 \leq x \leq 1, \tag{5.83}$$

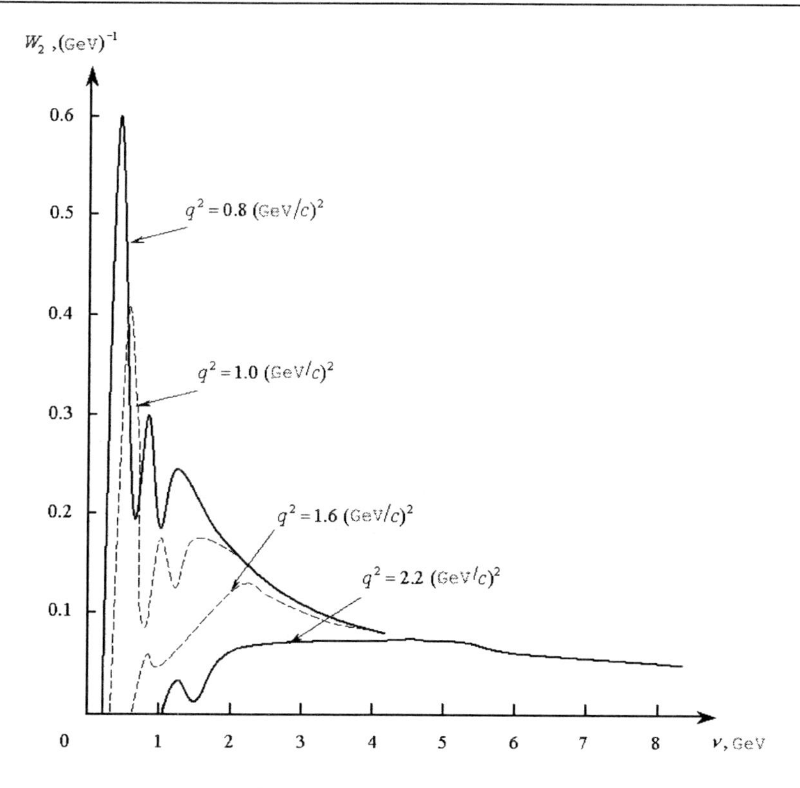

Figure 29. The $W_2(\nu, q^2)$-dependence of ν under different values of q^2.

which is connected with the hadron system invariant mass M_f by the relation

$$M_f^2 = m_p^2 - q^2(1/x - 1).$$

One logically passes to the dimensionless structure function too

$$\nu W_2(\nu, q^2) = G_2(x, q^2/m_p^2). \tag{5.84}$$

At $\nu \geq 4$ GeV^2 the quantity $G_2(x, q^2/m_p^2)$ becomes the function on the variable x only. In other words, when one fixes x, then the plot of $G_2(x, q^2/m_p^2)$ versus x is represented by the straight line being parallel to the ordinate. The experiments show that the other structure function has the analogous behavior

$$m_p W_1(\nu, q^2) = G_1(x, q^2/m_p^2), \tag{5.85}$$

where we have again passed to the dimensionless variable. Thus, in the deep inelastic region

$$| q^2 | \gg m_N^2, \qquad \nu \gg m_N, \tag{5.86}$$

the structure functions $G_{1,2}(x, q^2/m_p^2)$ become independent from any scale or they are scale invariant.

Recall, that the scale invariance, the scaling, is the invariance of physical theory with respect to space-time transformations

$$\mathbf{x} \to \rho \mathbf{x}, \qquad t \to \rho t, \tag{5.87}$$

where $\rho > 0$ is a numerical parameter of transformation. In quantum theory the transformations (5.87) are supplemented by ones

$$\mathbf{p} \to \rho \mathbf{p}, \qquad E \to \rho E. \tag{5.88}$$

Physical quantities are changed in accordance with their dimensionalities under the scale transformations. So, an electromagnetic field vector potential and a current are transformed by the law

$$\mathbf{A} \to \rho^{-1}\mathbf{A}, \qquad \mathbf{j} \to \rho^{-3}\mathbf{j}.$$

It is evident that dimensionless quantities are scale invariants. The particles masses also fall into this category. When the masses or other dimension quantities, which are not changed under the scale transformations, do not enter into motion equations or boundary conditions, then the corresponding theories are scale invariant. Free Lagrangians of the photon field and the gluon one possess the scale invariance. Clearly, in the real world where gravitational, weak, electromagnetic and strong interactions are switched on, the scaling does not take place. The absence of the scale invariance is caused in the first place by the fact that for physical particles the relation must be fulfilled

$$E^2 = m^2 + \mathbf{p}^2.$$

It is obvious that this relation is not invariant with respect to the transformations (5.88). On the other hand, there are no reasons which would obstruct exhibiting the scaling in Nature. As we saw, the scale invariance of the dimensionless structure functions, taking place in the deep inelastic *ep*-scattering, is one of such examples. Even before the first experiments for studying the inclusive electron-proton scattering, J. Bjorken predicted the scaling of the functions $G_{1,2}(x, q^2/m_p^2)$ in the deep inelastic region, that is,

$$\left.\begin{array}{l} G_1(x, q^2/m_p^2) \to F_1(x), \\ G_2(x, q^2/m_p^2) \to F_2(x), \end{array}\right\} \tag{5.89}$$

under

$$|\, q^2 \,| \to \infty, \qquad \nu \to \infty$$

and x is fixed. On this reason the phenomena of the structure functions scale invariance is named by Bjorken scaling. More later experiments, fulfilled on electrons, muons and neutrinos beams, displayed that Bjorken scaling is not exact (true, violation is insignificant and may be considered as a correction to the basic effect). We shall not go into causes of scaling violation since that is beyond the framework of the book (successive explanation could be obtained within the QCD only). At the given stage the more important thing for us is to understand conclusions which follow from the scale invariance.

The behavior of the functions $G_{1,2}(x, q^2/m_p^2)$ is greatly distinguished from the corresponding behavior of the elastic formfactors $G_{E,M}(q^2)$. Whereas $G_{E,M}(q^2)$ sharply fall down with the increase of $|\, q^2 \,|$, at $|\, q^2 \,| \to \infty$ the functions $G_{E,M}(q^2)$ do not depend on q^2 at all. This looks like that the proton would not have the electromagnetic formfactors in the deep inelastic scattering region. In other words, in this case the proton behaves as a point particle. However, the proton is not the point particle and its size defined from the elastic ep-scattering is far from small $\sim$ 1 Fermi. The only reasonable explanation of the experiments on ep-scattering resides in the fact that the electric charge inside the proton is concentrated in several points, that is, in particles entering into the composition of the proton. All this may be formulated by a somewhat different way. The presence of the scaling means that such dimensional parameters as the proton mass m_p and the corresponding length $\sim 1/m_p$ do not play any significant dynamical role in the deep inelastic scattering processes, that is, there is no distances scale in this case for the proton. Explain aforesaid on the example of an atom. In an atom, aside from its own size, there is one more distances, scale, the size of the atom nuclear. Thanks to uncertainty relation, the scale of distances is inversely proportional to the scale of energies. Existence of two scales in an atom leads to the fact that processes, taking place with it at low and high energies, are distinguished from each other by the radical way. When the distances are bigger than the nuclear size R_N, the system is approximately described by Coulomb potential. However, when the distances have the order of R_N or are smaller than R_N, Yukawa potential works. In that case when one scale is available, the qualitative change of the processes character is not in progress under energy increase. To put it otherwise, for the proton in particular, and for hadrons in general, the second scale is equal to infinity in the case of the inclusive scattering. This means, if hadron really represent the composite particle, then particles, entering to hadron, should have the negligibly small size, that is, they should be point or, what is the same, structureless. At collisions with these hadron blocks the electrons can be often scattered through large angles just as α-particles were scattered in Rutherford experiments when they were finding their way into atoms nuclei.

5.4. Parton Model

To prove the scale invariance of the structure functions $G_{1,2}(x, q^2/m_p^2)$ Bjorken used the hadron currents algebra. One naturally expected the appearance of the model which could explain the scaling from the point of view of the hadron structure. With this in mind, in 1969 R. Feynman suggested the parton model, the brief content of which we shall give below.

Extended nucleon (and any hadron as well) does not represent formation which is smeared by the continuous way in the space, but it consists of weakly confined point particles (partons). The scattering of the high energy electron occurs on partons incoherently, that is, the electron is elastically scattered on one of partons, not affecting others. Every nucleon participates in the inelastic reaction by only one parton (active parton), which transfers the fraction x_i (i is the index of parton kind) of the hadron four-momentum P_μ. Thus, for the ith parton we have

$$\mathbf{p}_i = x_i \mathbf{P}, \qquad E_i = x_i E, \tag{5.90}$$

When one denotes the partons number density, the parton distribution function, by $f_i(x_i)$, then $f_i(x_i)dx_i$ defines the number of partons with the four-momentum $x_i P_\mu$ in the range from x_i to $x_i + dx_i$. However, under fulfillment (5.90) the parton mass proves to be a variable quantity which seems strange at least. The situation is clearing up in the reference system where the time component of the vector q_μ is equal to zero (such system really exists since $q^2 < 0$). To find it we consider the system K which moves as related to the laboratory system with the velocity v being parallel to the vector $\mathbf{q}$. Using Lorentz transformation we obtain

$$q_0 = \frac{q'_0 - v \mid \mathbf{q}' \mid}{\sqrt{1-v^2}}, \tag{5.91}$$

where we have supplied the quantities in the laboratory system by the prime. When one chooses the velocity v to be equal to

$$v = \frac{q'_0}{\mid \mathbf{q}' \mid}, \tag{5.92}$$

then from Eq. (5.91) follows that we have achieved our goal — $q_0 = 0$. It is clear that in the system K the nucleon momentum is equal to

$$\mid \mathbf{P} \mid = \frac{m_N v}{\sqrt{1-v^2}}. \tag{5.93}$$

Using (5.92) and taking into consideration $q'_0 = \nu$, we obtain

$$\mid \mathbf{P} \mid = \frac{m_N \nu}{\sqrt{Q^2}} \tag{5.94}$$

and

$$\mathbf{P}^2 = \frac{m_N^2 \nu^2}{Q^2} = \frac{m_N \nu x}{2}, \tag{5.95}$$

where we have passed to the positive quantity $Q^2 = -q^2$ for the reasons of convenience. In the deep inelastic region the relations (5.86) take place, therefore,

$$\mid \mathbf{P} \mid \gg m_N \qquad \mathbf{P} \to \infty. \tag{5.96}$$

Thus, in the Lorentz system where $q_0 = 0$ and the relation (5.96) is fulfilled (the system of infinity momentum (SIM)), one may escape the question about a variable mass of a parton, if one assumes both m_i and m_N being equal to zero. The notion of partons has sense only in the reference system where a nucleon moves with the relativistic velocity. This circumstance is the reflection of the already known fact, that only the high energy virtual photon ($\mid q^2 \mid \geq 1-2$ GeV^2) may discern a parton. In the SIM the momentum transverse component (as related to the direction of the nucleon motion) of a parton appears to be negligibly small. Really, in the rest system of a nucleon the mean square longitudinal and transverse parton momenta are equal each other. It is clear, that in the SIM, whose velocity as related to the laboratory system is close to the light velocity ($v = \nu/\sqrt{\nu^2 - q^2} \approx 1$) the longitudinal parton momentum is much more bigger than the transverse parton momentum. The transition to the SIM has one more advantage, namely, it sheds light on a parton behavior in a nucleon. In this system the interactions acts frequency of partons with each other decreases, by virtue of the relativistic delay of the time. Thus, in the short time interval between interactions of the virtual photon with the parton, the parton behaves as a nearly free particle. Scattered partons (active partons) and the initial hadrons residues, which did not take part in interactions (the set of passive partons), turn into final hadrons thanks to strong interactions. The final hadrons produce two jets, one in the direction of a scattered parton, and other in the direction of an initial parton momentum. A hadron jet represents the set of hadrons having small (the order of 300 GeV) transverse momenta relative to the motion of the parent particle. The jets existence already on its own serves as evidence of weakness of hadron matter interaction on the small distances. Indeed, if the hadron matter produced what amounts to dense high excited cluster, then the isotropic configuration with a large value of an average transverse momentum (the order of an collision energy) would be natural for outgoing secondary particles.

The parton model proves to be extremely fruitful. First and foremost, with its help one managed to prove the scaling behavior of the structure functions $G_{1,2}(x, q^2/m_p^2)$. This model also gives an opportunity to establish the parameters of partons participating in electromagnetic interaction. Evidently, it would be rather attractive to identify the partons with the quarks. From quantum laws, quarks in hadrons can exist both in real and in virtual states. We agree to call the real quarks by the valence quarks[1]. So, three valence quarks enter into baryons whereas mesons consist of valence quark-antiquark pairs. Just valence quarks define additive quantum numbers of hadrons (electric charge, strange, baryon charge and so on). Thanks to the uncertainty relation, quark-antiquark pairs could be supplemented to valence quarks on a short time. One natural calls the quarks forming the sea of virtual quark-antiquark pairs by the sea quarks. For the belief to the quark-parton model to be strengthened, the electric charges and the spins of quarks should be measured. As a consequence of detail measurements and comparing the experiments on various hadron targets they managed to select the contributions coming from different kinds partons and define partons electric charges. It appeared that they coincide with the quarks electric charges, namely, are equal to $2/3$ and $-1/3$. Experiments unambiguously defined that the spin of the charged partons is equal to $1/2$.

To investigate the electron-nucleon scattering gives the opportunity to define the nucleon momentum fractions, which are transferred by quarks and antiquarks, when the nucleon moves with the big velocity. The quarks and antiquarks contributions to the nucleon momentum could be expressed through the quarks and antiquarks distribution functions

$$\sum_i \int dx x[f_i(x) + \overline{f}_i(x)] = 1 - \varepsilon, \tag{5.97}$$

where we have took into account the contributions coming both from valence and from sea quarks. If one assumes that the quarks and antiquarks are the sole pretenders on the partons role, then ε must take the value 0. In the end of 60s of XX century the experiments on probing the nucleons by virtue of the electrons were fulfilled at SLAC (Stanford Linear Accelerator Center) and some years later (in 1973) the similar raying of nucleons with the help of the neutrino beams was carried at CERN (Conseil Europeen pour la Recherche Nucleaire). These experiments gave

$$\varepsilon \approx 0.5.$$

This result means that nearly 50% of the nucleon momentum are transferred by partons which do not take part both in electromagnetic and weak interactions. In the QCD only the gluons, which are the carriers of the strong interaction between quarks, may pretend on the role of such particles.

[1]Such quarks are also named by the block or constituent quarks.

At present we discuss the question about the quarks masses. The particle mass could be exactly determined on its energy only for the free particle. Since the free quarks are not discovered up till now, then one assigns the precise meaning to their masses with difficulty. The quark mass problem is very similar to the problem concerning the electron mass in the solid state physics. The electron, when it is moving in a solid, behaves as a particle with the "effective" mass m_{ef}, which is significantly distinguished from its true mass m. Moreover, m_{ef} may depend on the motion features, because in the reality the masses difference $\Delta m = m_{ef} - m$ is caused by interaction of the electron with objects surrounding it. In this sense the masses of all the quarks are "effective", because they are defined in the processes in which the quarks are interacting with other particles. For this reason the quarks masses values, found under analyzing the energy levels location of the bound states (quarkoniums), may appreciably differ from those values which correspond to the weak decay of the quarks. They distinguish the current and constituent (block) quarks masses. Since, in the quantum field theory the interactions are formulated on the language of currents and potentials, it is natural to call the quarks entering into Lagrangians by the current quarks. Thus, the current masses concern to naked quarks and they do not take into account the contributions coming from their gluons and quark-antiquarks fur coats. In the QCD the current quark mass proves to be depended on the momentum transferred to the quark and be decreased with the momentum growth. So, at the scale ~ 2 GeV the current quarks masses are confined into the intervals

$$\left.\begin{array}{c} 1.5\,\text{MeV} < m_u < 5\,\text{MeV}, \qquad 3\,\text{MeV} < m_d < 9\,\text{MeV}, \\ 60\,\text{MeV} < m_s < 170\,\text{MeV}. \end{array}\right\} \tag{5.98}$$

Thanks to the fur coats contribution, the block masses exceed the corresponding current masses on 300 GeV approximately. From (5.98) it becomes clear that the success of the $SU(2)$-symmetry is basically caused by a closeness of the masses of the u- and d-quarks. The symmetry with respect to the $SU(3)$-group, which includes the more heavier s-quark, is already violated much stronger. However, on the irony of fate, just the approximate $SU(3)$-symmetry found the exact symmetry status in the same quark walk of life, provided that it is being used to describe interactions between quarks.

5.5. Color

So, late in the (19)60s, the hypothetical quarks have been acquiring the status of objects the reality of which, while indirectly, manifests in experiments. However, the quark model has a lot of unresolved problems, as before. One of problems is connected with the quarks statistics. As an example, we consider the Ω^--hyperon entering into the $3/2^+$ baryon decuplet. Its total wave function is the product of three wave functions which express the dependence on the space variables, the spin and the unitary spin, that is,

$$\Psi_\Omega = \Psi(\mathbf{r})\Phi(\mathbf{S})\Theta(\mathbf{S}^{(un)}), \tag{5.99}$$

where the quark filling of the Ω^--hyperon may be schematically represented by the following way

$$\Theta(\mathbf{S}^{(un)}) = | s\uparrow s\uparrow s\uparrow >, \tag{5.100}$$

(the arrow on the quark symbol defines the direction of the spin projection). It is clear that $\Theta(\mathbf{S}^{(un)})$ is completely symmetric with respect to any transposition of the s-quarks. In the case of three spins parallelism $\Phi(\mathbf{S})$ is also symmetric. Since the symmetry character of the wave function space part is determined by the factor $(-1)^L$ (for the $3/2^+$-decuplet, $L=0$), then $\Psi(\mathbf{r})$ proves to be symmetric under the three quarks transposition as well. Thus, the total wave function of three s-quarks system appears symmetric. However, the quarks have the half-integer spin, consequently, they obey to Fermi-Dirac statistics and Pauli principle is valid for them. As a result, the total wave function Ψ_Ω must be antisymmetric. To prove Pauli theorem about the connection of the spin with the statistics is based on such fundamental concepts of the quantum theory as the microcausality and locality. The refusal from this theorem were tantamount to Aurora volley announcing the October revolution beginning in Russia. But the evolutions way always was more preferable than the mysterious ways of revolutions. The painless solution of the conflict with the statistics proves to be possible under introducing the new discrete variable. This variable, called color, is assigned to all the quarks independently of the flavor. By the example of the Ω^--hyperon, it is clear that the minimum number of the new variable values should be equal to 3. We are restricted by three values for the color degree of freedom and shall designate them as R (red), G (green) and B (blue). To introduce the color allows to put three quarks into one and the same quantum mechanical state inside a hadron. For Pauli principle to be fulfilled, the baryon wave functions must be antisymmetrized on the color variables . So, the Ω^--hyperon wave function part, connected with the unitary spin, has the form

$$\Theta(\mathbf{S}^{(un)}) = \frac{1}{\sqrt{6}} \sum_{i,j,k=1}^{3} \varepsilon_{ijk} s_i s_j s_k, \tag{5.101}$$

where the coefficient $\frac{1}{\sqrt{6}}$ is related with the normalization, and not entirely convenient indices R, G, B are simply replaced by 1,2,3.

When we are talking about the kinematic aspects of the quark systems, three internal degrees of freedom, the colors, are conveniently considered as the eigenvalues of the color spin operator (Fig. 30). It should be stressed that by the definition the color spin operator has non-zero values for quarks and gluons only.

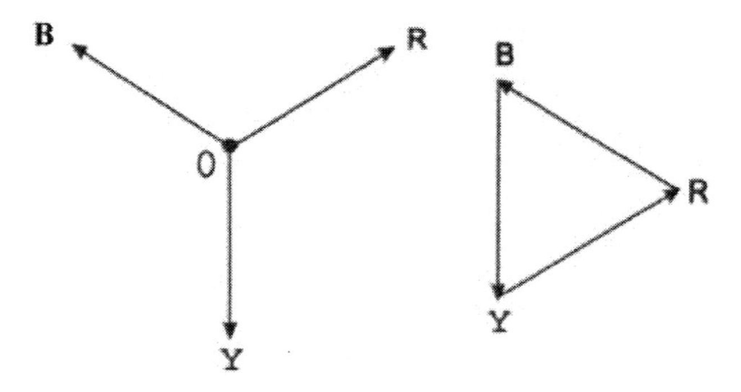

Figure 30. The color spin operators.

If the hypothesis of the quarks confinement is valid, the color spin belongs to the number of the hidden degrees of freedom. This, in its turn, means that all the particles, which are being observed in the free state, must have the color spin being equal to zero. As it follows from Fig. 30 there are two ways to obtain a colorless (or white) hadron. The first way is to mix three different colors (or anticolors) while in the second way every color is mixed with the corresponding anticolor. Then, for any baryons, whose wave function with consideration both for flavor and for color will be written in the form

$$B_{\alpha\beta\gamma}=\frac{1}{\sqrt{6}}\sum_{i,j,k=1}^{3}\varepsilon_{ijk}q_{\alpha i}q_{\beta j}q_{\gamma k}, \tag{5.102}$$

the color spins are mutually compensated and the total baryon color equals zero. Analogously, the mesons prove to be colorless because they consist of mixture of color and corresponding anticolor in equal proportion

$$M^{\alpha}_{\beta}=\frac{1}{\sqrt{3}}\sum_{i=1}^{3}\overline{q}^{\alpha i}q_{\beta i}. \tag{5.103}$$

Let us express the hadron colorless in the strict mathematical language. Assume that a quark changes its color, to say, it goes from one color state to another. Moreover, we shall consider that this new state is the linear combination of all the possible old states (for the sake of simplicity, we omit the flavor indices of the quarks)

$$q'_i=\sum_{i=1}^{3}U_{ij}q_j. \tag{5.104}$$

Now we require hadrons to have one and the same view both in the old and in new variables, i. e. hadrons should be the invariants of the transformation (5.104)

$$\sum_{i=1}^{3}q'_i\overline{q}'_i=\sum_{i=1}^{3}q_i\overline{q}_i, \tag{5.105}$$

$$\sum_{i,j,k=1}^{3}\varepsilon_{ijk}q'_iq'_jq'_k=\sum_{i,j,k=1}^{3}\varepsilon_{ijk}q_iq_jq_k. \tag{5.106}$$

As the antiquarks $\overline{q}$ are transformed by the complex conjugated matrices U^*, then from (5.105) follows

$$\sum_{i=1}^{3}U_{ij}U^*_{im}=\delta_{jm},$$

that is, the transformation (5.105) is performed by the unitary matrices. In its turn Eq. (5.106) leads to the condition

$$\det U=1 \tag{5.107}$$

which means that the transformation (5.105) is special. So, the transformations in question consist the special unitary group in three-dimensional color space, the $SU(3)$-group, that is, the explicit form of (5.104) is as follows

$$q'_j=[\exp(i\alpha_aT_a)]_{jk}q_k, \tag{5.108}$$

where $a = 1,2,...8$, α_a are the real parameters and $T_a = \lambda_a/2$ are the group generators. Further in order to distinguish this group from the group on flavors we shall be talking about it as the color group $SU(3)_c$. The observed hadrons are invariant with respect to the transformations of the $SU(3)_c$-group or, what is the same, are the color singlets. The color spin introduction liquidates not only the conflict with the statistic, it also forbids the existence of bound qq-states (of the type RB, GR, and so on) on the strength of the postulate: *only the colorless hadrons are observable*. Thus, the color scheme explains the exceptional role of the quark combinations $qqq, \overline{qqq}$, and $q\overline{q}$ in Nature.

Now we address to the dynamical aspects of the color hypothesis. It is evident that when we are considering the quarks dynamics the quark interactions should be taken into account. In this case, the role of the color degree of freedom is most of all similar to the role of the electric or gravitational charges and, let they forgive us for this terminological liberty, we shall be speaking about not the color spin but about the color charge. The quantum field theory describing interactions between quarks, the QCD, is built on the ground of localizing the $SU(3)_c$-symmetry by analogy with the quantum electrodynamics in which the Abelian gauge $S(1)$-group is localized. We shall be speaking of it in the sixth chapter. Now we are sufficient to know that the local gauge invariance leads to the conclusion about the existence of the massless gluons octet, the carriers of interaction between quarks. Since the gluons are connected with the color quarks, they are the color charge carriers. The gauge invariant Lagrangian of the QCD has the form

$$L_{QCD} = -\frac{1}{4}G^a_{\mu\nu}(x)G^{\mu\nu}_a(x) + \overline{q_k}(x)[i\gamma_\mu D^\mu_{\overline{k}j} - m_q\delta_{\overline{k}j}]q_j(x), \qquad (5.109)$$

where $a = 1,2,..8$,

$$G^a_{\mu\nu}(x) = \partial_\mu G^a_\nu(x) - \partial_\nu G^a_\mu(x) + g_s f_{abc} G^b_\mu(x) G^c_\nu(x),$$

$$D^\mu_{\overline{k}j} = \delta_{kj}\partial_\mu + ig_s G^\mu_{\overline{k}j}(x), \qquad G^\mu_{jk}(x) = \frac{1}{2}G^a_\mu(x)(\lambda_a)_{jk},$$

$G^\mu_{\overline{k}j}$ is four-dimensional potential of the gluon field in the point x (its components represent the Hermitian 3×3-matrices in the color space). From (5.109) follows that the two color gluons bear the color and anticolor charge. It is clear that the combination depleted of the color charge $g_0 = R\overline{R} + B\overline{B} + G\overline{G}$ is the color singlet. Exchange of this singlet changes by no means the color state of the quark, and consequently, g_0 can not pretend on the role of particle bearing the color interaction between the quarks. The color parts of the wave functions for residuary eight gluons can be established by just the same manner as the wave functions unitary parts of the baryon octet were found. Having fulfilled the replacements

$$u \to R, \qquad d \to G, \qquad s \to B,$$

in Fig.21 and in Eqs. (5.3), (5.4), we obtain the following expressions for the gluons wave functions color parts

$$\left.\begin{array}{llll} g_1 = R\overline{G}, & g_2 = R\overline{B}, & g_3 = G\overline{R}, & g_4 = G\overline{B}, \\ g_5 = B\overline{R}, & g_6 = B\overline{G}, & g_7 = \sqrt{\frac{1}{2}}\left(R\overline{R} - G\overline{G}\right), & \\ & g_8 = \sqrt{\frac{1}{6}}\left(R\overline{R} + G\overline{G} - 2B\overline{B}\right). & & \end{array}\right\} \qquad (5.110)$$

As the gluons have the color charges they can interact with each other. For this reason the QCD is the nonlinear field theory or, to put this another way, the QCD includes the Feynman diagrams with the vertices which describe emitting or absorbing the gluon caused by the gluon. The inclusion of the gluons contribution into vacuum polarization allows to explain the quarks behavior specificity at large transfers of the momentum. With penetrating in the gluon fur coat which surrounds the quark, the quark color charge is being decreased. This means that in the limit of infinitely small distances separating the quarks, the color interaction between them is switched off and they behave very similar to free particles (asymptotic freedom). In the case of the deep inelastic scattering off electrons on protons the quarks, which exhibit themselves as partons, are in the protons just in the same condition.

To build the successive quantum theory of quark-gluons interactions we are needed to use the so-called interaction representation in which the quarks and gluons are described by the free equations of motion. It is apparent that in the QCD such an operation is not absolutely lawful because the quarks and the gluons are not observed in the free states. The quarks in hadrons might be considered as free particles at small distances, and only in this case it is lawful to use the QCD, based on the perturbation theory methods (perturbative QCD). With the growth of the distances between the quarks the effective coupling constant of the quarks increases and, as a result, the perturbative QCD has not ceased to work. By now the quarks confinement has not received the final understanding within the QCD, that is, it has been remaining only the hypothesis confirmed by experiments.

Let us briefly discuss some models which explain the confinement of quarks and gluons inside hadrons. So, one may account for the confinement by that the hadrons in the color states are much heavier than those in the colorless states and, for this reason, the latter are not observed in up-to-date experiments. The analogy with atoms helps to understand this idea. Let the neutral (nonionized) atoms correspond to the white hadrons while the charged ions do to the hadrons color states. It is clear that the ions have larger energy and are going to come back in the neutral atom state. Such a tendency is explained by the fact that the electromagnetic interaction, which could be approximately described by Coulomb potential $V_c \sim \alpha_{em}/r$ in the atom case, acts between the electric charges. One naturally assume that there exists color interaction to hinder flying the quark out the hadron. This idea has found embodiment in the enough descriptive bag model. The simplest variant, the MIT bag model (Massachusetts technology Institute), is based on the assumption: *a hadron represents the bag with the sharp borders to hinder flying out all color objects*. The hadron system is described by the Lagrangian function

$$L = \int d\mathbf{r}[\mathcal{L}_{QCD} - f(\mathbf{r})], \tag{5.111}$$

where $f(\mathbf{r})$ defines the walls pressure and, consequently, makes provision for the confinement both quarks and gluons. If the quarks become widely separated then the gluon fields, propagating between the quarks, are stretched into straight lines and the bag takes the form of the tube. In the case of interaction of quarks with antiquarks the picture looks the most simple. When one forgets about the nonlinearity for the time being, then the system $q\overline{q}$ would be similar to the electric dipole whose field lines distribution in the space is displayed in Fig. 31 a.

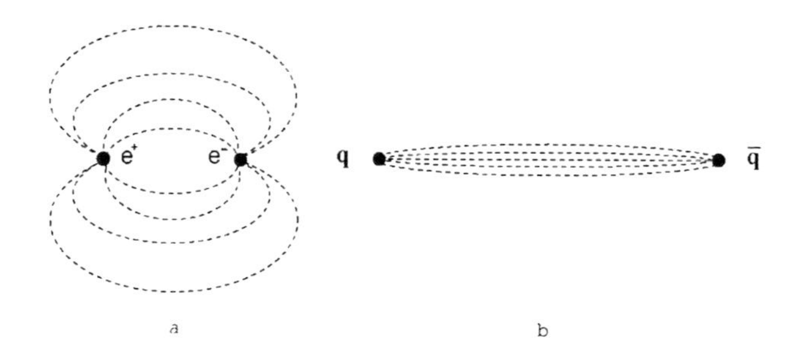

Figure 31. The distributions of the field lines in the $q\overline{q}$ system.

The presence of interaction between gluons results in compressing these lines into the color field tube, the gluonguide (Fig. 31 b). The interaction structure is such that the tube cross section area S is kept constant under the growth of the distance between the members of the pair $q\overline{q}$ (r). But the field line numbers depend only on the sources charges. Consequently the field strength in the gluonguide is being kept constant and the field energy increases proportionally to the tube volume Sr. Since the cross section area S is fixed the energy of the system $q\overline{q}$ is linearly increasing with the growth of r. It means that any process, in which the energy finite amount is transferred to this system, is not able to separate the quark from the antiquark. In reality the flux tube can not be stretched infinitely since the production of the quark-antiquark pair from vacuum has become energetically more profitable which corresponds to the transition of the one-hadron state into two-hadron state. In the first approximation the effective potential of interaction between q and $\overline{q}$ has the form

$$V^c(r) = -\frac{4}{3}\frac{\alpha_s}{r} + Ar + V_0, \tag{5.112}$$

where A and V_0 are positive constant quantities, that is, it looks like a whirlpool (Fig. 32).

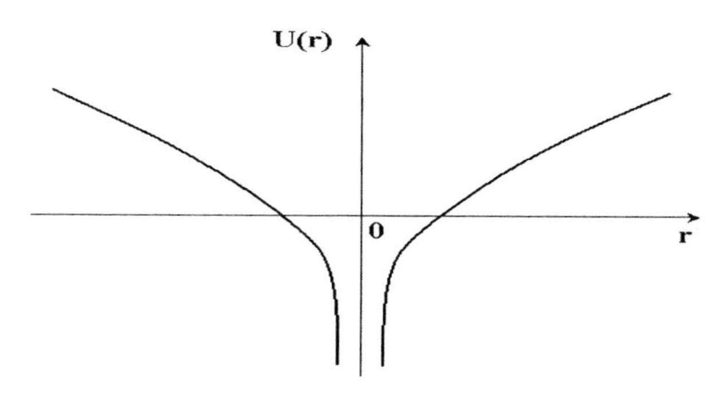

Figure 32. The effective potential of interaction between q and $\overline{q}$.

The first Coulomb-like term $\sim 1/r$ corresponds to the one-gluon exchange (the analog of the one-photon exchange in electrodynamics), and the linear term $\sim r$ providing the quark confinement inside the hadron is explained by the contribution of multigluons exchanges. In the next approximation the relativistic corrections, the spin-spin and spin-orbit interactions, are taken into account in the expression for the potential $V^c(r)$. Thus $V^c(r)$ does not decrease with the distance to the source but it increases as one moves farther and farther away from the source. Just the same reasons can be applied to any composite color singlet. If one tries to disjoin some color part of this system (say, a qq-pair in a baryon) from others, its energy will linearly increase resulting in the confinement of color components.

One more sufficiently perspective attempt of the confinement explanation is the so-called Wilson lattice theory. In lattice theories one is assumed that the space and time do not make up a continuum but represent the points discrete set which resemble the crystal lattice (most commonly, the cubic one). The quarks are placed in the lattice sites and the field lines of the gluons field connect lattice arbitrary site with its nearest neighbors only. Further one is assumed that the energy of interaction between quarks or between quarks and antiquarks is proportional to a length of a string to connect them. One may draws uncounted set of strings between two points of a such space-time. In the lattice theory the quantum mechanics average is fulfilled on these strings. Integrals appearing in the process may be analytically calculated in the region of so called strong coupling when the lattice step is much more bigger than the typical scale of the quantum fluctuations of the gluon fields ($\sim 10^{-13}$ cm). The pass to the continuous space-time is realized by the way of decreasing the lattice step. In this case since the lattice sites are merged with each other, then the problem of calculating the large multiplicity integrals appears. The problem is usually resolved with Monte Carlo method. As this method could be applied to the finite multiplicity integrals the lattices with the finite number of the sites along everyone of four axes are only considered. In the long run one may show that under the certain conditions the quarks confinement taking place in the strong coupling region is persisting under decreasing the lattice step as well.

Since the gluons bear the color they can not exist in free states as well. If the QCD is true, one should expect the existence of hadrons containing only the gluons. The most simple bound gluon state is two gluons forming the color singlet. Of course, one may build the color singlets from three and more gluons. Hadrons of such a kind are called gluonium or glueball. Different glueballs may be discriminated by the spin and the mass only. From the theoretical point of view the gluonium identification seems to be very difficult because it is impossible to point the decays or other properties of the gluonium which would certainly discriminate the gluonium from the quarkonium having the same quantum numbers. On the other hand, calculations show that the most intensive gluonium formation must occur in those reactions and decays in which the gluons rather than the quarks are produced at small distances. Examples are found in decays of heavy mesons ψ and Υ. However, up to date the gluonium has been discovered. The numerical calculations on a computer being done within the QCD allow to obtain the definite predictions concerning the masses of the most light glueballs. In doing so the typical masses scale proves to be of order of 1.5 GeV.

One such a confirmation of existence of the quarks and gluons is the processes of detecting the hadron jets. To understand it we recall how elementary particles are observed in Wilson chamber. The track left by the particle is not manifestation of the particle itself at all. It is the result of interaction between the particle and matter filling the chamber. This interaction leads to the production of great numbers of ions along the particle trajectory. It is evident that the interaction distorts the elementary particle motion. For high energy particles (only such particles have left the track) such distortions (for example, the track jitter in a transverse plane) are negligibly small and we could state the trajectory of itself particle is being observed directly. Analogously, the multiparticle cluster of fast hadrons with small transverse momenta, the hadron jet, is the track for a parton which is outgoing from the deep inelastic scattering region. The jet is not only the visualization way of the "free" quark or gluon but the form of their existence as well. Here a vacuum, a structure of which is set by the QCD, plays the role of an environment filling the track chamber. In this vacuum, just the same as in the QED vacuum, small-scale fluctuations (SCF) are present thanks to the asymptotic freedom phenomenon. Apart from the SCF, gluon fields long-wavelength fluctuations (LWF) caused by nonlinear character of the QCD exist. At this time, the notions of "small" or "large" are determined from the viewpoint of a parton, that is, the LWF are realized on the distances of the quark Universe radius ($\sim$ 1 Fermi). The LWF correspond to the really strong interquarks interaction. Just they play the role of "matter" with which a high-energy parton interacts. The average transverse momentum $< p_T >$ of creating hadrons just conforms to the LWF scale:$< p_T >\approx 300$ MeV ≈ 1 Fermi^{-1}. Thus the hadron jet created by the quark or the gluon may be identified with the itself quark or gluon in just the same sense as the drops chain in Wilson chamber is considered by the particle trajectory.

In 1975, basing upon the results of analyzing the e^+e^--annihilation into hadrons on the electron-positron storage ring SPEAR, the Stanford research group announced on the discovery of the quark jets. It has appeared that when the jet energy grows the average angle of the jet spread decreases, that is, hadrons are increasingly gathered round the direction of flying away q and $\overline{q}$. At the jet energy of the order of 18 GeV the hadrons which constitute the jet occupy only 2% of the total solid angle. Investigations have also shown that the correspondence *quark*$\leftrightarrow$*jet* has the universal character. This means the composition of hadrons in jet (relationships between p, n, π, K and so on) and hadrons distribution on momenta do not depend on what concrete reaction the given flavor quark, the jet primogenitor, is produced.

The first indirect manifestations of gluon jets (1979) were connected with researching the decays of the Υ-meson[1] into hadrons. According to the QCD the three gluons must be produced at annihilation of the $b\overline{b}$-pair, that is, the decays in question must have the three-jets nature. True enough, in the case of the Υ-mesons it is difficult to observe three gluon jets directly because the gluons energy is too small as yet. For this reason two indirect methods are used. The essence of the former is as follows. Usually, at large energies in the result of the electron-positron annihilation two quark jets in the opposite directions are created. If one increases the energy E_{cm} to an extent that the Υ-meson begins to be born, the situation will be changed. In the final state instead of the quark-antiquark pair three gluons will be created, their momenta being allocated now on the whole space. As a consequence, at $E_{cm} \geq m_\Upsilon$ the two-strings structure of the created hadrons, which existed at $E_{cm} < m_\Upsilon$, should disappear. Just that effect was registered at DESY (Deutsches Electronen-Synchrotron).

The second method of the indirect observing the gluon jets was based on the events kinematics. At the electron-positron pair annihilation the Υ-meson is created in the rest. Consequently, the total momentum of the created gluons must be equal to zero as well. But three the three-dimensional vectors the sum of which equals zero must lay in the same plane. With the good precision this should be also fulfilled for the particles momenta of

[1]This meson represents the bound state of two quarks b and $\overline{b}$ to be dealt in the next section.

the final hadron state. Of course, this plane is varied from one decay to the another. The analysis of the Υ-decays showed the final hadrons momenta do lay in the same plane.

In the same year, a little bit later, under increasing the PETRA (Positron-Electron Tandem Ring Accelerator) energy up to 20 GeV the bremsstrahlung of the gluon by the quark was observed in the following process

$$e^+ + e^- \to q + \overline{q} + g.$$

The additional hadrons created by the bremsstrahlung gluon should lead to the azimuthal asymmetric thickening of one of the quark jets. With the energy increasing and the enhancement of the statistics set one managed to select the gluon from the basic jet and measure the distributions both on energy and on the angle of flying out the gluon jet. Measurements have shown the new jet behaves similar to the bremsstrahlung photon in the reaction

$$e^+ + e^- \to \mu^+ + \mu^- + \gamma,$$

that is, just as the particle with the spin 1 should behave.

Once one has managed to "see" the quark, the next task was to count the quarks number (with allowance made for the color and the flavor). For this purpose the process of the e^+e^--annihilation into hadrons appeared the most suitable

$$e^+ + e^- \to q + \overline{q} \to 1\,\text{jet} + 2\,\text{jet}. \qquad (5.113)$$

The quark and the antiquark which are produced at the reaction second stage can not exist in the free states since they are the color objects. They extract the color quark-antiquark pair suitable to them from a vacuum and recombine with this pair into colorless hadrons which fly apart as two jets directed oppositely. At sufficiently high energies the contributions coming from created quark-antiquark pairs with different colors and flavors are noncoherent, that is, the total cross section of the reaction (5.133) is presented in the form of the cross sections sum over quarks of all the colors and flavors. The cross section obtained may be compared with that of the e^+e^- annihilation into other point particle, muons,

$$e^+ + e^- \to \mu^+ + \mu^-. \qquad (5.114)$$

Both the quarks and charged leptons are described by the Dirac equation. However, mass and charge entering into this equation are quite different for quarks and leptons. In the case of high energies one may neglect their masses to get the following expression for the ratio of the cross sections of the reactions (5.113) and (5.114)

$$R = \frac{\sigma_{e^+e^-\to hadrons}}{\sigma_{e^+e^-\to\mu^+\mu^-}} = \frac{\sum_i \sigma_{e^+e^-\to q_i\overline{q}_i}}{\sigma_{e^+e^-\to\mu^+\mu^-}} = \sum_i \left(\frac{e_{q_i}}{e}\right)^2. \qquad (5.115)$$

If the quarks did not have the color degree of freedom this ratio would be[1]

$$R = \left(\frac{e_u}{e}\right)^2 + \left(\frac{e_d}{e}\right)^2 + \left(\frac{e_s}{e}\right)^2 = \frac{4}{9} + \frac{1}{9} + \frac{1}{9} = \frac{2}{3}. \qquad (5.116)$$

[1]Running ahead we stress, it is valid only when energies are smaller than the threshold of creating hadrons which consist of more heavier quarks, i. e. $E_{e^-e^+} < 3$ GeV.

The experiment gave three times as much value. But such trebling must appear with due regard for the color degrees of freedom. Really, every quark-antiquark pair of the given flavor may be created in three different color states.

The origin of ordinary strong interactions between real colorless hadrons (for example, between nucleons in the atom nuclear) is explained by interacting the charged structural components entering into hadrons, that is, much the same way as the origin of the electromagnetic nature chemical couplings between the electrically neutral atoms and molecules in matter. In particular, the π-mesons exchange between nucleons[2] may be connected with the quark-antiquark pair production inside a nucleon and a subsequent conversion of these pairs into the π-mesons which can fly out of the nucleons because they are colorless objects.

Thus we have managed to build all the hadrons, which were known by 1975, with the help of the u_i, d_i, and s_i-quarks. And the understanding illusion visited us once again:*all matter in the Universe consists of combinations of nine quarks and four leptons (electron, electron neutrino, muon, and muon neutrino).* It seemed that the problem of matter structure was close to the completion. It only remained for us to make more precise the properties of the fields describing the weak and strong interactions. True enough one "but" was again, namely, the so called quark-lepton symmetry.

Stable matter composes of the electrons, the u and d-quarks. We have every reason to believe that the electron neutrino is a stable particle as well. We call this totality by the first generation of the quarks and leptons. Further we introduce a new quantum number S^W with the same algebra as the ordinary spin has and consider the representation with the weight $1/2$. Let us place the first generation particles into the weak isospin doublets

$$\begin{pmatrix} \nu_e \\ e^- \end{pmatrix}, \quad \begin{pmatrix} u \\ d \end{pmatrix}. \tag{5.117}$$

Notice, the charges differences of the neutrinos and charged leptons are equal to those of the *up* and *down* quarks while the charges algebraic sum of the quarks (with allowance made for trebling on the color) and leptons is equal to zero. There exists the second lepton generation too

$$\begin{pmatrix} \nu_\mu \\ \mu^- \end{pmatrix}, \tag{5.118}$$

whereas the second quark generation is kept unfilled

$$\begin{pmatrix} ? \\ s \end{pmatrix}. \tag{5.119}$$

If one assumes that the quark-lepton symmetry is the immovable law of Nature, the existence of the quarks of a new flavor is needed.

5.6. c-Quark and $SU(4)$-Symmetry

In the autumn of 1974 in Brookhaven National Laboratory the group under the supervision of C. Ting began investigating the process of the electron-positron pair production at the

[2]Such a treatment of nuclear forces hold good when the distance between nucleons exceeds 8×10^{-14} cm.

collisions of the protons with the helium target (pp-collisions) in the mass region between 2 and 4 GeV. It was discovered that majority of the e^-e^+-pairs has the mass being approximately equal to 3.1 GeV, that is, the pronounced maximum takes place in the cross section of the process

$$p + Be \to e^+ + e^- + X, \tag{5.120}$$

where we have designated the nondetectable particles plurality by symbol X. If this maximum corresponds to the true resonance then it should be present in the cross sections of some other reactions. And really, at about the same time the group working on facility SPEAR (B. Richter as a supervisor) discovered this resonance under investigating the processes

$$e^+ + e^- \to, \qquad e^+ + e^- \to e^+ + e^-, \qquad e^+ + e^- \to \mu^+ + \mu^-. \tag{5.121}$$

The both groups simultaneously reported about the discovery of the new particle with the mass ~ 3.1 GeV. Since it was highly difficult to share the palm, the new particle was called the double name J/ψ (the symbol J means Ting's name in Chinese and the name ψ was proposed by the SPEAR Collaboration). The name J/ψ has been saved not only as a tribute of respect to its path-breakers but also due to the play on words, J/ψ= gi/psi=gipsy, which has been so amusing the romantic soul of a physicist.

The particle J/ψ has the spin 1, the negative parity and it is a long-liver by the microworld standards, that is, it has the anomalously small decay width $\Gamma \approx 70$ keV whereas ordinary resonances have $\Gamma = 100 - 200$ MeV. To clarify the true nature of the new particle was being proceeded about three years. Amongst the working hypotheses which have appeared, there was even such one:J/ψ *is not a hadron but the long-awaited neutral intermediate boson, one of carriers of weak interaction.* With the passage of time larger and larger arguments arise which begin to turn the scale in favour of the hypothesis:J/ψ *consists of a quark and a antiquark, $c\bar{c}$, with a new flavor.* From the simplest estimation $m_c \simeq m_{J/\psi}/2 = 1.55$ GeV it follows at once that the c-quarks should be much more heavier than the u, d and s-quarks. To match the theory and experiments one should make an assumption that the c-quark is the carrier of the new quantum number, the charm $= 1$ for c-quark and $= -1$ for $\bar{c}$). Then J/ψ represents the particle with the hidden charm and, on its structure, is very similar to φ-meson which is constituted of the strange quark and antiquark (the particle with the hidden strange). This resemblance has served as a key to understand the small decay width of the discovered particle. In Figs. 33a and 33b the quark diagrams corresponding to the φ-meson decay are displayed.

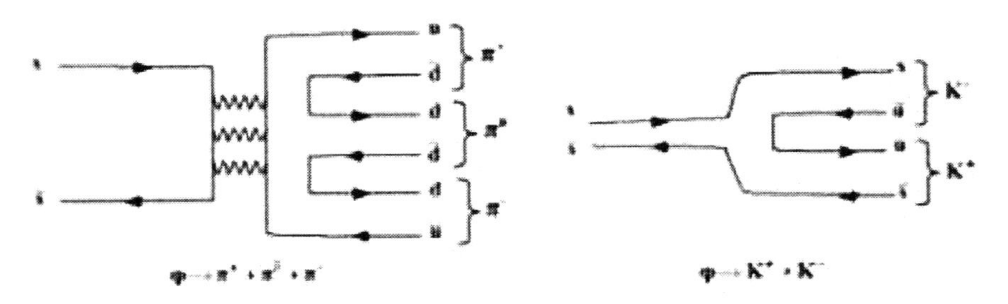

Figure 33. The quark diagrams of the φ-meson decay.

The distinction from Feynman diagrams resides in the fact that here the quarks, when $t = \pm\infty$, are not free particles since hadrons hold them captive as before. Besides, in these diagrams strong interactions between quarks are not usually displayed. Just the same as in the case of Feynman diagrams the arrow directed backwards the time corresponds to the antiparticle. In Fig. 33a the s-quarks entering into the φ-meson composition get over to the K-mesons composition. In the second case (Fig. 33b) the strange quarks are annihilated and instead of them the pairs of the u- and d-quarks appear. The experiments show that the φ-meson predominantly decays through the channel $K+\overline{K}$ (the relative probability, the branching, $Br \approx 84\%$) and very reluctantly does through the channel $\pi^+ + \pi^- + \pi^0$ ($Br \approx 15\%$). At this example we see how to work the approximate semi phenomenological rule by Okubo — Zweig — Iizuka (OZI) which assumes the systematization of relative amplitudes of hadrons interaction reactions depending on a topology of quark diagrams to display these processes. The largest degree of suppression is present in the diagrams where the quarks and antiquarks lines going out of one and the same hadron are connected with each other and represent the block which is not related with the rest of the diagram. In this case the quark-antiquark pair belonging to one and the same hadron disappears. The process where the same quark and antiquark pass into the different hadrons of the final state is an alternative to such a process. All this shows once again the behavior uncommonness of the quarks. In processes of inclusive scattering the created quark-antiquark pairs behave in such a way as though they beforehand deduce in what groups on two (mesons) or three (baryons) they should be unified. The OZI rule may be also understood as the manifestation of the specific quark "thinking", namely, the quarks choose the possible variant of the future hadron prison already at their creating. The OZI rule, as applied to the J/ψ-particle, means J/ψ would predominantly decay through the particles containing the c-quark and the light u- and d-quarks (Fig. 34b).

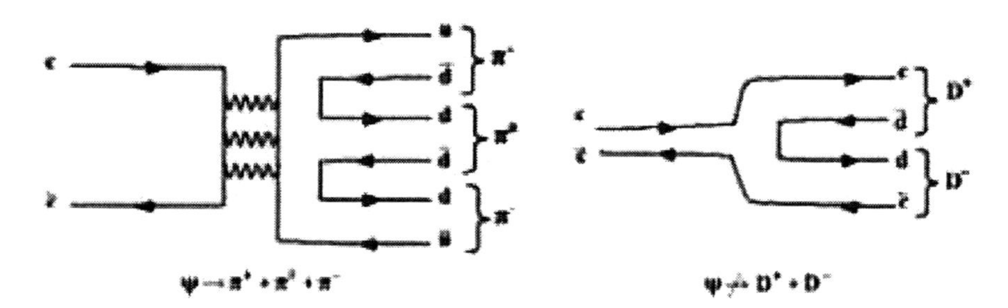

Figure 34. The quark diagrams of the J/ψ-particle decay.

These particles having the explicit charm are called the D-meson. However, by a lucky chance, the decay

$$J/\psi \to D^+ + D^- \qquad (5.122)$$

proved to be forbidden energetically ($2m_D > m_{J/\psi}$). The decays on π- and K-mesons were allowed energetically but they were connected with the annihilation of the c-quarks and, as a result, greatly suppressed by virtue of the OZI rule (Fig. 34a). Thus, the small value of the J/ψ-meson decay width served as the indication of the existence of the new kind of the quark c.

About a week later after the J/ψ-meson discovery, at SPEAR the narrow resonance ψ' placed at slightly more high energy was detected. With further increasing an energy one had found some resonances both with the spin 1 and with the spin 0 in the neighborhood of 4 GeV. If one displays the mass spectrum of these particles (we call them the ψ-particles) graphically then this spectrum will resemble the atom spectral lines picture (Fig. 35).

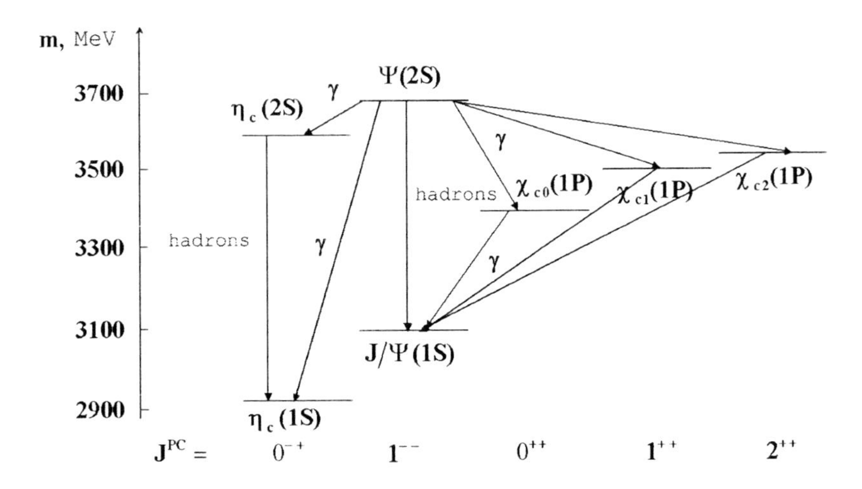

Figure 35. The charmonium spectrum.

According to the contemporary concepts, this mass spectrum can be related to the energy levels spectrum of the system $c\overline{c}$. By analogy with the well known positronium the bound system $c\overline{c}$ is called the charmonium. However, unlike the positronium where the binding forces have electromagnetic nature, the charmonium owes its existence to strong interaction. The particles denoted by a prime represent the radial excitations of the underlying states. To classify the charmonium levels we make use of the spectroscopic designations $^{2S+1}L_J$. Of course, this is a nonrelativistic classification but the nonrelativistic approach is lawful because of the c-quark large mass (by now experiments give the current quark mass values in the interval from 1.15 to 1.35 GeV).

From the data concerning only ψ-particles it was impossible to make the final conclusion in favour of the existence of the c-quark. The hypothesis about the charm quark is ultimately confirmed only after the discovery of hadrons with the explicit charm (1976): mesons

$$D^+ = c\overline{d}, \qquad D^0 = c\overline{u},$$

$$D^- = d\overline{c}, \qquad \overline{D}^0 = u\overline{c},$$

$$F^+ = c\overline{s}, \qquad F^- = s\overline{c},$$

and baryons

$$\Lambda^+ = cdu, \qquad \Sigma_c^+ = udc, \qquad \Sigma_c^{++} = uuc, \ldots$$

These discoveries furnished the genuine triumph of the quark theory of hadrons structure.

Since the mass of the discovered fourth quark was much more bigger than the masses of the u-, d- and s-quarks it was already hard to say of that the quarks are the manifestation of the approximate $SU(4)$-symmetry (flavor symmetry) hadrons would have. Notice, the $SU(2)$-symmetry violation is $\sim 1\%$ while the $SU(2)$-symmetry violation is $10 \div 20\%$. According to the contemporary point view a violation of a $SU(n_f)$-symmetry is caused by the quark masses difference, namely, the more this difference the stronger the violation of the corresponding $SU(n_f)$-symmetry.

5.7. b and t-Quarks

Soon after the J/ψ-meson discovery in 1975 one more charged lepton, a τ-lepton, was detected in the independent experiments series. The direct confirmation of existing the ν_τ-neutrino will have taken place much more later (2001). However, even in 1975, physicists majority basing upon the idea of the Nature unity was considering that the τ-lepton also has the neutrino satellite. Thus, the quark-lepton symmetry which had been rebuilt with the c-quark discovery was violated again. The new lepton pair should correspond to a new quark pair. The first confirmation of this hypothesis happened in 1977. In Fermi laboratory (FERMILAB) at the proton accelerator with the energy up to 400 GeV a new particle with the mass 9.45 GeV and $J^P = 1^-$ was discovered. It was called Υ-particle (upsilon). Subsequently in DESY (Hamburg) and in Cornell University (USA) at investigation of the electron-positron collisions the existence of the Υ-particle was confirmed and its excited states (Υ', Υ'' and so on) were detected. Since the properties of the Υ-particle have been very similar to those of the J/ψ-meson, one assumed that it is the bound state of the quark-antiquark pair of the fifth kind, $\Upsilon = b\overline{b}$. As the particles belonging to the Υ family (we call it by upsilonium) have the masses around 10 GeV then the mass of the quarks entering into them must be by no means smaller than 5 GeV (according to the contemporary data the mass of the current c-quark lays in the interval from 4 to 4.4 GeV). The investigations of the Υ-meson properties have shown the b-quark has $I = 0$ and $Q = -1/3$. Physists were going on the already well rolled road for the small width of the Υ-meson decay to be explained. The new quantum number, the beauty $b = 1$, which is conserved in strong interaction only, was assigned to the b-quark. Then the OZI rule suppressed decays into hadrons which do not contain b-quarks and Nature, for its part, in order not to lead into temptation researchers, gave orders in such a way that the decays into a pair of mesons with the explicit beauty proves to be forbidden energetically (the most light beauty mesons $B^+ = u\overline{b} and B^0 = d\overline{b}$ have the mass 5270 MeV). As we know the quantity

$$R = \frac{\sigma_{e^+e^- \to hadrons}}{\sigma_{e^+e^- \to \mu^+\mu^-}}$$

is the sensitive tool to define the quark flavors number. Between thresholds of a $q_i\overline{q}_i$-pairs production the quantity R is constant and when the next i threshold had been achieved it was abruptly increased on the value $3Q_i^2$. In Fig. 36 we display the experimentally measured values of R^{exp} as a function of energy in the center of mass system.

The resonant levels of the charmonium and upsilonium are put on the stepped monotonous behavior. Upon subtracting the resonant contributions R^{exp} is well matched

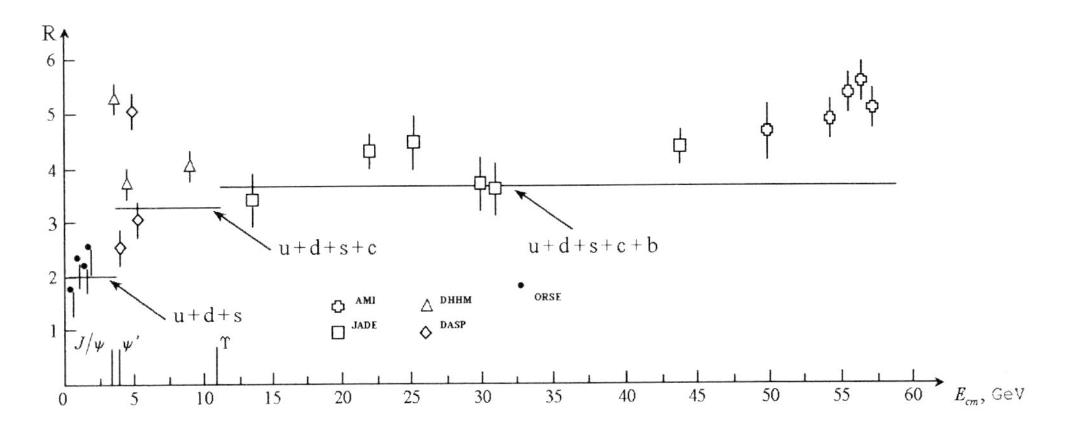

Figure 36. Experimental values of $R = \sigma_{e^+e^-\to hadrons}/\sigma_{e^+e^-\to\mu^+\mu^-}$ as a function of energy in the CMS.

with the experimental value which is the powerful argument in favour of the b-quark existence.

After the b-quark discovery the hunting season is opening on its partner on the weak isospin doublet which was called t-quark. Its name originates from the first letter of the word *top*, though one may consider that the origin is the word *truth*. At first, the indirect evidences for the existence of the sixth quark were obtained. Among those, the results of investigating the process

$$e^+ + e^- \to b + \overline{b} \tag{5.123}$$

at the CERN electron-positron collider called LEP (Large Electron Positron) were the most convincing. To study the asymmetry in the scattering of the b-quark confirmed the b-quark does represent the member of the $SU(2)$ doublet of the weak isospin, as the electroweak interaction model by Weinberg, Salam and Glashow (WSG) predicted. Recall, by that time the WSG theory has gone through the basic stage of its experimental checkout. Moreover, the detailed measurements of the properties of the $W^\pm$- and Z-bosons which were fulfilled at LEP, SLC (Stanford Linear Collider, e^+e^--collider), CERN $Sp\overline{p}S$ ($p\overline{p}$-collider) and FERMILAB Tevatron ($p\overline{p}$-collider) gave the unquestionable proofs in favour of the existence of the heavy quark connected with the b-quark by means of electroweak interaction. In April 1994 the first experiments, which directly pointed the t-quark existence, appeared. They were carried out at FERMILAB Tevatron which represents the six-kilometer storage ring where the protons and antiprotons having the energy 1.8 TeV in the system of center of mass are rotated in opposite directions. The $p\overline{p}$ collisions occur in two points of the ring where two detectors system, CDF and D$\oslash$ detectors, are placed. The collider luminosity [1] was being constantly increased and in 1994 it reached the value $L = 10^{31}$ cm$^{-2}\cdot$s^{-1} which allowed to detect the t-quark. At looking for the t-quarks one is guided by the fact that the cross sections both of their productions and of their consequent decays depend on the t-quark mass value. The first to raise the low bound on m_t to $\sim$ 175 GeV was CDF Collaboration. Already in 1994 they observed twelve events with $m_t \sim$ 175 GeV which correspond

[1]The luminosity L is the collisions number in second on unit section, that is, L being multiplied by the given process cross section σ expressed in cm^2 gives the number of the corresponding events in second.

to the selection criterions for the t-quark. But the expected background constituted roughly six events and, although the interpretation of the obtained results as the t-quark observation offered the most probable, the statistical providing of the result was insufficient to recognize it as the t-quark discovery. And only in February 1995, in one and the same day, both Collaborations, CDF and D⊘ sent off the reports about the t-quark detection to the press.

According to the QCD, at energy in the center of mass system $E = 1.8$ TeV, the $t\bar{t}$ pair production in $p\bar{p}$ collisions mainly occurs at the cost of the subprocesses

$$q+\overline{q} \to t+\bar{t}, \qquad g+g \to t+\bar{t}, \tag{5.124}$$

where we designated the gluons by the symbol g. We choose the axis z along the proton beam direction and shall study such variables of final products of a reaction as the transverse momentum relative to z axis p_T and connected with it the transverse energy E_T. The standard model (SM) predicts that the t-quark decays through the channel $t \to W^+b$ just about always. Distribution on the transverse momentum for one of the decay products has the peculiarity which allows not only to detect such decays but to define the mass of the unstable particle as well. This method based on merely kinematic arguments became very popular after it had been used under discovering the W-boson in $p\overline{p}$ collisions (in that case, the electrons and the positrons with very high transverse momenta, $p_T \sim 40$ GeV, were detected). At sizeable excess of the decaying particle mass m_i over the decay product masses, the events concentration maximum proves to be observed close to the transverse momentum value of one of the decay product k_T being approximately equal to $m_i/2$. However, this method of the t-quark identification did not work since the W-boson was the unstable particle with a very small life time while the b-quark served as a source of producing the hadron jets. The W-boson decays into $q\overline{q}'$ pairs ($u\overline{d}$ or $\overline{s}$) with the probability $2/3$ and does into one of three lepton families ($\bar{l}\nu_l$) with the probability $1/3$. Thus, there are 6 high-energy point fermions which could be either charged and neutral leptons or quarks giving rise to the production of the hadron jets. It was very difficult to select the decays with the τ-leptons from the hadron background and, for this reason, they were not taken into consideration. High background (signal-to-noise correlation was smaller than 10^{-4}) made an inclusion in analysis of final states with six hadron jets impossible. As a result the following channels decay of the $t\bar{t}$-pair were available to observe

$$t\bar{t} \to e^+\nu_e b e^- \overline{\nu}_e \overline{b}, \qquad (1/81), \tag{5.125}$$

$$t\bar{t} \to \mu^+\nu_\mu b \mu^- \overline{\nu}_\mu \overline{b}, \qquad (1/81), \tag{5.126}$$

$$t\bar{t} \to e^\pm\nu_e b \mu^\mp \overline{\nu}_\mu \overline{b}, \qquad (2/81), \tag{5.127}$$

$$t\bar{t} \to e^\pm\nu_e b q\overline{q}' \overline{b}, \qquad (12/81), \tag{5.128}$$

$$t\bar{t} \to \mu^\pm\nu_\mu b q\overline{q}' \overline{b}, \qquad (12/81), \tag{5.129}$$

where the numbers in the square brackets denote the branching predicted by the SM. First three dilepton channels prove to be the most clear but they have very small statistics. Last two channels, *lepton+hadronjets*, have high statistics (close to 30% on the total decay width $t\bar{t}$) but suffer from extremely large background. At $p\overline{p}$ collisions the total cross section of the $t\bar{t}$-pair production is the function not only the parton distributions inside the proton and the antiproton but the mass of the t-quark as well. So, at $m_t = 100, 155$ GeV it has the values $\sim 10^2$ and 10 pb respectively. It should be stressed that these values are very small by strong interaction standards. When $E_{\text{c.m.s.}} = 1.8$ TeV the $p\overline{p}$ collisions also lead to the production of hadron jets and lepton pairs in the final state at the cost of the production of W- and Z-gauge bosons in virtual states. It has been just these events which constitute the main background for the $t\overline{t}$-pair production. To compare we give the values of the competed cross sections 1) for the W-production — $\sigma_W \sim 20$ nb; 2) for the Z-production — $\sigma_Z \sim 2$ nb; 3) for the WW-production — $\sigma_{WW} \sim 10$ pb; 4) for the WZ-production — $\sigma_{WZ} \sim 5$ pb.

Let us consider the criterions used by CDF and D∅ Collaborations to select the signal from the background. The heavy quark pair production ($m_t = 173.8 \pm 5.2$ GeV) and its consequent decay generates the final states with the more large average energy than the background events. To investigate the distribution of the final hadrons and leptons on the transverse momenta proves to be useful in this case as well. High energy electrons, muons and hadrons were recorded with the help of different detectors and were easily distinguishable from each other. As for the neutrino is concerned the disbalance of the total transverse energy E_T or the missing transverse momentum p_T testify to its existence in the final state. It is evident that the quantity

$$H_T = \sum_{i=1}^{N_p} E_T^i$$

where the summation is realized on all hadron jets and basic electron clusters by which are meant a leptons plurality including even if one electron, is very useful for the kinematic analysis. At the investigation of the final states *lepton+hadron jets*, two methods were used by D∅ Collaboration to select the signal from the background. The first based upon the kinematic analysis (KA) resided in the demand $H_T > 200$ GeV and in the presence of at least four hadron jets with $E_T > 15$ GeV. The second method was connected with the b-quark identification (B tagging-out) through the decays

$$b \to \mu^- + \overline{\nu}_\mu + X, \qquad b \to \mu^+ + \nu_\mu + X. \tag{5.130}$$

Under studying the data on *lepton+hadron jets* CDF Collaboration used already other methods. One of them was also connected with the b-quarks and was received the name:"tagging-out of the second vertex (TSV)". The tracks of charged products of the b-quarks decay are detected in the drift chamber. Only the tracks with $p_T > 1.5$ GeV (decays of the b-quarks which produced from the $t\overline{t}$-pairs and the W-bosons are their sources) were of interest. The b-quark decay point in the silicon stripped detector was rebuilt by the method extrapolations. It allowed to select the signal from a background by means of a comparison of a different events intensity. The second method, received the name "tagging-out of soft leptons (TSL)", was based on detecting the low-energy ($p_T \sim 2$ GeV) muons and electrons near hadron jets.

In Table 5.1 we give the expected number background events and the number of the events connected with the t-quark production at the observations results by CDF and D∅ Collaborations.

As it follows from Table 5.1, in all channels both Collaborations observed the sizeable excess of the signal over a background that undoubtedly testifies the $t\overline{t}$ pairs production in

Table 5.1.

Sampling	Background	Signal
dileptons (CDF)	1.3 ± 0.3	6
dileptons (D⊘)	0.65 ± 0.15	3
leptons+jets (D⊘ KA)	0.93 ± 0.5	8
leptons+jets (D⊘ B-tagging-out)	1.21 ± 0.26	6
leptons+jets (CDF TSV)	6.7 ± 2.1	27
leptons+jets (CDF TSL)	15.4 ± 2.3	23

the $p\overline{p}$ collisions.

To account for the experimental data one should assign to the t-quark a new quantum number t (the top or the truth) being equal to 1 which along with the strange, the charm, and the beauty is conserved in strong interaction but is not conserved in electromagnetic and weak interactions. It was natural to wait for the existence of a quarkonium consisting of the t-quark and t-antiquark, the toponium. However, by now the toponium has been discovered. It is not inconceivable that the Nature presented us only two kinds of the simplest "quark" atoms with the great numbers of the energy levels. Recall, that in the Atom Physics the hydrogen atom was such a present. But in the Nuclear Physics, mildly speaking, the Nature already was a little bit miserly and constrained itself by the deuteron only which has got none of the excited levels.

Introducing the b- and t-quarks extends the flavor symmetry of strong interaction up to the $SU(6)$ group. However, thanks to the sharp gradation of the quark masses this symmetry is strongly violated. On the other hand, since up to the present hadrons containing the constituent t-quark have not been discovered then for practical calculations we could successfully use the $SU(5)$ flavor symmetry whose the violation degree is much less ($m_b \ll m_t$). The flavor symmetry plays a double role, it not only defines the classification of hadrons on various multiplets but also establishes series of dynamical relations between amplitudes of different processes of hadrons interaction. In the low energy region ($E \leq m_0$, where m_0 is an average mass value in a multiplet) strongly violated flavor symmetry possesses a weak predictive force and is unlikely applicable for a practical use. However, at high energies ($E \gg m_0$) it becomes the useful tool of an investigation.

5.8. Looking for Free Quarks

Up to now we considered the constituent quarks model, according to which practically the whole hadron mass is confined in the quarks, and assumed that the quarks dynamics is completely defined by the QCD. However, there is nothing to prevent us to conjecture that the negative results on the quarks production are caused only by the insufficient energy of contemporary accelerators to create the quarks in a free state. This, in its turn, means that at the hadrons production the giant quark mass is eaten by such a huge binding energies. And only the fact of the free quarks existence can prove this statement.

To look for the free quarks is being carried on two directions. The former is a geophysical-chemical approach which is based on the assumption: the stable quarks occur in matter surrounding us. In this case the negative charged quarks will be captured by nuclei and form either quark atoms or quark ions with the nonintegral charge. It is evident that the created compounds should have specific physical and chemical properties. Thus the experiments of this approach are aimed for discovering the characteristic manifestation of the fractionally charged particles existence: the lowered ionization constituting 1/9 or 4/9 of the integer charged particle ionization; an unusual value of e/m in mass-spectroscopic experiments; an anomalous behavior of a levitating matter grain in an electrostatic field; a nonstandard position of spectral lines in quark atoms and so on. The quarks were being sought in terrestrial matter, in lunar soil, in meteorites. To look for quark atoms in solar matter with the help of spectroscopic methods was also carried out. It is natural that the most reliable constraints on the free quarks existence have been obtained during searching for the quarks in the stable matter of the Earth. Different variants of experiments lead to upper limit values of a possible quarks concentration in a matter which lay in the interval from 5×10^{-15} to $5\times10^{-28}\ \frac{\text{quarks}}{\text{nucleon}}$.

The second direction includes in itself the attempts of detecting the quarks in cosmic rays and accelerators directly. In experiments on accelerators the free quarks are not discovered up to the masses 250 GeV in $p\overline{p}$ collisions (CDF, 1992) and 84 GeV in e^-e^+ collisions (LEP, DELPHI, 1997) at the production cross sections higher than 10 and 1 pb respectively. Experiments on detecting the quarks in collisions of cosmic rays with particles in the upper atmosphere layers, which were fulfilled in a wide interval of energies and, consequently, did not have severe constraints on the mass of created quarks, have not also brought to success and have set the constraint on the quarks flux from cosmos: $<2.1\times10^{-15}\ \frac{\text{quarks}}{\text{cm}^2\cdot\text{sr}\cdot\text{s}}$ (KAM2, 1991).

Meanwhile, if the free quarks existence is not forbidden in principle then they would be created on early stages of the Universe evolution when the temperature was very high, say $kT > 2m_q$. At such a temperature the quarks are in the state of the thermodynamic equilibrium with other fundamental particles (number of created quarks is equal to that of annihilating quarks). At $kT\sim m_q$ the equilibrium is violated: the quarks production reactions have been switched off and the quarks start to burn away. This burning away takes place at the cost of the reactions of the kind

$$q+\overline{q}\to \text{mesons},\qquad q+\overline{q}\to\overline{q}+\text{baryon}. \tag{5.131}$$

Since the reactions (5.131) are exothermic then their cross sections tend to constant values whose sum we denote by σ_0. If one assumes that the cross section of the quark destruction has the typical nuclear scale, say $\sigma_0\sim m_\pi^{-2}$, one may show the quark-to-proton-concentration ratio in the present Universe must be

$$\frac{n_q}{n_p}\sim 10^{-12}.$$

This number is greater than the gold abundance on the Earth, but the quark Klondike has never been opened to date. One would assume the quarks are unstable particles and all this relic quark sea has had time to disappear by now. However on the strength of the electric charge conservation law, at least one of the quarks must be stable and should live till the present day. So, either the free quarks are really absent in Nature or their production cross section has as minimum the atom scale rather than the nuclear one.

It should be noted that some experimental groups make reports concerning the free quarks observations every now and then. So, the Stanford University group investigated a behavior of a niobium ball $\sim 10^{-4}$g which levitated in the nonuniform magnetic field. They observed the cases when the electric charge of the ball was equal to $\pm e/3$. As this takes place, the corresponding quarks concentration has the order of $\approx 10^{-20}\,\frac{\text{quarks}}{\text{nucleon}}$. However these data are not confirmed by other investigations yet. And the result is included in a category of the reliable one if and only if it is independently obtained by several different groups using different experimental methods.

Physics development shows the Nature gives answers to correctly posed questions only. There is no sense in the question which tormented ancient scholastics: "How many angels could be put in a sword tip?" The question, physicist tortured themselves at the very outset of Quantum Theory, appeared to be senseless:"What is the electron — the particle or the wave?" May be, when we are trying to detect the quarks in a free state we are in the analogous situation? It is not expected that the quarks represent the specific kind of quasiparticles, field quanta, which describe collective oscillations of corresponding freedom degrees of a hadron. We faced such formations in other regions of physics and before. Among these there are: the magnon, the quantum of the spin oscillations in magneto-ordered systems; the plasmon, the quantum of the charges density oscillations in conductive mediums; the phonon, the quantum of the elastic oscillations of the atoms or molecules in crystal lattice. At switching off interaction similar particles are pulled down into compound parts and stop their existence. For example, the phonon decays and turns into plurality of independent motions of particles which constitute a crystal. Then the quarks have the sense only as dynamical essences inside hadrons in just the same way as the phonons which can not exist outside a crystal. However, be it as it may, looking-for the free quarks is continued. The problems of their detection on accelerators of next generation, LHC (Large Hadron Collider), NLC (Next Liner Collider), FMC (First Muon Collider) and so on, are intensively discussed.

Chapter 6

Standard Model

6.1. Abelian Gauge Invariance and QCD

According to Noether's theorem, l dynamical invariants, that is, l being conserved in time combinations of field functions and their derivatives, correspond to every finite-parametric continuous transformation of coordinates and field functions under which an action variation is turned into zero. So, the momentum conservation law follows from the invariance with respect to space translations, the energy conservation law does from the invariance with respect to time translations, and the angular momentum conservation law does from the invariance with respect to space rotations.

With the exception of the above mentioned dynamical invariants connected with the symmetry of Minkowski space-time, in particle physics one also introduces dynamical invariants caused by symmetries of a physical system with respect to transformations in abstract spaces. Such invariants are called internal quantum numbers while we name the corresponding symmetries by nongeometric (internal or dynamical) symmetries. A good example illustrating the connection

$$\textit{internal symmetry} \rightarrow \textit{invariance} \rightarrow \textit{conservation law}$$

is the electric charge conservation law.

So, there is an arbitrary field described by N-component complex functions $\psi_k(x)$ $\psi_k^\dagger(x) = (\psi_k^*)^T$ (k=1,2,....N). From the Lagrangian reality condition of this field follows that the Lagrangian $L(x)$ must contain bilinear combinations of field functions and their derivatives only of the kind

$$\psi_k^\dagger(x)A^{kl}\psi_l(x), \qquad \partial^\mu\psi_k^\dagger(x)B^{kl}_{\mu\nu}\partial^\nu\psi_l(x), \tag{6.1}$$

$$\partial^\mu\psi_k^\dagger(x)C^{kl}_\mu\psi_l(x), \qquad \psi_k^\dagger(x)C'^{kl}_\nu\partial^\nu\psi_l(x), \tag{6.2}$$

where A^{kl}, $B^{kl}_{\mu\nu}$, C^{kl}_μ, C'^{kl}_ν are the quantities being independent on x, and the indices k and l may have tensor or matrix dimensions. For example, the Lagrangian of the free electron-positron field is given by the expression

$$L = \frac{i}{2}(\overline{\psi}\gamma^\mu\partial_\mu\psi - \partial_\mu\overline{\psi}\gamma^\mu\psi) - m\overline{\psi}\psi$$

and we have

$$A^{kl}=-m(\gamma^0)^{kl},\qquad B^{kl}_{\mu\nu}=0,\qquad C'^{kl}_{\mu}=-C^{kl}_{\mu}=\frac{i}{2}(\gamma^0\gamma^\mu)^{kl},$$

where k and l are the matrix indices.

From Eqs. (6.1) and (6.2) it follows that the field functions are defined with an accuracy of an arbitrary phase factor. This means that Lagrangian of the system is kept invariant under the transformations

$$\left.\begin{array}{l}\psi_k(x)\to\psi'_k(x)=U(\alpha)\psi_k(x)=\exp(i\alpha)\psi_k(x),\\ \psi^\dagger_k(x)\to\psi'^\dagger_k(x)=U^\dagger(\alpha)\psi^\dagger_k(x)=\exp(-i\alpha)\psi^\dagger_k(x),\end{array}\right\}\tag{6.3}$$

where α is an arbitrary constant number. To put this another way, physical reality corresponding to the descriptions in terms of the old $(\psi_k(x))$ and new $(\psi'_k(x))$ field functions is the same.

The transformations $U(\alpha)$ generate the one-parametric group of the local gauge transformations which is also called the gauge transformations group of the first kind. The group $U(\alpha)$ is unitary, that is,

$$U(\alpha)U^\dagger(\alpha)=I,$$

where I is the unit matrix. Since all its elements commute with each other it is Abelian. Let us consider the electron-positron field and carry out an infinitesimal transformation of the field functions

$$\left.\begin{array}{l}\psi'(x)=(1+i\delta\alpha)\psi(x)=\psi(x)+\delta\psi(x),\\ \overline{\psi}'(x)=(1-i\delta\alpha)\overline{\psi}(x)=\overline{\psi}(x)+\delta\overline{\psi}(x).\end{array}\right\}\tag{6.4}$$

The theory invariance means that the Lagrangian variation turns into zero under the transformations (6.4), that is, it takes place

$$\begin{aligned}\delta L&=\frac{\partial L}{\partial\psi}\delta\psi+\frac{\partial L}{\partial\overline{\psi}}\delta\overline{\psi}+\frac{\partial L}{\partial(\partial_\mu\psi)}\delta(\partial_\mu\psi)+\frac{\partial L}{\partial(\partial_\mu\overline{\psi})}\delta(\partial_\mu\overline{\psi})=\\ &=i\alpha\left[\frac{\partial L}{\partial\psi}\psi-\frac{\partial L}{\partial\overline{\psi}}\overline{\psi}\right]+i\alpha\left[\frac{\partial L}{\partial(\partial_\mu\psi)}\partial_\mu\psi-\frac{\partial L}{\partial(\partial_\mu\overline{\psi})}\partial_\mu\overline{\psi}\right]=\\ &=i\alpha\left[\frac{\partial L}{\partial\psi}-\partial_\mu\left(\frac{\partial L}{\partial(\partial_\mu\psi)}\right)\right]\psi-i\alpha\left[\frac{\partial L}{\partial\overline{\psi}}-\partial_\mu\left(\frac{\partial L}{\partial(\partial_\mu\overline{\psi})}\right)\right]\overline{\psi}+\\ &\qquad+i\alpha\partial_\mu\left[\frac{\partial L}{\partial(\partial_\mu\psi)}\psi-\frac{\partial L}{\partial(\partial_\mu\overline{\psi})}\overline{\psi}\right]=0.\end{aligned}\tag{6.5}$$

The expressions in the square brackets are equal to zero on the strength of Lagrange-Euler equations for ψ and $\overline{\psi}$. Thus the Lagrangian invariance with respect to (6.4) leads to a current conservation

$$\partial_\mu j^\mu=0,$$

where

$$j_\mu=i\alpha\left[\frac{\partial L}{\partial(\partial_\mu\psi)}\psi-\frac{\partial L}{\partial(\partial_\mu\overline{\psi})}\overline{\psi}\right].$$

For an arbitrary field with k degrees of freedom the current of the kind

$$j^{\mu}(x)=i\alpha\left(\frac{\partial L(x)}{\partial[\partial_{\mu}\psi_k(x)]}\psi_k(x)-\frac{\partial L(x)}{\partial[\partial_{\mu}\psi_k^{\dagger}(x)]}\psi_k^{\dagger}(x)\right) \tag{6.6}$$

satisfies the continuity equation. Integrating the continuity equation over the three-dimensional volume and using Gauss theorem we arrive at the conservation law of the corresponding charge

$$Q=\int j^0(x)d\mathbf{r}=\text{const}. \tag{6.7}$$

It is evident that the same will take place for the current being equal to const $\times\, j^{\mu}$ as well, if the current j^{μ} satisfies the continuity equation. In the method we have used a charge measurement unit is not fixed. This could be done with the help of additional physical assumptions only. Supposing $\alpha=q$, where q is the electric charge of particles corresponding to a wave field, we come to the electric charge conservation law

$$Q_{em}=\int j^0_{em}(x)d\mathbf{r}=iq\int\left(\frac{\partial L(x)}{\partial[\partial_0\psi_k(x)]}\psi_k(x)-\frac{\partial L(x)}{\partial[\partial_0\psi_k^{\dagger}(x)]}\psi_k^{\dagger}(x)\right)d\mathbf{r}=\text{const}. \tag{6.8}$$

To gain a better understanding of consequences of the gauge transformation (6.3) we make it into a geometrical form. For the sake of simplicity we consider an one-component field $\phi(x)$ (such fields describe spinless particles). The field functions $\phi(x)$ and $\phi^*(x)$ could be represented in the form

$$\phi(x)=\frac{\phi_1(x)+i\phi_2(x)}{\sqrt{2}},\qquad \phi^*(x)=\frac{\phi_1(x)-i\phi_2(x)}{\sqrt{2}}, \tag{6.9}$$

where $\phi_1(x)$ and $\phi_2(x)$ are real quantities. Then the gauge transformations (6.3) will show up as follows

$$\left.\begin{array}{l}\phi_1'(x)+i\phi_2'(x)=\exp(i\alpha)[\phi_1(x)+i\phi_2(x)],\\ \phi_1'(x)-i\phi_2'(x)=\exp(-i\alpha)[\phi_1(x)-i\phi_2(x)].\end{array}\right\} \tag{6.10}$$

Since Eq. (6.10) could be rewritten as

$$\begin{pmatrix}\phi_1(x)\\ \phi_2(x)\end{pmatrix}'=\begin{pmatrix}\cos\alpha & \sin\alpha\\ -\sin\alpha & \cos\alpha\end{pmatrix}\begin{pmatrix}\phi_1(x)\\ \phi_2(x)\end{pmatrix}, \tag{6.11}$$

then it is evident that the gauge transformations (6.3) may be treated as rotations of a vector $\phi(x)=(\phi_1(x),\phi_2(x))$ about an angle α. On the group-theoretic slang the above mentioned means that the $U(1)$ group is locally isomorphic to the orthogonal rotations group $SO(2)$ in the two-dimensional space of the real functions $\phi_1(x)$ and $\phi_2(x)$. Since $\alpha=$ const, then this transformation must be one and the same in all the points of the space-time continuum, i.e. it is a "global" gauge transformation. In other words, when in the internal space of the field $\phi(x)$ the rotation on the angle α is fulfilled in one point, the same rotation must be simultaneously fulfilled in all other points. If the conserved quantity, not being the source of a physical field, be connected with the invariance with respect to this transformation, there would be no occasions for the trouble. However the electric charge produces the electromagnetic field in the space. Moreover the propagation velocity of electromagnetic interaction does not equal to infinity but it has the finite value. To avoid the conflict with the short-range interaction (or the field concept) we are forced to localize the gauge transformation $U(1)$. This means the phase α must be different for various points of the space-time, i. e. $\alpha = \alpha(x)$. The corresponding transformation is called the "local" gauge transformation or the gauge transformation of the second kind.

Let us examine physical consequences of the theory invariance with respect to the local gauge transformations. By way of example we consider the electron-positron field $\psi(x)$ with the Lagrangian rewritten in the form

$$L = i\overline{\psi}(x)\gamma^{\mu}\partial_{\mu}\psi(x) - m\overline{\psi}(x)\psi(x). \tag{6.12}$$

Under the local gauge transformation of the $U(1)$ group the field function transformation law is given by the expression

$$\psi(x) \to \psi'(x) = \exp[i\alpha(x)]\psi(x). \tag{6.13}$$

We see, that, thanks to the presence of the derivative, the Lagrangian (6.12) is not invariant under this transformations

$$\partial_{\mu}\psi(x) \to \exp[i\alpha(x)][\partial_{\mu}\psi(x) + i\psi(x)\partial_{\mu}\alpha(x)]. \tag{6.14}$$

The invariance of Lagrangian will be ensured, if one introduces a new derivative in such a way that the derivative of the field function is transformed just the same manner as the field function itself, that is,

$$D_{\mu}\psi(x) \to \exp[i\alpha(x)]D_{\mu}\psi(x). \tag{6.15}$$

The relation (6.15) will be fulfilled under condition

$$D_{\mu} = \partial_{\mu} + igA_{\mu}(x), \tag{6.16}$$

where at the local transformations (6.13) the introduced vector field $A_{\mu}(x)$ must behave in the following manner

$$A_{\mu}(x) \to A_{\mu}(x) - g^{-1}\partial_{\mu}\alpha(x). \tag{6.17}$$

We call the new derivative D_{μ} by a covariant derivative. Now our system, apart from the fermion field, includes the vector field $A_{\mu}(x)$ too. Consequently the Lagrangian (6.12) should be supplemented by the free vector field Lagrangian which, in its turn, must not violate the local gauge invariance and must be relativistic covariant. If we also demand the fulfillment of the superposition principle, we can be dealing with only a quantity of the kind

$$aF_{\mu\nu}(x)F^{\mu\nu}(x) + b\tilde{F}_{\mu\nu}(x)\tilde{F}^{\mu\nu}(x), \tag{6.18}$$

where

$$F_{\mu\nu}(x) = \partial_{\mu}A_{\nu}(x) - \partial_{\nu}A_{\mu}(x), \qquad \tilde{F}_{\mu\nu}(x) = \frac{1}{2}\varepsilon_{\mu\nu\lambda\sigma}F^{\lambda\sigma}(x).$$

Now to obtain the QED Lagrangian we sufficiently identify g with the electron electric charge e and $A_{\mu}(x)$ with the electromagnetic field potential. In so doing we should choose the coefficients in (6.18) by the following way: $a = -1/4$, $b = 0$. Thus the Lagrangian invariance with respect to the local gauge transformations group produces not only the electric charge conservation but it also leads to harmony with special relativity theory. Appearance of the gauge boson corresponding to the gauge group $U(1)$, the electromagnetic interaction carrier, the photon, is one more consequence of the localization of this group. It should be stressed that adding the mass term $m^2 A_\mu A^\mu/2$ to the total QED Lagrangian is forbidden by the gauge invariance. The conservation of the invariance demands that the corresponding gauge boson, the interaction carrier, must be massless.

Thus, imposing the natural requirement of the local gauge invariance on the free fermion Lagrangian, we arrive at the QED. Then, if one is distracted apart some arbitrariness in a choice of free fields Lagrangians the above mentioned may be thought as the strong argument in favor of that the local gauge invariance represents the principle laying at the heart of a theory of any interaction.

6.2. Nonabelian Gauge Invariance and QCD

Thus the electromagnetic field appears as the compensated field which ensures the charged fields invariance with respect to the local one-parametric group $U(1)_{em}$. In 1954 C. N. Yang and R. L. Mills investigated the local generalization of non-Abelian three-parametric group $SU(2)$. As a result they came to recognize that in this case the local gauge invariance of the theory already demands introducing three-parametric compensated field. Obvious generalization of this fact lies in a statement: in the case of n-parametric local gauge group the theory invariance demands introducing n-parametric compensated field. But since these gauge fields were massless they led to long-range forces which are absent in Nature. In this connection the Yang-Mills theory first has attracted purely academic interest and the prototype of the future theory of strong interaction one can be made out in it in no way. There were not such notions as the quark and the gluon. They appeared a decade later and, in the beginning, they were not connected with a mathematical apparatus of non-Abelian theories in any way. Still ten years were necessary in order that the synthesis of these two ideas led to the QCD formulation.

The QCD is based on developing the idea stated in **6.1**. But now, in place of the $U(1)_{em}$ gauge group, we are dealing with the phase transformations group of the color quarks fields, the $SU(3)_c$ group. We shall consider for simplicity that the quarks have one flavor only (one flavor approximation). Then the free quarks Lagrangian is given by the expression

$$L_0 = \overline{q_k}(x)(i\gamma_\mu \partial^\mu - m)q_k(x), \tag{6.19}$$

where $k = R, B, G$. Let us investigate the consequences of the invariance of L_0 with respect to the local gauge transformations of the non-Abelian group $SU(3)_c$

$$q'_k(x) = U_{kj}(x)q_j(x) = \{\exp[i\alpha_a(x)T_a]\}_{kj}q_j(x), \qquad \overline{q}'_k(x) = \overline{q}_j(x)U^\dagger_{kj}(x), \tag{6.20}$$

where $T_a = \lambda_a/2$ and λ_a are the Gell-Mann matrices (4.144).

So, our task is to ensure the local $SU(3)_c$ invariance of the Lagrangian L_0. Pass in (6.20) to infinitesimal transformations

$$q'_k(x) = [\delta_{kj} + i\alpha_a(x)(T_a)_{kj}]q_j(x). \tag{6.21}$$

In this case the derivative of the quark field function is transformed by a law

$$\partial_\mu q'_k(x) = [\delta_{kj} + i\alpha_a(x)(T_a)_{kj}]\partial_\mu q_j(x) + i(T_a)_{kj}q_j(x)\partial_\mu\alpha_a(x) \tag{6.22}$$

and violates the invariance of L_0. To rescue the situation we introduce eight gauge fields $G^a_\mu(x)$ and build covariant derivatives

$$D^\mu_{kj}(x) = \delta_{kj}\partial^\mu + ig_s(T_a)_{kj}G^a_\mu(x), \tag{6.23}$$

where $g_s \equiv g_{SU(3)_c}$. Further we replace the ordinary derivatives with the covariant ones in L_0

$$L_0 = \overline{q_k}(x)[i\gamma^\mu D^\mu_{\overline{k}j}(x) - m\delta_{\overline{k}j}]q_j(x) =$$

$$= \overline{q_k}(x)(i\gamma_\mu\partial^\mu - m)q_k(x) - g_s[\overline{q_k}(x)\gamma_\mu(T_a)_{\overline{k}j}q_j(x)]G^{a\mu}(x). \tag{6.24}$$

The invariance of the Lagrangian will be provided under a condition

$$\overline{q_k}'(x)D^{\mu\prime}_{\overline{k}j}(x)q'_j(x) = \overline{q_k}(x)D^\mu_{\overline{k}j}(x)q_j(x),$$

what, in its turn, gives

$$D^{\mu\prime}_{kj}(x) = U_{ki}(x)D^\mu_{il}(x)U^\dagger_{jl}(x). \tag{6.25}$$

By analogy with the QED we demand that the transformation law of the gauge fields $G^a_\mu(x)$ has the form

$$G^{a\prime}_\mu(x) = G^a_\mu(x) - \frac{1}{g_s}\partial_\mu\alpha_a(x). \tag{6.26}$$

However, in this case, the last quantity in (6.24) is not the invariant with respect to the $SU(3)_c$ transformations. Really, taking into account the algebra of the λ_a matrices we obtain

$$[\overline{q_k}(x)\gamma_\mu(T_a)_{\overline{k}j}q_j(x)]' = [\overline{q_k}(x)\gamma_\mu(T_a)_{\overline{k}j}q_j(x)] + i\alpha_b(x)\overline{q_k}(x)\gamma_\mu(T_aT_b -$$

$$-T_bT_a)_{\overline{k}j}q_j(x) = [\overline{q_k}(x)\gamma_\mu(T_a)_{\overline{k}j}q_j(x)] - f_{abc}\alpha_b(x)(\overline{q_k}(x)\gamma_\mu(T_c)_{\overline{k}j}q_j(x)),$$

From the obtained expression follows that the gauge invariance of the Lagrangian (6.19) will be restored if we replace the transformation law (6.19) with

$$G^{a\prime}_\mu(x) = G^a_\mu(x) - \frac{1}{g_s}\partial_\mu\alpha_a(x) - f_{abc}\alpha_b(x)G^c_\mu(x). \tag{6.27}$$

Now we should supplement the Lagrangian L_0 by that of the free gauge bosons L_G. Thanks to the presence of the last term in Eq. (6.27), the field tensor $G^a_\mu(x)$ has more complicated form than its analog in the QED. It is not difficult to show that the gauge invariance will be ensured by the following choice of L_G

$$L_G = -\frac{1}{4}G^a_{\mu\nu}(x)G^{\mu\nu}_a(x), \tag{6.28}$$

where

$$G^a_{\mu\nu}(x) = \partial_\mu G^a_\nu(x) - \partial_\nu G^a_\mu(x) - g_s f_{abc}G^b_\mu(x)G^c_\nu(x). \tag{6.29}$$

Thus, using only the requirement of the Lagrangian invariance with respect to the local gauge group $SU(3)_c$, we have obtained the total Lagrangian of the color quarks q_k and the vector gluons G_μ

$$L=\overline{q_k}(x)(i\gamma_\mu\partial^\mu - m)q_k(x) - g_s[\overline{q_k}(x)\gamma_\mu G^\mu_{\overline{k}j}(x)q_j(x)]-$$

$$-\frac{1}{4}G^a_{\mu\nu}(x)G^{\mu\nu}_a(x), \qquad (6.30)$$

where $G^\mu_{\overline{k}j}(x)=(\lambda_a)_{\overline{k}j}G^{a\mu}(x)/2$, The number of the quark field phases, we can change by arbitrary way, is equal to eight. Consequently to compensate all the phases changes we need eight gluons as well. Since the introduction of the gluon mass term leads to the violation of the local gauge invariance the carriers of strong interaction, the gluons, are massless.

It is obvious from the form of L_G that in the QCD the kinetic energy of the gluons is not already purely kinetic, since it contains interaction between the gluons ($\sim g_sG^a_\mu G^b_\nu G^c_\lambda$ and $\sim g_s^2G^a_\mu G^b_\nu G^c_\lambda G^d_\sigma$). Thus in the QCD the Feynman diagrams include vertices in which only the gluons are met. In other words, the gluons possess nonlinear self-interaction, and this circumstance is caused by that they have themselves the color charge. To define Lagrangian of the theory or, what is the same, th establish the evolution equation is the basic milestone under a new theory production. The QCD appearance has sharply changed the situation in the strong interaction theory. By now the QCD is the sole serious candidate which claims for describing the hadrons structure and hadrons interaction processes. Many important questions of the QCD have been already resolved and the obtained theoretical results are being used under interpretation and description of the experimental data. However the QCD is in the making as yet. At "large" distances ($\geq 10^{-13}$ cm) the nonlinearity leads to such forces between the quarks and gluons which do not allow to appear the quarks and gluons in free states. Just the treatment of effects connected with the large distances is the QCD stumbling-stone. The basic unsolved problems of the QCD are related with it.

The gauge fields, which are introduced to provide the local non-Abelian gauge invariance of the theory, now are called Yang-Mills fields and the equations they satisfy in the free case

$$\partial^\nu G^a_{\mu\nu}(x)+g\varepsilon_{abc}G^{b\nu}(x)G^c_{\mu\nu}(x)=0, \qquad (6.31)$$

where ε_{abc} are structural constants of the local gauge group to be under consideration, are called Yang-Mills equations.

6.3. Spontaneous Symmetry Breaking. Higgs Mechanism

The QCD is not the sole descendant of the Yang-Mills theory. The electroweak interaction theory was also built on the basis of the non-Abelian gauge theory. However the way of its production was already more complex. The point is that gauge fields are long-range by its nature. This fact immediately leads to the zero mass of the interaction carriers as it takes place both in the QCD and in the QED. Weak interaction exists only at very small distances and the weak interaction carriers, the $W^\pm$- and Z-bosons, must have a huge mass. So, it was necessary to combine two incompatible things, namely, the local gauge invariance and the non-zero mass of the $W^\pm$- and Z-bosons. It appeared that one needs to use the idea of a spontaneous symmetry breaking to solve this problem.

An evolution of any physical system is defined by two factors: 1) a form of Lagrangian (or Hamiltonian), and 2) initial conditions. As an example, we consider a smooth surface in the form of peaked hat resting on a horizontal ground (Fig. 37).

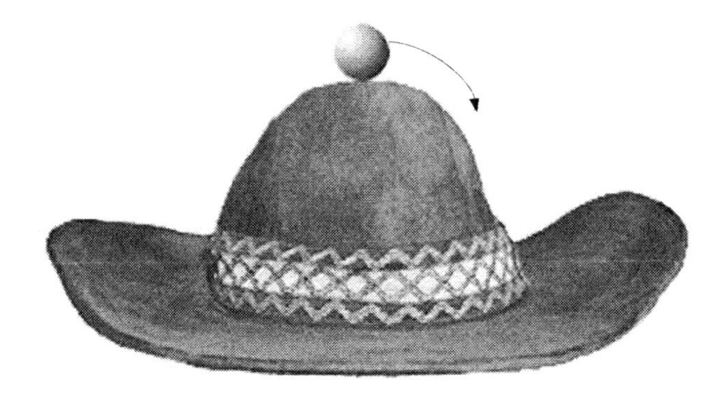

Figure 37. The spontaneous symmetry violation.

We place a ball on the hat top and we shall consider that gravitational force is a sole force acting on the system. It is clear that the system possesses the explicit symmetry with respect to rotations about a horizontal axis passing through the hat center. However the system is not stable. Really, if we move the ball out of position, that is, change the initial condition, it will be rolled down and the system symmetry will be violated. When having stopped on the fixed place of the hat brim the ball sets the selected direction from the central axis. The system has found a stability at the cost of the symmetry violation. Since the state with the violated symmetry has more low energy the ball prefer to roll on the hat brim. In the stable configuration the initial rotational symmetry of gravitational force exists as before but now it exists in the hidden form. The observed system state does not reflect the symmetry of the interaction which is present in the system.

In the quantum field theory two analogous conditions define a character of a symmetry manifestation: 1) a form of Lagrangian (or Hamiltonian); 2) a vacuum form. In the QCD and the QED both Lagrangian and the vacuum state were invariant under the symmetry transformations. In such cases we say the symmetry of the theory has been not violated. In 1960 Y. Nambu and J. Goldstone showed that there exist theories whose Lagrangian is invariant with respect to symmetry transformations while a vacuum is not invariant. In this case we use the term "*spontaneous symmetry breaking*".

We investigate the spontaneous breaking of the local gauge symmetry by the example of the $U(1)$ group. Consider the world which consists of charged scalar particles only and is defined by Lagrangian

$$L \equiv T - V(\varphi) = [\partial_\mu\varphi(x)]^*[\partial^\mu\varphi(x)] - \mu^2\varphi^*(x)\varphi(x) - \lambda[\varphi^*(x)\varphi(x)]^2 \qquad (6.32)$$

with $\lambda > 0$. When $\mu^2 > 0$, the Lagrangian will describe a self-interacting (according to the law $[\varphi^*(x)\varphi(x)]^2$) scalar field with the mass μ^2. In this case the value $\varphi(x) = 0$ corresponds to the vacuum (the minimum of $V(\varphi)$. In other words, the average on the vacuum of $\varphi(x)$ turns into zero ($< 0|\varphi(x)|0 >= 0$). However we wish to study the case with $\mu^2 < 0$. If one introduces the field functions according to the relation

$$\varphi(x) = \frac{\varphi_1(x) + i\varphi_2(x)}{\sqrt{2}},$$

the Lagrangian (6.32) takes the form

$$L = \frac{1}{2}\{[\partial_\mu\varphi_1(x)]^2 + [\partial_\mu\varphi_2(x)]^2 - \mu^2[\varphi_1^2(x) + \varphi_2^2(x)] - \lambda[\varphi_1^2 + \varphi_2^2(x)]^2/2\}. \qquad (6.33)$$

From this writing it is evident that the minima of the potential $V(\varphi)$ lie on the circle of the radius v in the plane φ_1, φ_2 (see, Fig. 38)

$$\varphi_1^2 + \varphi_2^2 = v^2, \qquad v^2 = -\frac{\mu^2}{\lambda}. \qquad (6.34)$$

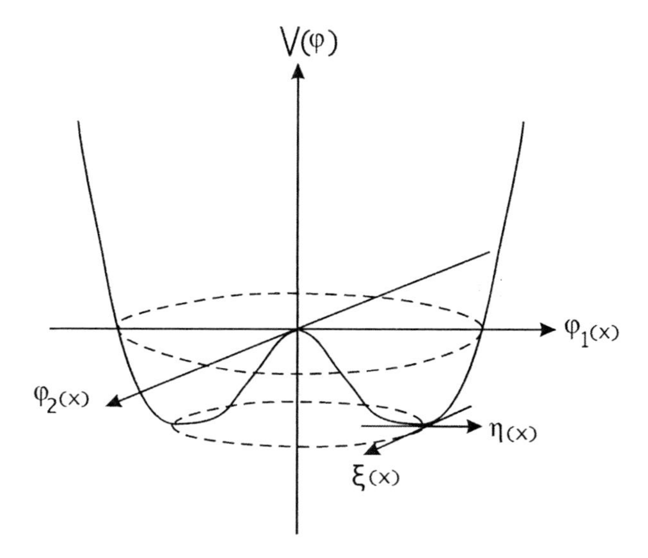

Figure 38. The potential $V(\varphi)$ of the complex scalar field.

This means the constant scalar field, the so called scalar field vacuum condensate, exists in the vacuum. This quantity, that is, the energy shift of the ground state, can not be measured directly since the quantity under test is the difference between the given energy and the vacuum energy.

In order to make Lagrangian (6.33) to be invariant with respect to the local gauge transformations of the $U(1)$ group

$$\varphi'(x) = \exp[i\alpha(x)]\varphi(x),$$

we introduce the covariant derivative

$$D_\mu(x) = \partial_\mu + ieA_\mu(x), \qquad (6.35)$$

where $A_\mu(x)$ is transformed by the law

$$A'_\mu(x) = A_\mu(x) - \frac{1}{e}\partial_\mu\alpha(x). \qquad (6.36)$$

Having supplemented the obtained Lagrangian with the free gauge bosons Lagrangian we arrive at

$$L = [\partial_\mu - ieA_\mu(x)]\varphi^*(x)[\partial^\mu + ieA^\mu(x)]\varphi(x) - \mu^2\varphi^*(x)\varphi(x) -$$

$$-\lambda[\varphi^*(x)\varphi(x)]^2 - \frac{1}{4}F_{\mu\nu}(x)F^{\mu\nu}(x). \tag{6.37}$$

Within elementary particle physics in the majority of instances we can not obtain exact solutions. We most commonly have to use the perturbation theory series expansion and calculate fluctuations close to a minimum energy. If we try to carry out the expansion in a neighborhood of the unstable point $\varphi = 0$, the perturbation theory series will not converge. The correct activity method is to carry out the expansion in a neighborhood of the minimum of the potential $V(\varphi)$, that is, in a neighborhood of the stable vacuum. For the minimum we choose the point $\varphi_1 = v, \varphi_2 = 0$. Notice, the vacuum is not already invariant with respect to the $U(1)$ group, i. e. the symmetry appears to be spontaneously broken. To expand L in a neighborhood of the vacuum we introduce real fields $\eta(x)$ and $\xi(x)$ which describe quantum fluctuations around this minimum

$$\varphi(x) = \frac{\eta(x) + i\xi(x) + v}{\sqrt{2}}. \tag{6.38}$$

To substitute (6.38) into the Lagrangian (6.37) leads to the expression

$$L' = \frac{1}{2}\{[\partial_\mu\xi(x)]^2 + [\partial_\mu\eta(x)]^2\} - v^2\lambda\eta^2(x) + \frac{1}{2}e^2v^2A_\mu(x)A^\mu(x) +$$

$$+evA_\mu(x)\partial^\mu\xi(x) - \frac{1}{4}F_{\mu\nu}(x)F^{\mu\nu}(x) + L_{int}, \tag{6.39}$$

where L_{int} denotes the terms describing interactions between the $\eta(x)$ and $A_\mu(x)$ fields. From the Lagrangian (6.39) follows

$$m_\xi = 0, \qquad m_\eta = \sqrt{2\lambda v^2}, \qquad m_A = ev. \tag{6.40}$$

In summary, we have attained the goal to be sought. Our gauge bosons have found the mass. However, as this takes place, the other problem connected with occurrence of a massless scalar particle has appeared. Such particles are called Goldstone bosons. Let us gain an understanding of the situation. Having given the mass to the field $A_\mu(x)$, we thereby increased the number of polarization degrees of freedom from 2 to 3, because now the field $A_\mu(x)$ can have the longitudinal polarization too. But the simple shift of the field variables which is given by Eq. (6.38) can not create new degrees of freedom in any way. It is obvious that not all the fields entering into L' correspond to the physical particles. It is beyond doubt that just the Goldstone boson brings about the suspicion. Let us show that it does not really belong to the physical sector. Since the theory is gauge invariant we can carry out any gauge transformation (fix the gauge[1]) and the physical contents of the theory is keeping invariable. The approximate equality

$$\varphi(x) = \frac{\eta(x) + i\xi(x) + v}{\sqrt{2}} \approx \frac{\eta(x) + v}{\sqrt{2}} \exp\left[\frac{i\xi(x)}{v}\right] \tag{6.41}$$

[1]The gauge which allows to exclude the Goldstone boson is called *unitary gauge*.

which is true to lowest order in $\xi(x)$ can suggest the required transformation form. It is clear that we should introduce new fields of the kind

$$\varphi'(x) = \exp\left[-\frac{i\xi(x)}{v}\right]\varphi(x) = \frac{\eta(x)+v}{\sqrt{2}},$$

$$A'_\mu(x) = A_\mu(x) + \frac{1}{ev}\partial_\mu\xi(x).$$

Further, taking into account

$$D_\mu\varphi(x) = \exp\left[\frac{i\xi(x)}{v}\right]\left[\partial_\mu\varphi'(x) + ieA'_\mu(x)\varphi'(x)\right] =$$

$$= \exp\left[\frac{i\xi(x)}{v}\right]\frac{1}{\sqrt{2}}\{\partial_\mu\eta(x) + ieA'_\mu(x)[\eta(x)+v]\},$$

and

$$|D_\mu\varphi(x)|^2 = \frac{1}{2}|\partial_\mu\eta(x) + ieA'_\mu(x)[\eta(x)+v]|^2, \qquad F_{\mu\nu} = \partial_\mu A'_\nu(x) - \partial_\nu A'_\mu(x),$$

we rewrite the Lagrangian (6.32) in the following form

$$L'' = \frac{1}{2}|\partial_\mu\eta(x) + ieA'_\mu(x)[\eta(x)+v]|^2 - \frac{\mu^2}{2}[\eta(x)+v]^2 -$$

$$-\frac{\lambda}{4}[\eta(x)+v]^4 - \frac{1}{4}F_{\mu\nu}(x)F^{\mu\nu}(x) = L_0 + L_{int}, \tag{6.42}$$

where

$$L_0 = \frac{1}{2}[\partial_\mu\eta(x)]^2 - \frac{\mu^2}{2}\eta^2(x) - \frac{1}{4}[\partial_\mu A'_\nu - \partial_\nu A'_\mu]^2 + \frac{1}{2}(ev)^2 A'_\mu(x)A^{\mu\prime}(x),$$

$$L_{int} = \frac{1}{2}e^2 A'_\mu(x)A^{\mu\prime}(x)\eta(x)[\eta(x)+2v] - \lambda v^2\eta^3(x) - \frac{\lambda}{4}\eta^4(x).$$

To sum up, the field $\xi(x)$ disappears from the Lagrangian, that is, the seeming additional degree of freedom connected with the Goldstone boson proved spurious in practical situations. Obviously, this degree of freedom corresponds to the liberty of the choice of the gauge transformation. The Lagrangian describes a mere two interacting massive particles, the vector gauge boson $A_\mu(x)$ with the mass $M = ev$ and the scalar particle $\eta(x)$ with the mass $m = \sqrt{2}\mu$ (the Higgs boson). Having absorbed the Goldstone boson the massless gauge field became massive. The number of degrees of freedom has been conserved since the massive vector particle with three spin projections appeared from the massless vector field, having two spin states, and from the massless scalar particle. In the field theory this phenomenon was discovered in 1964 and gave the name *Higgs mechanism*.

6.4. Weinberg-Salam-Glashow Theory

The symmetry $SU(2)_L \times U(1)_Y$ was first proposed by S. L. Glashow (1961) and then it was expanded to include the massive vector bosons (W and Z) by S. Weinberg and A.Salam (1967-1968). At building the electroweak interaction theory by Glashow, Weinberg, and Salam one may single out four basic stage. It should be stressed that the analogous division takes place for any version of an electroweak theory as well.

Firstly, one should choose a gauge group G in such a way that it includes all the necessary vector particles. The gauge symmetry $SU(2)_L \times U(1)_Y$ [1] is the base of the WSG theory. Here $SU(2)_L$ represents a group of the weak isospin while $U(1)_Y$ does a group of the weak hypercharge (the term "weak" is used in order to stress the difference from the corresponding characteristics of strong interaction). Thus the weak isospin $\mathbf{S^W}$ and the weak hypercharge Y^W (more precisely, $Y^W/2$) are generators of the gauge transformations of $SU(2)_L$ and $U(1)_Y$ respectively. The weak hypercharge is assigned to every field so that the analog of the formula by Gell-Mann and Nishijima is fulfilled

$$Q = S_3^W + \frac{Y^W}{2}. \tag{6.43}$$

The unbroken local $SU(2)_L \times U(1)_Y$ symmetry demands the existence of four massless vector bosons. Three of them W^1, W^2, W^3, represent gauge bosons of the non-Abelian group $SU(2)_L$ and their interaction is characterized by the gauge constant g. B describes the gauge field of the Abelian group $U(1)_Y$ and its interaction is determined by the gauge constant g'.

In the second stage one should choose the representation of the symmetry group for the matter particles (leptons and quarks). Since the theory will describe weak processes which, as it is well known, do not conserve the parity, the theory must be explicitly mirror-asymmetrical from the beginning. This asymmetry is realized as follows. The left-hand components of the fermions

$$\psi_L(x) = \frac{1}{2}(1+\gamma_5)\psi(x)$$

form the weak isospin doublets with respect to the $SU(2)_L$ group

$$\begin{pmatrix} \nu_{eL} \\ e_L^- \end{pmatrix}, \begin{pmatrix} \nu_{\mu L} \\ \mu_L^- \end{pmatrix}, \begin{pmatrix} \nu_{\tau L} \\ \tau_L^- \end{pmatrix}, \qquad S^W = \frac{1}{2}, \ Y^W = -1 \tag{6.44}$$

$$\begin{pmatrix} u_L \\ d_L \end{pmatrix}, \begin{pmatrix} c_L \\ s_L \end{pmatrix}, \begin{pmatrix} t_L \\ b_L \end{pmatrix}, \qquad S^W = \frac{1}{2}, \ Y^W = \frac{1}{3}, \tag{6.45}$$

while the right-hand components of all fermions excepting the neutrinos

$$\psi_R(x) = \frac{1}{2}(1-\gamma_5)\psi(x)$$

represent the weak isospin singlets

$$e_R^-, \ \mu_R^-, \ \tau_R^-, \qquad S^W = 0, \ Y^W = -2,$$

[1] We used the designation $SU(2)_{EW} \times U(1)_{EW}$ for this group before.

$$u_R,\ c_R,\ t_R, \qquad S^W = 0,\ Y^W = \frac{4}{3}, \tag{6.46}$$

$$d_R,\ s_R,\ b_R, \qquad S^W = 0,\ Y^W = -\frac{2}{3}.$$

The absence of the neutrino singlets in the WSG theory was connected with the fact that the neutrino was considered the massless particles by the time of this theory production.

At the $SU(2)_L \times U(1)_Y$ global transformations the transformation law of the left-hand and right-hand components of the field $\psi(x)$ has the form

$$\psi'_L(x) = \exp[i(\alpha\cdot\mathbf{S}^W + \frac{\beta Y^W}{2})]\psi_L(x), \qquad \psi'_R(x) = \exp[\frac{i\beta Y^W}{2}]\psi_R(x). \tag{6.47}$$

We begin our consideration with the leptons. The lepton sector of the WSG theory is described by the Lagrangian

$$L_l = i\sum_{l=e,\mu,\tau}[\overline{\psi}_{lL}(x)\gamma^\mu\partial_\mu\psi_{lL}(x) + \overline{\psi}_{lR}(x)\gamma^\mu\partial_\mu\psi_{lR}(x)]. \tag{6.48}$$

To introduce the mass terms into the Lagrangian (6.48) directly

$$m_l[\overline{\psi}_{lL}(x)\psi_{lR}(x) + \overline{\psi}_{lR}(x)\psi_{lL}(x)]$$

violates the gauge invariance. This makes us to use the mechanism of the mass generation at the cost of the spontaneous symmetry breaking not only for the weak interaction carriers but for leptons as well. For this purpose the Lagrangian of the Yukawa type which describe interactions between leptons and Higgs fields is used

$$L_Y = -\sum_{l=e,\mu,\tau} f_l[\overline{\psi}_{lL}(x)\psi_{lR}(x)\varphi(x) + \varphi^\dagger(x)\overline{\psi}_{lR}(x)\psi_{lL}(x)]. \tag{6.49}$$

In order to make the neutrino massive it is sufficient to introduce the neutrino singlets in the theory. Notice, there is no theoretical principle which allows to choose the Yukawa constants f_l and they unfortunately remain an arbitrary parameters of the theory.

In the third stage one needs to localize the gauge group in question, that is, carry out the replacement

$$\alpha \to \alpha(x), \qquad \beta \to \beta(x).$$

This, as it is known, demands the transition to the covariant derivatives and an introduction of the free gauge bosons Lagrangian. In the case of the $SU(2)_L \times U(1)_Y$ gauge group the covariant derivatives for the fields entering into the Lagrangian have the form

$$D_\mu = \partial_\mu - ig\mathbf{S}^W\cdot\mathbf{W}_\mu - ig'\frac{Y^W}{2}B_\mu. \tag{6.50}$$

Then, recalling that at $S^W = 1/2$ the matrices $\sigma_k/2$ are the generators of the $SU(2)$ transformations, we obtain for the fields $\psi_L(x)$ and $\psi_R(x)$

$$D_\mu\psi_{lL}(x) = \left[\partial_\mu - \frac{ig}{2}\sigma\cdot\mathbf{W}_\mu(x) + \frac{ig'}{2}B_\mu(x)\right]\psi_{lL}(x), \tag{6.51}$$

$$D_\mu \psi_{lR}(x) = \left[\partial_\mu + ig' B_\mu(x)\right] \psi_{lR}(x). \tag{6.52}$$

The free gauge bosons Lagrangian is determined by the expression

$$L_0 = -\frac{1}{4} W^a_{\mu\nu}(x) W^{a\mu\nu}(x) - \frac{1}{4} B_{\mu\nu}(x) B^{\mu\nu}(x),$$

where $W^a_{\mu\nu}(x)$ is a tensor of a non-Abelian field

$$W^a_{\mu\nu}(x) = \partial_\mu W^a_\nu(x) - \partial_\nu W^a_\mu(x) + g\varepsilon_{abc} W^b_\nu(x) W^c_\mu(x), \qquad a,b,c = 1,2,3,$$

$B_{\mu\nu}(x)$ is a tensor of an Abelian field

$$B_{\mu\nu}(x) = \partial_\mu B_\nu(x) - \partial_\nu B_\mu(x).$$

To pass to the covariant derivative leads to the appearance of two basic interactions in the total Lagrangian: interaction of the weak currents isotriplet $\mathbf{J}^\mu(x)$ with three vector bosons $\mathbf{W}^\mu(x)$

$$g\mathbf{J}^\mu_l(x) \cdot \mathbf{W}_\mu(x) = g\overline{\psi}_{lL}(x)\gamma^\mu \mathbf{S}^W \cdot \mathbf{W}_\mu(x) \psi_{lL}(x), \tag{6.53}$$

and interaction of the weak hypercharge current $j^{Y\mu}_l(x)$ with fourth vector boson $B_\mu(x)$

$$\frac{g'}{2} j^{Y\mu}_l(x) B_\mu(x) = g' \overline{\psi}_l(x) \gamma^\mu \frac{Y^W}{2} \psi_l(x) B_\mu(x), \tag{6.54}$$

where

$$\psi_l(x) = \psi_{lL}(x) + \psi_{lR}(x).$$

In the fourth stage we must give the mass both to the weak interaction carriers and to the leptons. For this purpose we shall make use of the mechanism of the spontaneous symmetry breaking on the chain

$$SU(2)_L \times U(1)_Y \to U(1)_{em}.$$

The supplementary two-component complex scalar Higgs field $\varphi(x)$ (four degrees of freedom) is introduced

$$\varphi(x) = \frac{1}{\sqrt{2}} \begin{pmatrix} i\Phi_1(x) + \Phi_2(x) \\ H(x) - i\Phi_3(x) \end{pmatrix} = \begin{pmatrix} \varphi^+(x) \\ \varphi^0(x) \end{pmatrix}, \qquad S^W = 1/2,\ Y^W = 1, \tag{6.55}$$

where $\mathrm{Im}\,\Phi_i(x) = 0$ and $\mathrm{Im}\,H(x) = 0$. The Lagrangian describing the Higgs fields doublet, exclusive of the kinetic energy, contains the potential energy of self-interaction as well

$$L_H = |D_\mu \varphi(x)|^2 - V(\varphi) \tag{6.56}$$

where

$$V(\varphi) = -\mu^2 \varphi^\dagger(x)\varphi(x) - \lambda[\varphi^\dagger(x)\varphi(x)]^2 \tag{6.57}$$

with $\lambda > 0$ and $\mu^2 < 0$. The spontaneous symmetry breaking is realized by the shift of the neutral Higgs field component on the real constant $v = \sqrt{-\mu^2/\lambda}$

$$\varphi(x) = \frac{1}{\sqrt{2}} \begin{pmatrix} i\Phi_1(x) + \Phi_2(x) \\ H(x) - i\Phi_3(x) + v \end{pmatrix} = \varphi'(x) + \xi_0, \tag{6.58}$$

where

$$\xi_0 = \frac{1}{\sqrt{2}} \begin{pmatrix} 0 \\ v \end{pmatrix}$$

and $< 0|\varphi'(x)|0 >= 0$. Then the parametrization of fluctuations close to ξ_0 in the lowest order of Φ takes the form

$$\varphi(x) = \frac{1}{\sqrt{2}} \exp[i\sigma \cdot \Phi(x)/v] \begin{pmatrix} 0 \\ v + H(x) \end{pmatrix}. \tag{6.59}$$

Notice, at any choice of $\varphi(x)$ the symmetry violation inevitably results in the appearance of the mass on the corresponding gauge bosons. But when the invariance both of the Lagrangian and of the vacuum with respect to some gauge transformation subgroup is conserved then the gauge bosons connected with these subgroups are kept massless. Under the choice

$$< 0|\varphi(x)|0 >= \frac{1}{\sqrt{2}} \begin{pmatrix} 0 \\ v \end{pmatrix}$$

with $S^W = 1/2$, $S_3^W = -1/2$ and $Y^W = 1$ both the $SU(2)_L$ and $U(1)_Y$ gauge symmetries are violated. Since the generators of the groups $U(1)_{em}$, $SU(2)_L$ and $U(1)_Y$ satisfy the relation (6.43) then

$$Q\xi_0 = 0, \tag{6.60}$$

or

$$\xi_0' = \exp[i\alpha(x)Q]\xi_0 = \xi_0. \tag{6.61}$$

Thus both the final Lagrangian and the vacuum are invariant with respect to the $U(1)_{em}$ group transformations what ensures the zero-mass photon.

As a result of the shift in $|D_\mu\varphi(x)|^2$ the terms which are bilinear on the components of W_μ^a and B_μ appear. They give the contribution to the mass matrix of the gauge bosons

$$\left| [-\frac{ig}{2}\sigma \cdot \mathbf{W}_\mu(x) - \frac{ig'}{2}B_\mu(x)]\varphi(x) \right|^2 =$$

$$= \frac{1}{8} \left| \begin{pmatrix} gW_\mu^3(x) + g'B_\mu(x) & g[W_\mu^1(x) - iW_\mu^2(x)] \\ g[W_\mu^1(x) + iW_\mu^2(x)] & -gW_\mu^3(x) + g'B_\mu(x) \end{pmatrix} \begin{pmatrix} 0 \\ v \end{pmatrix} \right|^2 =$$

$$= \frac{g^2v^2}{8} [(W_\mu^1(x))^2 + (W_\mu^2(x))^2] + \frac{v^2}{8} [gW_\mu^3(x) - g'B_\mu(x)]^2. \tag{6.62}$$

As it follows from Eq. (6.62) the fields $W_\mu^3(x)$ and $B_\mu(x)$ prove to be mixed. This is no surprise, since they have the identical quantum numbers. For diagonalization of the last term in (6.62) we pass to a new basis

$$\begin{pmatrix} Z_\mu \\ A_\mu \end{pmatrix} = \frac{1}{\sqrt{g^2 + g'^2}} \begin{pmatrix} g & -g' \\ g' & g \end{pmatrix} \begin{pmatrix} W_\mu^3 \\ B_\mu \end{pmatrix} = \begin{pmatrix} \cos\theta_W & -\sin\theta_W \\ \sin\theta_W & \cos\theta_W \end{pmatrix} \begin{pmatrix} W_\mu^3 \\ B_\mu \end{pmatrix}, \tag{6.63}$$

where

$$\tan\,\theta_W = \frac{g'}{g}.$$

In the new basis the term $v^2[gW^3_\mu(x)-g'B_\mu(x)]^2/8$ takes the form

$$\frac{1}{2}m_Z^2 Z_\mu(x)Z^\mu(x),$$

where

$$m_Z=\frac{v\sqrt{g^2+g'^2}}{2}. \tag{6.64}$$

Further crossing to complex self-conjugate fields

$$W^\pm_\mu=\frac{W^1_\mu\mp iW^2_\mu}{\sqrt{2}},\qquad W_\mu\equiv W^-_\mu,\ W^*_\mu\equiv W^+_\mu \tag{6.65}$$

and taking into account (6.64) we rewrite Eq.(6.62) in the form

$$m_W^2 W^*_\mu(x)W^\mu(x)+\frac{1}{2}m_Z^2 Z_\mu(x)Z^\mu(x), \tag{6.66}$$

where

$$m_W=\frac{gv}{2}. \tag{6.67}$$

Thus the weak interaction carriers have acquired the mass whereas the photon is kept massless. The massless gauge bosons had two polarization states. After they had acquired the mass the number of their polarization states increased on one. They borrow these three additional degrees of freedom from the Higgs bosons. However the Higgs field had four degrees of freedom. What destiny has the last Higgs component? It appeared that the remained Higgs boson becomes massive and passes into the physical particles sector (the physical Higgs boson).

Substituting (6.58) into (6.57) one may convinces that the field $H(x)$ has get the mass $m_H^2=2\lambda v^2$ while the fields $\Phi_i(x)$ have been kept massless, that is, they represent the Goldstone fields. To eliminate the Goldstone bosons we use the parametrization of the field $\varphi(x)$ in the form (6.59). We carry out the $SU(2)_L$ gauge transformation in the total Lagrangian written in terms of $\varphi(x)$

$$\varphi'(x)=\mathcal{U}(x)\varphi(x)=\frac{1}{\sqrt{2}}\begin{pmatrix}0\\ v+H(x)\end{pmatrix},$$

$$\frac{\sigma_k}{2}W'_{k\mu}(x)=\mathcal{U}(x)\left(\frac{i}{g}\partial_\mu+\frac{\sigma_k}{2}W_{k\mu}(x)\right)\mathcal{U}^{-1}(x),$$

$$\psi'_L(x)=\mathcal{U}(x)\psi_L(x),\qquad B'_\mu(x)=B_\mu(x),\qquad \psi'_R(x)=\psi_R(x)$$

where

$$\mathcal{U}(x)=\exp[-i\sigma\cdot\Phi(x)/v].$$

The transformation law for $W^\mu_k(x)$ follows from the Lagrangian invariance with respect to $\mathcal{U}(x)$, what is ensured by the condition

$$\mathcal{U}^{-1}(x)D'_\mu(x)\mathcal{U}(x)\psi_L(x)=D_\mu(x)\psi_L(x).$$

It is not difficult to show the fields $\Phi_1(x), \Phi_2(x)$ and $\Phi_3(x)$ have not been already contained in the final Lagrangian. To sum up, three Goldstone bosons are eliminated by the gauge transformation from the theory (to be gauged) while the liberated three degrees of freedom cross into the transverse components of the $W^{\pm}$- and Z-bosons which became massive.

The physical Higgs boson is not isolated from the rest of the part of the model. It interacts both with the leptons and with the gauge bosons. The corresponding Lagrangian is given by the expression

$$L_H = -\sum_l f_l \bar{l}(x) l(x) H(x) + \frac{g^2}{4}\left[W_\mu^*(x)W^\mu(x) + \right.$$

$$\left. + \frac{1}{2\cos^2\theta_W} Z_\mu(x) Z^\mu(x)\right]\left[H^2(x) + 2vH(x)\right]. \tag{6.68}$$

For the massive neutrinos, as we told before, we should introduce the neutrino singlets in the theory. This leads to the appearance of the following term in L_H

$$-\sum_l f_{\nu_l} \bar{\nu}_l(x) \nu_l(x) H(x).$$

The leptons also get the masses in the result of the shift of the field φ on the constant (6.58)

$$m_l = \frac{f_l v}{\sqrt{2}}, \qquad m_{\nu_l} = \frac{f_{\nu_l} v}{\sqrt{2}}. \tag{6.69}$$

As it follows from Eqs. (6.68) and (6.69) the coupling constants describing interaction of the Higgs bosons with the W- and Z-bosons are proportional to gm_W and gm_Z respectively, that is, they are much more larger than the coupling constants determining interaction between the Higgs bosons and the fermions.

Let us consider the sum of the terms (6.53) and (6.54) which describe interaction between the leptons and the gauge bosons of the $SU(2)_L \times U(1)_Y$ group. Taking into account the explicit form of the Pauli matrices we may present interaction of the weak currents isotriplet $\mathbf{J}_\mu(x)$ with three $\mathbf{W}^\mu(x)$- bosons in the form

$$g\mathbf{J}_l^\mu(x) \cdot \mathbf{W}_\mu(x) =$$

$$= \frac{g}{2}(\bar{\nu}_{lL}(x), \bar{l}_L(x))\gamma^\mu \begin{pmatrix} 0 & W_\mu^1(x) - iW_\mu^2(x) \\ W_\mu^1(x) + iW_\mu^2(x) & 0 \end{pmatrix} \begin{pmatrix} \nu_{lL}(x) \\ l(x) \end{pmatrix} +$$

$$+ \frac{g}{2}(\bar{\nu}_{lL}(x), \bar{l}_L(x))\gamma^\mu \begin{pmatrix} W_\mu^3(x) & 0 \\ 0 & -W_\mu^3(x) \end{pmatrix} \begin{pmatrix} \nu_{lL}(x) \\ l_L(x) \end{pmatrix} =$$

$$= \frac{g}{\sqrt{2}}\left[\bar{\nu}_{lL}(x)\gamma^\mu l_L(x) W_\mu^*(x) + \bar{l}_L(x)\gamma^\mu \nu_{lL}(x) W_\mu(x)\right] + gJ_l^{3\mu}(x) W_\mu^3(x), \tag{6.70}$$

where

$$gJ_l^{3\mu}(x)W_\mu^3(x) = \frac{g}{2}[\bar{\nu}_{lL}(x)\gamma_\mu \nu_{lL}(x) W_\mu^3(x) - \bar{l}_L(x)\gamma_\mu l_L(x) W_\mu^3(x)]. \tag{6.71}$$

The interaction of the weak hypercharge current $j_l^{Y\mu}(x)$ with the fourth vector boson $B_\mu(x)$, in its turn, may be written as

$$\frac{g'}{2}j_l^{Y\mu}(x)B_\mu(x) = -\frac{g'}{2}\overline{\psi}_{lL}(x)\gamma^\mu\psi_{lL}(x)B_\mu(x) - g'\overline{\psi}_{lR}(x)\gamma^\mu\psi_{lR}(x)B_\mu(x) =$$

$$= -\frac{g'}{2}[\overline{\nu}_L(x)\gamma^\mu\nu_{lL}(x) + \bar{l}_L(x)\gamma^\mu l_L(x) + 2\bar{l}_R(x)\gamma^\mu l_R(x)]B_\mu(x). \qquad (6.72)$$

Now we should unify the last term in (6.70) with (6.72). It is evident at a glance that the choice of the basis in the form (6.63) provides the absence of the electromagnetic interaction of the neutrino. Taking into consideration Eq. (6.63) we obtain without trouble

$$gJ_l^{3\mu}(x)W_\mu^3(x) + \frac{g'}{2}j_l^{Y\mu}(x)B_\mu(x) = -\frac{g'g}{\sqrt{g^2+g'^2}}\left[\bar{l}_L(x)\gamma^\mu l_L(x) + \right.$$

$$\left. + \bar{l}_R(x)\gamma^\mu l_R(x)\right]A_\mu(x) + \left[\frac{g^2+g'^2}{2\sqrt{g^2+g'^2}}\overline{\nu}_{lL}(x)\gamma^\mu\nu_{lL}(x) - \right.$$

$$\left. -\frac{g^2-g'^2}{2\sqrt{g^2+g'^2}}\bar{l}_L(x)\gamma^\mu l_L(x) + \frac{g'^2}{\sqrt{g^2+g'^2}}\bar{l}_R(x)\gamma^\mu l_R(x)\right]Z_\mu(x). \qquad (6.73)$$

Since Eq. (6.43) takes place then the following relation should be fulfilled

$$(j_\mu)_{em} = J_\mu^3 + \frac{1}{2}j_\mu^Y,$$

that is, the electromagnetic interaction $-j_{em}^\mu(x)A_\mu(x)$ must be included into interactions (6.53) and (6.54). Then it becomes clear that the first term in (6.73) describes the electromagnetic interaction of the charged leptons, and the multiplier $gg'/\sqrt{g^2+g'^2}$ is nothing more nor less than the electric charge

$$|e| = \frac{gg'}{\sqrt{g^2+g'^2}}. \qquad (6.74)$$

Taking into consideration this circumstance and incorporating Eqs. (6.70), (6.73) we obtain the final expression for the interaction Lagrangian of the gauge bosons with the leptons

$$L_G = \frac{g}{2\sqrt{2}}\left[\overline{\nu}_l(x)\gamma^\mu(1+\gamma_5)l(x)W_\mu^*(x) + \bar{l}(x)\gamma^\mu(1+\gamma_5)\nu_l(x)W_\mu(x)\right] -$$

$$-e\bar{l}(x)\gamma^\mu l(x)A_\mu(x) + \frac{g}{4\cos\theta_W}\{\overline{\nu}_l(x)\gamma^\mu[1+\gamma_5]\nu_l(x) +$$

$$+\bar{l}(x)\gamma^\mu[4\sin^2\theta_W - 1 - \gamma_5]l(x)\}Z_\mu(x). \qquad (6.75)$$

Since all the terms in the Lagrangian (6.75) represent the quantities of the type *current×potential*, then the weak interaction caused by the exchanges of the W- and Z-bosons is commonly called an interaction of the charged and neutral currents respectively.

To develop the electroweak interactions theory for the quarks is performed in full analogy with the above mentioned scheme for the leptons. There is the following correspondence between the quarks and the leptons

$$\begin{pmatrix} \nu_{eL} \\ e_L^- \end{pmatrix} \leftrightarrow \begin{pmatrix} u_L^\alpha \\ d_L^\alpha \end{pmatrix}, \qquad \begin{pmatrix} \nu_{\mu L} \\ \mu_L^- \end{pmatrix} \leftrightarrow \begin{pmatrix} c_L^\alpha \\ s_L^\alpha \end{pmatrix}, \qquad \begin{pmatrix} \nu_{\tau L} \\ \tau_L^- \end{pmatrix} \leftrightarrow \begin{pmatrix} t_L^\alpha \\ b_L^\alpha \end{pmatrix}, \tag{6.76}$$

$$\left.\begin{array}{lll} \nu_{eR} \leftrightarrow u_R^\alpha, & e_R^- \leftrightarrow d_R^\alpha, & \nu_{\mu R} \leftrightarrow c_R^\alpha, \\ \mu_R^- \leftrightarrow s_R^\alpha, & \nu_{\tau R} \leftrightarrow t_R^\alpha, & \tau_R^- \leftrightarrow b_R^\alpha, \end{array}\right\} \tag{6.77}$$

where α is the color quark index[1]. But what actually happens is that, instead of the d-, s- and b-quarks, their linear combinations enter into singlets and doublets of the weak isospin. We shall designate them by symbols d', s' and b'. They are connected with the unprimed quarks by the relations

$$q'^d = \begin{pmatrix} d' \\ s' \\ b' \end{pmatrix} = \mathcal{M}^{CKM} q^d = \mathcal{M}^{CKM} \begin{pmatrix} d \\ s \\ b \end{pmatrix} =$$

$$= \begin{pmatrix} c_{12}c_{13} & s_{12}c_{13} & s_{13}e^{-i\delta_{13}} \\ -s_{12}c_{23} - c_{12}s_{23}s_{13}e^{i\delta_{13}} & c_{12}c_{23} - s_{12}s_{23}s_{13}e^{i\delta_{13}} & s_{23}c_{13} \\ s_{12}s_{23} - c_{12}c_{23}s_{13}e^{i\delta_{13}} & -c_{12}s_{23} - s_{12}c_{23}s_{13}e^{i\delta_{13}} & c_{23}c_{13} \end{pmatrix} \begin{pmatrix} d \\ s \\ b \end{pmatrix}, \tag{6.78}$$

where $\mathcal{M}^{CKM}$ is the Cabibbo — Kobayashi — Maskawa matrix (CKM), $c_{ij} = \cos\theta_{ij}^{CKM}$, $s_{ij} = \sin\theta_{ij}^{CKM}$, i, j are generations indices being equal to $1, 2$ and 3, θ_{ij}^{CKM} are mixing angles and a phase multiplier $e^{\pm i\delta_{13}}$ describes the CP parity violation. So in Eqs. (6.76) and (6.77) one should make the replacement

$$q_i^d \to q_i'^d = \mathcal{M}_{ij}^{CKM} q_j^d.$$

The CKM matrix is unitary. Its elements could be determined from the weak decays of hadrons and from the experiments on the deep inelastic scattering of the neutrino by hadrons. Since the matrix $\mathcal{M}^{CKM}$ is the product of three noncommuting matrices of the rotations in three planes [12], [13] and [23] of the abstract space then the parametrization (6.78) is not unique [2].

In place of Eqs. (6.53) and (6.54) we have for the quarks

$$g\mathbf{J}_q^\mu(x)\cdot\mathbf{W}_\mu(x) + \frac{g'}{2} j_q^{Y\mu}(x)B_\mu(x) = g\overline{\psi}_{qL}(x)\gamma^\mu\mathbf{S}^W\cdot\mathbf{W}_\mu(x)\psi_{qL}(x) +$$

$$+ g'\overline{\psi}_q(x)\gamma^\mu\frac{Y^W}{2}\psi_q(x)B_\mu(x). \tag{6.79}$$

Passing to the mass eigenstates basis (W, W^*, Z, γ) from the gauge basis (W^1, W^2, W^3, B) with the help of formulas (6.63), (6.65) and using the relation (6.79), we obtain the following expression for the Lagrangian describing the interaction between the quarks and the gauge bosons

$$L_q = \frac{g}{\sqrt{2}}\left[\overline{q}_{iL}^u(x)\gamma^\mu\mathcal{M}_{ik}^{CKM}q_{kL}^d W_\mu(x) + \overline{q}_{kL}^d(x)\gamma^\mu\mathcal{M}_{ki}^{CKM*}q_{iL}^u W_\mu^*(x)\right] +$$

[1] As the electroweak interactions do not change the quark flavor further the flavor index will be neglected.

[2] We have used the parametrization accepted in Review of Particle Physics.

$$+\frac{g}{4\cos\theta_W}\left[\overline{q}_i^u(x)\gamma^\mu\left(1-\frac{8}{3}\sin^2\theta_W-\gamma_5\right)q_i^u(x)-\overline{q}_i^d(x)\gamma^\mu\left(1-\frac{4}{3}\sin^2\theta_W-\right.\right.$$

$$\left.\left.-\gamma_5\right)q_i^d(x)\right]Z_\mu(x)+\frac{e}{3}\left[2\overline{q}_i^u(x)\gamma^\mu q_i^u(x)-\overline{q}_i^d(x)\gamma^\mu q_i^d(x)\right]A_\mu(x). \qquad (6.80)$$

We define the Yukawa Lagrangian for the quarks by analogy with the lepton case. Then, after spontaneous symmetry breaking we get the Lagrangian determining the interaction of the quarks with the physical Higgs boson in the form

$$L=-\sum_i f_{q_i}\overline{q}_i(x)q_i(x)H(x),$$

where the Yukawa constant f_{q_i} defines the mass of the q_i-quark

$$m_{q_i}=\frac{f_{q_i}v}{\sqrt{2}}.$$

The Fermi theory was a predecessor of the WSG theory. This theory most advantageously managed with the description of weak interaction in the low energy region. Then, according the correspondence principle, at low energies the WSG theory must reproduce the results of the former theory. We take advantage this circumstance to define the linkage between the parameters of the new and old theories. It may be beneficial to recall some essential points of the Fermi theory.

By the early (19)30s it was established that beta-radioactivity of all nuclei is caused by a pair of fundamental reactions where the proton and neutron are interconverted

$$n\to p+e^-+\overline{\nu}_e, \qquad (6.81)$$

$$p\to n+e^++\nu_e. \qquad (6.82)$$

The electron and the antineutrino or the positron and the neutrino which appear as a result of the beta-decay are created because they do not exist in the radioactive nuclear. This phenomenon is analogous to the process of the photon emitting by the electron when it transits with the one orbit on the other orbit located close to the nuclear. E. Fermi used this analogy and already known mathematical apparatus of the quantum theory of electromagnetic interaction in his weak interaction theory proposed in 1934. The interaction Lagrangian by Fermi has the form of the product *current×current*

$$L_F=\frac{G_F}{\sqrt{2}}j_\mu^\dagger(x)j^\mu(x). \qquad (6.83)$$

The weak current entering into (6.63) is built out of the wave functions of the particles which are pairwise unified: neutron — proton, electron — electron antineutrino and so on. So, for the process (6.83) the one current is nucleon and it transfers the neutron to the proton. The other current is lepton and it creates the pair, the electron and the electron antineutrino. These currents belong to the class of the charged currents since they change the electric charge of the particles to interact. In the both currents the charge is increased on $|e|$: the positive charged proton arises from the neutral neutron while the electron antineutrino does from the electron. The interaction (6.83) gave the name *four-fermion contact interaction*[1]. The structure of the charged weak currents has been finally established in the middle of (19)50s. Analysis of experiments being done by that time led to the conclusion that the weak current represents the sum of the vector V and the axial vector A ($V+A$ structure). Thus the lepton part of the current is defined by the expression

$$j_l^\sigma(x) = \overline{e}(x)\gamma^\sigma(1+\gamma_5)\nu_e(x) + \overline{\mu}(x)\gamma^\sigma(1+\gamma_5)\nu_\mu(x) + \overline{\tau}(x)\gamma^\sigma(1+\gamma_5)\nu_\tau(x). \quad (6.84)$$

After determination of the composite nucleons structure the quarks occupy the nucleons place in the charged weak current nucleon part

$$j_q^\sigma(x) = \overline{d}'(x)\gamma^\sigma(1+\gamma_5)u(x) + \overline{s}'(x)\gamma^\sigma(1+\gamma_5)c(x) + \overline{b}'(x)\gamma^\sigma(1+\gamma_5)t(x). \quad (6.85)$$

If one is constrained by the first order of perturbation theory on the weak interaction constant G_F under calculations then the Fermi theory gives a fine accordance with an experiment. However the corrections of the high orders on G_F represent the integrals which become infinite at the large energies, that is, physically meaningless. Consequently the Fermi theory must be suitably reconstructed. The refusal of a locality of interaction is the most evident way. In other words, the analogy between quantum electrodynamics and weak interaction theory should be deeper, namely, weak interaction is also carried by gauge bosons.

To connect the WSG theory parameters with the Fermi constant G_F we consider the muon decay through the channel

$$\mu^- \to e^- + \overline{\nu}_e + \nu_\mu. \quad (6.86)$$

The corresponding diagrams in the momentum representation for the Fermi theory and the WSG theory are displayed in Figs. 39 and 40.

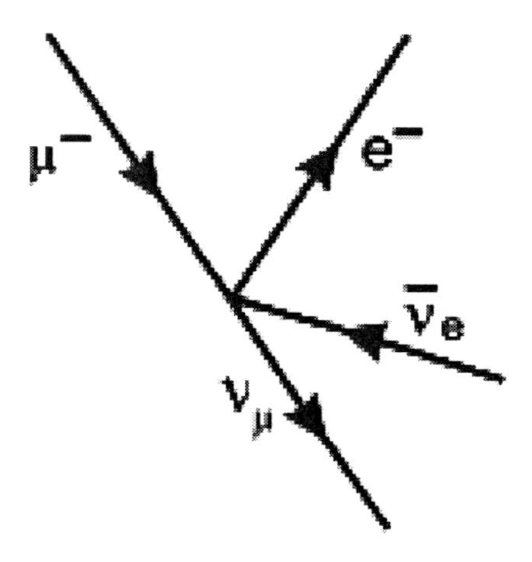

Figure 39.

[1]Since four fermion wave functions enter into the Lagrangian we call the interaction by four-fermion. As the interaction takes place in one and the same point x we name it by contact.

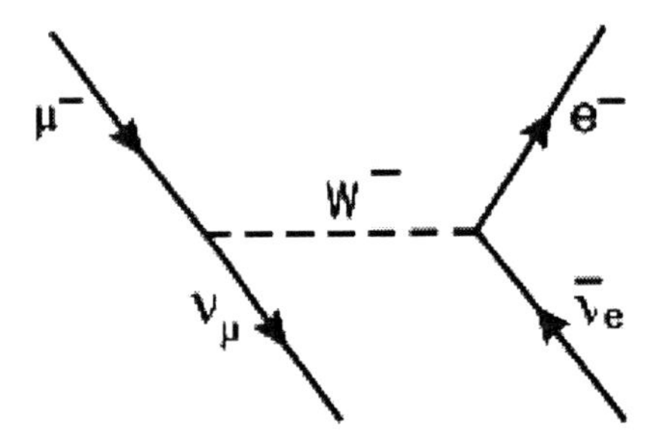

Figure 40.

In the WSG theory the amplitude of the decay (6.86) is given by the expression

$$A_{WSG} = \frac{g^2}{8}\sqrt{\frac{m_\mu m_{\nu_\mu} m_e m_{\nu_e}}{E_\mu E_e E_{\nu_e} E_{\nu_\mu}}}\overline{\nu}_\mu(p_{\nu_\mu})\gamma^\sigma(1+\gamma_5)\mu(p_\mu)\left(\frac{g_{\sigma\beta} - q_\sigma q_\beta/m_W^2}{m_W^2 - q^2}\right)\times$$

$$\times\overline{e}(p_e)\gamma^\beta(1+\gamma_5)\nu_e(p_{\nu_e}), \tag{6.87}$$

where $q = p_\mu - p_{\nu_\mu}$ and the expression standing in the parenthesis describes the propagation of the virtual W^- boson, i.e. is the propagator of this particle. In the Fermi theory the decay amplitude has the form

$$A_F = \frac{G_F}{\sqrt{2}}\sqrt{\frac{m_\mu m_{\nu_\mu} m_e m_{\nu_e}}{E_\mu E_e E_{\nu_e} E_{\nu_\mu}}}\overline{\nu}_\mu(p_{\nu_\mu})\gamma^\sigma(1+\gamma_5)\mu(p_\mu)\overline{e}(p_e)\gamma_\sigma(1+\gamma_5)\nu_e(p_{\nu_e}). \tag{6.88}$$

Under small q the expressions (6.87) and (6.88) must coincide. As in this case the W boson propagator is reduced to $g_{\sigma\beta}/m_W^2$, the required linkage is given by

$$\frac{g^2}{8m_W^2} = \frac{G_F}{\sqrt{2}}. \tag{6.89}$$

The constant g characterizes emitting and absorbing the $W^\pm$ bosons, much as e defines emitting and absorbing the photons. From (6.74) follows that $e > g$ and, therefore, weak interaction is in essence stronger than electromagnetic one. However, as it was shown, the weak processes amplitudes are proportional to g^2/m_W^2 at low energies[1]. So, owing to that the W bosons are very heavy, the weak interaction processes appear to be much orders of magnitude weaker than electromagnetic processes.

Not only do the WSG theory unify electromagnetic and weak interactions, but it also predicts the existence of new phenomena in weak interaction physics, the neutral currents. In 1973 the first reactions caused by the neutral currents were observed

$$\nu_\mu + p \to \nu_\mu + p + \pi^+ + \pi^-. \tag{6.90}$$

Information confirming the neutral currents existence also follows from experiments on observing the parity violation in atom physics. The interaction constant of the neutral currents proves to be approximately the same as that of the charged currents.

[1]Recall, that the separation on electromagnetic and weak interactions has a sense only at energies < 100 GeV.

The WSG theory predicts the linkage between the masses of the W- and Z-bosons as well. From the relations (6.64) and (6.67) follows

$$m_Z = \frac{m_W\sqrt{g^2+g'^2}}{g} = \frac{m_W}{\cos\theta_W}. \tag{6.91}$$

Using Eqs. (6.74) and (6.89), we obtain

$$m_W = \sqrt{\frac{\pi\alpha_{em}}{G_F\sqrt{2}}}\sin^{-1}\theta_W. \tag{6.92}$$

Having done three independent experiments one may define the constants G_F, $\sin\theta_W$, α_{em}, the knowledge of which allows to determine not only the masses of the W- and Z-bosons but the vacuum average v (vacuum expectation value) of the Higgs field as well

$$v = \frac{m_W\sin\theta_W}{\sqrt{\pi\alpha_{em}}}.$$

The Weinberg angle value may be found out of different experiments concerning nuclear physics, physics of weak interactions at low energies and high energy physics. By 1983 the results of $\sin^2\theta_W$ determination have become to be matched. The averaged value was given by

$$\sin^2\theta_W \approx 0.23. \tag{6.93}$$

Then substituting the values

$$\alpha_{em} \approx \frac{1}{137}, \qquad G_F = 1.17\times 10^{-5}\,\text{GeV},$$

and (6.93) into Eqs. (6.91) and (6.92) we obtain

$$m_W \approx 80\,\text{GeV}, \qquad m_Z \approx 91\,\text{GeV}. \tag{6.94}$$

It is evident that the discovery of the W- and Z-bosons would be the deciding step on the way of the WSG theory checkout. For these purposes the proton-antiproton collider was built in CERN. It began to operate in summer 1981. The direct production of the W-boson with the subsequent decay through the electron and electron antineutrino

$$u+\overline{d} \to W^+ \to e^- + \overline{\nu}_e \tag{6.95}$$

is displayed in Fig.41.

The cross section of the reaction (6.95) is a function of the colliding quarks energy and as soon as the energy in the center of mass system is approaching to m_W, the W-boson is exhibiting as a resonance. In the resonance region the cross section has the sharp maximum the hight and the width of which are predicted by the WSG theory. At the resonance the cross section value can be calculated by application of the Breit-Wigner formula (4.106).

It is beyond doubt that to directly observe the quark collisions is impossible since the quarks in the free states are unavailable for us. The proton-antiproton collisions are just the most changing. A monochromatic proton beam may be considered as the quark beam with the wide distribution on momenta, that is, when P is a proton momentum then x_iP

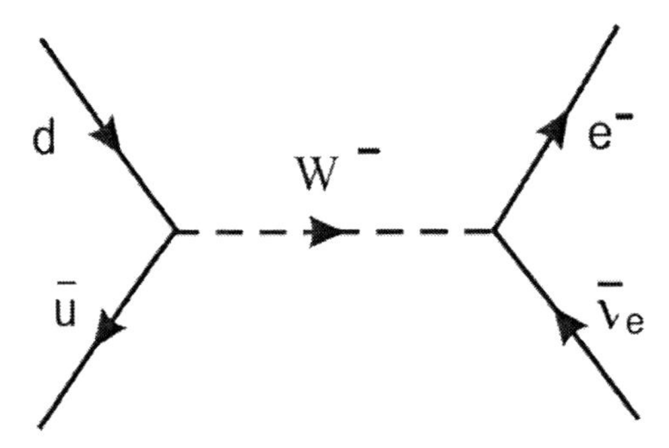

Figure 41. The Feynman diagrams for the process $u+\overline{d} \to W^{+*} \to e^- + \overline{\nu}_e$.

is a momentum of an i-quark ($0 \leq x_i \leq 1$). An antiproton beam looks analogously. In the process of the W production the quark picks for itself the antiquark with the suitable momentum. In the center of mass system the energy of the impacts of the quark with the antiquark $s_{q\overline{q}}$ is connected with $s_{p\overline{p}}$ by the relation

$$s_{q\overline{q}} = s_{p\overline{p}} x_q x_{\overline{q}}.$$

The distribution functions of the quarks both in the proton and in the antiproton are such that in order to provide the right correlation between the proton quark and the antiproton antiquark the following condition must be fulfilled

$$x_q \approx x_{\overline{q}} \geq 0.25. \tag{6.96}$$

Thus there is one wide region of optimal energies for the $p\overline{p}$ impacts at the given W boson mass. For $m_W = 80$ GeV it is given by

$$400 \leq \sqrt{s_{p\overline{p}}} \leq 600 \text{ GeV}.$$

In actual fact the following processes are investigated at CERN $p\overline{p}$ collider

$$p+\overline{p} \to W^{\pm} + X, \tag{6.97}$$

where X is arbitrary hadrons plurality. To detect the $W^{\pm}$ bosons was carried out through the lepton decays

$$W^+ \to e^+ + \nu_e, \qquad W^- \to e^- + \overline{\nu}_e. \tag{6.98}$$

Such processes are represented by diagrams which include the elements both of the quarks diagrams and of the Feynman ones. So the diagram pictured in Fig.42 corresponds to the process

$$p+\overline{p} \to W^- + X \to e^- + \overline{\nu}_e + X. \tag{6.99}$$

To obtain the cross section of the reaction (6.99) one should integrate the cross section of the reaction (6.95) at the resonance over the distributions of the quarks in the proton and the antiproton.

Since the W boson mass is large then charged leptons $l^{\pm}$, appearing under the W boson decay, have a large transverse momentum. Thus the detection of $l^{\pm}$ is not caused any

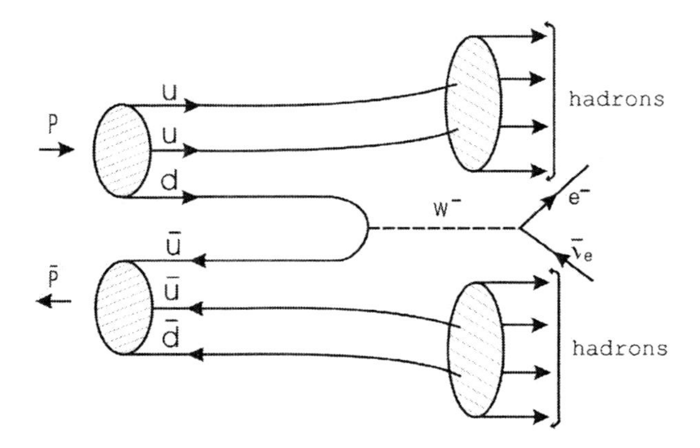

Figure 42. The diagrams for the process $p+\overline{p} \to W^{-*}+X \to e^- + \overline{\nu}_e + X$.

troubles. The neutrino recording was based on purely kinematic reasons. For this purpose the special detector was used. Its sensitivity relatively to all charged or neutral particles which were created in the impact process, was uniform in the whole volume on the whole solid angle. Since interactions are observed in the center of mass system any appreciable disbalance of a momentum signals about the presence of one or more numbers of noninteracting particles (presumably, neutrinos). Calorimeters are ideally suited for the role of such detectors because their energy registration efficiency may be done to be sufficiently homogeneous for different particles to hit in them. Notice, in the intrinsic frame of reference of the W-boson the momentum carried away by the neutrino p_ν is equal to $m_W/2$, i.e. it is very large.

In January of 1983 two independent Collaborations UA1 and UA2 working at CERN $p\overline{p}$ collider presented the first results on detecting the W bosons in the reactions (6.97), (6.98).

In June of 1983 the group UA1 has informed about the observation of the first five cases of the creation and the decay of the Z bosons. In August of the same year the group UA2 has detected the eight analogous events. The Z-bosons are created in the reaction

$$p+\overline{p} \to Z+X \tag{6.100}$$

and are detected through the decays

$$Z \to e^- + e^+, \qquad Z \to e^- + e^+ + \gamma, \qquad Z \to \mu^- + \mu^+. \tag{6.101}$$

These experiments have led not only to the discovery of the W- and Z-bosons, they have also shown that their properties are exactly described by the WSG theory. It should be noticed that the masses values (6.94) are approximate rather than precise. To obtain the precise values of the gauge bosons masses we must take into consideration the interaction of particles with the vacuum or, what is the same, incorporate the higher orders of the perturbation theory (radiative corrections) under calculating the cross sections. The radiative corrections (RC) influence on the values of α_{em} and $\sin^2\theta_W$ is especially significant. The calculations showed that the inclusion of the RC changes the gauge bosons masses values in the formulas (6.94) approximately on 5%.

Chapter 7

Fundamental Particles

The evolution of our notions about the matter structure, the four steps on the Quantum Stairway, may be schematically represented by Fig.43.

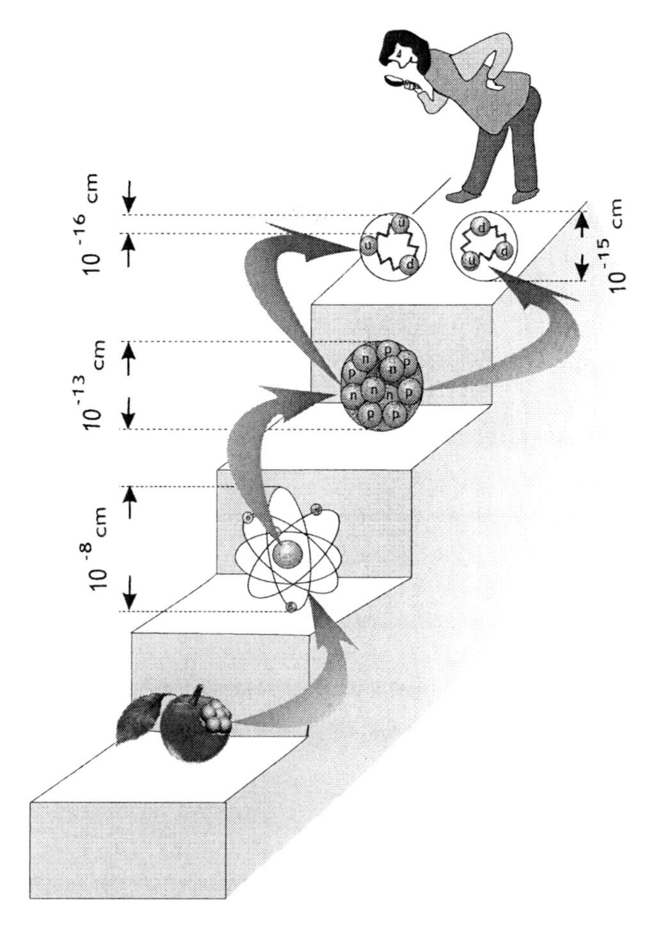

Figure 43. The Quantum Stairway.

Every stage of this Stairway presents a separate region of Physics and, consequently, the phenomena range described by it, is characterized first and foremost by particles velocities

Table 7.1.

q	Q	I	T	T_3	B	s	c	b	t
u	2/3	1/2	1/2	1/2	1/3	0	0	0	0
d	-1/3	1/2	1/2	-1/2	1/3	0	0	0	0
c	2/3	1/2	0	0	1/3	0	1	0	0
s	-1/3	1/2	0	0	1/3	-1	0	0	0
t	2/3	1/2	0	0	1/3	0	0	0	1
b	-1/3	1/2	0	0	1/3	0	0	-1	0

and a region size which is available for particles motions. An object having an intrinsic structure from the viewpoint of the higher stage is considered as a structureless one in all phenomena at any underlying stage. So an atom is supposed to be a point particle in Classical Physics, a nucleus — in Atom Physics, a nucleon — in Nuclear Physics.

Below in Table 7.1 we give the additive quantum numbers of the quarks

Before when we talked about the quark-lepton symmetry we were pointing the way of its storing, placing the quarks and the leptons into the weak isospin doublets. In so doing it appeared that the down quarks subject to the mixing. The absence of the similar mixing in the lepton sector would shake not only our belief in the quark-lepton symmetry, but it would also make us to refuse the belief in the Nature unity. The analogous phenomenon proves to be taken place in the lepton sector as well. But we have learned about it much later, in 2001 only. The experiments showed that the electron, muon and tau-lepton neutrinos are not the states with definite mass value but they represent the mixtures of the physical states ν_1, ν_2 and ν_3. The corresponding neutrino sector mixing matrix $\mathcal{M}^{NM}$ has the same form as the CKM matrix, that is,

$$\mathcal{M}^{NM} = \mathcal{M}^{CKM}(\theta_{ij}^{CKM} \rightarrow \theta_{ij}^{NM}),$$

where θ_{ij}^{NM} are neutrinos mixing angles. Moreover, the mixing angles of the lepton sectors are connected with those of the quark sectors by the relations

$$\theta_{12}^{NM} + \theta_{12}^{CKM} = \frac{\pi}{4}, \qquad \theta_{23}^{CKM} + \theta_{23}^{NM} = \frac{\pi}{4}, \qquad \theta_{13}^{NM} \sim \theta_{13}^{CKM}.$$

So, the following fermions are part of the composition of the quark-lepton matter

$$\begin{pmatrix} \nu_e \\ e^- \end{pmatrix}, \quad \begin{pmatrix} u \\ d' \end{pmatrix}^{\alpha} \tag{7.1}$$

$$\begin{pmatrix} \nu_\mu \\ \mu^- \end{pmatrix}, \quad \begin{pmatrix} c \\ s' \end{pmatrix}^{\alpha}, \tag{7.2}$$

$$\begin{pmatrix} \nu_\tau \\ \tau^- \end{pmatrix}, \quad \begin{pmatrix} t \\ b' \end{pmatrix}^{\alpha}. \tag{7.3}$$

The first generation plays the particular role. In effect all we see around us in the Nature consists of the first generation fermions. All the members of the second and the third generations are unstable, the exception is probably provided by the neutrinos. They appear only in accelerators and in phenomena produced by the cosmic rays. The fermions of the second and third generations played the important role in the early Universe, in the first instants of the Big Bang. In particular, the neutrino flavors number has defined the quantitative ratio between hydrogen and helium in the Universe. The second and third generations made an impact on the masses values of the first generation particles. In its turn, the ratio of the masses $m_u : m_d : m_e$ made possible engendering the life in the Universe. It is very surprising that to choose the first generation particles masses values the Nature has used the trick being sufficiently rare in its repertoire, the parameters fine-tuning. We shall intimate that this is the case.

Let us consider the simplest but the most important atom of our Universe, the hydrogen atom. The stability of the atom is governed by that the reaction

$$p + e^- \rightarrow n + \nu_e \tag{7.4}$$

is energetically forbidden since the masses of the electron and the proton which constitute it satisfy an inequality

$$m_e < \Delta m, \tag{7.5}$$

where $\Delta m = m_n - m_p \approx 1.3$ MeV (we have neglected the neutrino mass). It is clear that the fragile equilibrium expressed by Eq.(7.5) may be violated even by an insignificant change of m_u, m_d and m_e. Consequences of the hydrogen instability would be catastrophic. If the hydrogen that represents the main fuel for the Universe stars would be absent, then the ordinary stars did not exist and the Universe acquired an absolutely other appearance. Then in order to make our world stable, so to say with a store, why not increase the value of Δm in the relation (7.5). However, the other problem connected with the deuterium is waiting for us on this way. Its nucleus, the deuteron, possesses the most small binding energy $E_c \approx 2.24$ MeV. A guarantor of the deuteron stability is the fact that in it the decay of the neutron through the channel

$$n \rightarrow p + e^- + \overline{\nu}_e. \tag{7.6}$$

is energetically unprofitable. In this case the energy conservation law demands

$$m_p + m_n - E_c = m_p + m_p + m_e + T, \tag{7.7}$$

where T is the kinetic energy of the decaying particles. From the positivity of T follows that the decay is forbidden under condition

$$E_c + m_e > \Delta m. \tag{7.8}$$

Thus, if we made Δm too large and violated (7.8), then the deuterium would be unstable that led to its complete lack in the Nature. However the deuterium production is the first step in the chain of nuclear transformations tracing from the hydrogen to more heavy elements which were not in the early Universe. To summarize, in the case of the deuterium lack the routine way of producing the elements being heavier than the hydrogen would become impossible.

The small ratio of the electron and the proton masses is the cause of such an important phenomenon as the exact localization of the nucleus in the electrons cloud that, in its turn, defines a molecules architecture. Otherwise the stable configurations, characterizing the life itself and much in the world surrounding us, would not exist.

The processes of annihilating the electrons-positrons into hadrons have set the upper bound on the quarks size. By now this bound is $\leq 10^{-16}$ cm. Such a limitation exists for leptons as well. Within the accuracy of the modern experiment the particles entering into the both groups are considered as point particles, that is, at present we have all the reasons to believe that these particles are fundamental.

It is convenient to present the fundamental particles of the SM in the three-dimensional coordinate system Θ, in which the operator eigenvalues of the weak isospin projection on the third axis S_3^W are plotted along the z axis while the plane xy is used for the definition of the eigenvalues of the color spin operator $S^{(col)}$. Draw a regular-shaped triangular prism in Θ (Fig. 44).

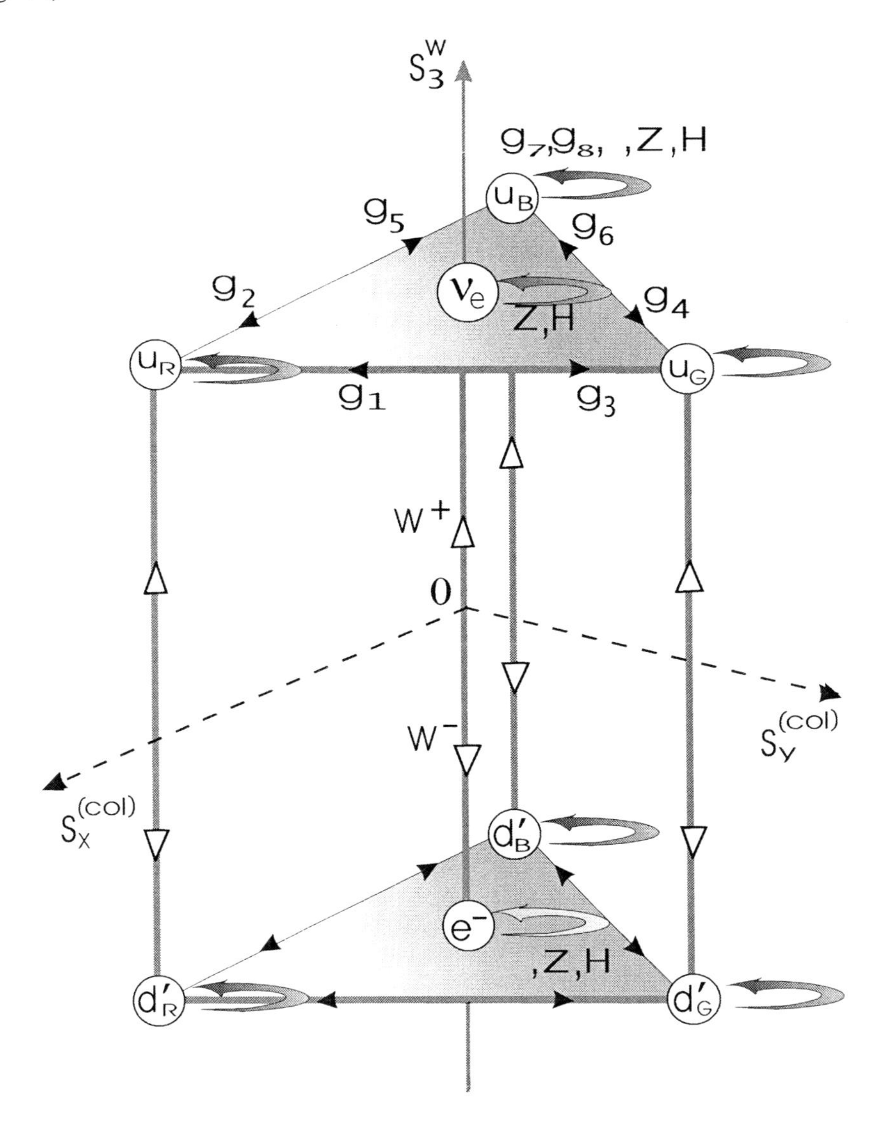

Figure 44. The fundamental particles of the SM.

Now we place three color states of the u quark in the vertices of the upper base (the points corresponding to $S_3^W = 1/2$) and three color states of the d' quark in the vertices of the low base ($S_3^W = -1/2$). The colorless leptons will be housed in the straight line $x = y = 0$: the electron neutrino in the center of the upper base while the electron in the center of the low base. Two analogous prisms will correspond to the fundamental fermions of the second and the third generations. The quarks and the leptons form the first group of the fundamental particles, the matter particles group. The next group, the interactions carriers group, constitute four gauge bosons of electroweak interaction W^-, W^+, Z, γ, eight gluons of the QCD g_i and an yet-undiscovered gravitational interaction carrier, the graviton G. We call this group by the gauge bosons group. Let us be distracted from the graviton existence and display the remaining interactions carriers in the form of the arrows, meaning the result of interactions between the bosons and the fermions, at the same figure. The emitting or the absorbing of the charged $W^{\pm}$ bosons leads to the transitions between the vertices of the fermion triangles with $\Delta S_3^W = \pm 1$, i.e. it produces the motion along the z axis. From eight existing gluons two are responsible for the processes without a change of the quarks color whereas six — for the processes changing the quarks color. The clockwise motions and counterclockwise motions along the perimeters of the equilateral fermion triangles correspond to the gluons to change the color.

We are coming now to displaying the gauge bosons, which do not effect the change of the electric charge, of the weak isospin, of the color and, consequently, do not alter the fermion position in the Θ space. Such a particles are the photon, the neutral gauge boson and two remaining gluons. Apart from the above mentioned fundamental particles, there is one more particle which has the common features with the both groups but still stands aside from them, i.e. it actually forms the third group of the fundamental particles. This is the Higgs boson having the zero-spin and electric charge being equal to zero. It does not enter into the matter composition and its interaction with all the fundamental fermions is not attached to the definite class. However, according the SM all the fundamental particles acquired their masses thanks to just the Higgs boson. It may be an echo of the spontaneous symmetry breaking which happened (if any) in the epoch of the early Universe [1]. Under emitting and absorbing the Higgs boson the fermion does not alter its position in the Θ space as well. Let us agree the bosons, whose interactions with the matter particles results in

$$\Delta Q = \Delta S_3^W = \Delta S^{(col)} = 0,$$

to display by the arrows which begin and end on the fermions.

At a sight in Fig.44 it is difficult to get rid of a temptation to declare the quarks and the leptons, which belong to the bases triangles, by the quark-lepton octet of some symmetry group. The fermions of the second and the third generations would also constitute the analogous octets. Having stood on this point of view we reduce all the plurality of the fundamental matter particles to three superparasitism each of them could be in eight states.

[1] In November of 2000 the reports about the Higgs boson observation with the mass being equal to 115 GeV came from two groups L3 and ALEE (LEP) which investigated the reaction

$$e^+ + e^- \to Z + H.$$

However two other Collaborations working at LEP, OPAL and DELPHI, did not confirm the results of their colleagues. In the year of 2001 LEP ceased the operation, leaving the Higgs boson history unwritten.

However such a simple and elegant scheme has its exotic. The term "multiplet" has a sense if and only if the transitions between all its states are allowed. But in our case only the *quark↔quark* and *lepton↔lepton* transitions exist whereas the *lepton↔quark* transitions are forbidden. This circumstance is a consequence of a lack of interconnections between the electroweak and the QCD fields within the SM. But if we choose the symmetry group, which allow to place all the fermions of each generation into a fundamental octet, and subsequently gauge this symmetry and spontaneously break it then we obtain frightened huge numbers both of the Higgs bosons and of the gauge bosons. It is clear that such a scheme of the Universe has hardly the right to the life, since, as Aristotle said, "Nature always realizes the best of possibilities".

The $SU(5)$ group is a minimum group of the GUT which includes the SM group as a subgroup. In this model one can not manage to place all the known fundamental fermions of each generation into one representation. However it could be done with the help of two representations, the quintet and the decuplet representations of the $SU(5)$ group. The quintet for the first generation has the form

$$(\overline{d}'_R, \overline{d}'_G, \overline{d}'_B, e^-, \nu_e),$$

while the corresponding decuplet is represented by an antisymmetric matrix

$$\frac{1}{\sqrt{2}}\begin{pmatrix} 0 & \overline{u}_B & -\overline{u}_G & -u_R & -d'_R \\ -\overline{u}_B & 0 & \overline{u}_R & -u_G & -d'_G \\ \overline{u}_G & -\overline{u}_R & 0 & -u_B & -d'_B \\ u_R & u_G & u_B & 0 & -e^+ \\ d'_R & d'_G & d'_B & e^+ & 0 \end{pmatrix}.$$

In so doing all the fermions fields are considered by the left-hand chiral fields, that is, the functions of all the fields are multiplied by the quantity $(1+\gamma_5)/2$. Evidently, if one displays the fermion multiplets of the $SU(5)$ group in the coordinate system Θ, then the obtained picture will not yet hold the same aesthetic appeal as it was in the case of Fig. 44. As the $SU(5)$ group has 24 generators then the corresponding gauge transformation is achieved by 24 gauge bosons. Twelve of them are the gauge bosons of the SM. The remaining bosons, the $X_i^{\pm}$ and $Y_i^{\pm}$ bosons ($i = 1,2,3$), have the masses $\sim M_{GU}$ and the charges $\pm 4e/3$, $\pm e/3$. There is also a great deal of physical Higgs bosons. For example, this number equals 16 in the version proposed by George and Glashow.

Under increasing the GUT group dimensionality, the number both of the physical Higgs bosons and of the gauge bosons grow. So in the $SO(10)$ group the number of the gauge bosons yet reaches 45.

On these examples we see one of possibilities — the achieved stage on the Quantum Stairway is the last one but the list of the fundamental particles may be wider. Increasing the list may occur both through the physical Higgs bosons and through the gauge bosons. There are also no reasons to believe that the number of the fermion generations may not be greater than three. The other possibility remains to be opened, namely, all the fundamental particles, or some of them at least, for example, the quarks and the leptons represent in fact an objects consisting of subarticles, peons. Then some new forces responsible for unifying the peons into the quarks and the leptons must exist. In this case the strong and the electroweak forces appear no more fundamental than the chemical or the nuclear forces and the attempt of building the GUT from the elementary quarks and leptons will be doomed to failure. Are the peons the last undivided matter blocks or is the climbing on the next stage of the Quantum Stairway waiting for us? It is clear only one that as early as the first century (B.C.) Lacertus said: "But without doubt, the known limit of breaking in pieces has been set".

It turned out so that at our climbing the third stage of the Quantum Stairway the QED was to be responsible for the matter structure. Just it became the first working model of the quantum field theory. Its basic properties — the relativistic covariance, the local gauge invariance, the convergence of perturbation theory series, changing the interaction constant with a momentum transferred — got those strings from which the Gobelin tapestry of the Modern Physics was weaved. It is difficult to foresee what aspect would assume the elementary particles theory if the carriers of interactions between the electrons were massive particles with the spin 3/2 rather than the photons. And we finish this Chapter by the words of the poet

Sad words "might be",
can't be forgotten
by mice and people too.

Chapter 8

Devices of Elementary Particle Physics

8.1. Particle Accelerators

From the times of the atomic nucleus discovery by Rutherford it has been clear that a study of the nuclei structures requires the production of accelerated beams. Natural sources of accelerated particles, radioactive elements, exhibit very low intensities, restricted energies, and they are absolutely uncontrollable. To produce high-energy particles, the development of special accelerating facilities was being started. Presently, giant accelerators symbolize modern elementary particle physics. The experiments involving elementary particles are impossible without the use of accelerators, and the progress of Physics is inconceivable without such experiments. During the XXth century typical accelerator energies were varying from a few eV to several TeV. In other words, the attainable energies were growing exponentially, being doubled every 2.5 years. It is obvious that such a rapid energy growth cannot be expected in the SIXTH century.

Modern accelerator facilities include three basic blocks:

(1) accelerator that effects the kinetic energy increase, formation and ejection of high-intensity particle beams;

(2) detector that comprises a system for registration of the interaction processes and analysis of the reaction products; besides, a particular section of the detector may be used as a target;

(3) equipment for input/output, storage and processing of the experimental information; unit for automatic control of the whole accelerating system.

According to the acceleration principle, accelerator facilities may be classed as synchrotron and linear accelerators; depending on the collision method, they may be subdivided into the stationary-target and colliding-beam machines. Both stable, e.g. electron, proton, neutrino, and unstable particles, e.g. muons, may be accelerated in the process.

The first accelerator designed by P. Van de Graafian in 1931 was of the high-voltage linear type. It represented a combination of the high-voltage source (generator) and acceleration vacuum tube, where the charged particles, moving between the generator poles, acquired an energy corresponding to the voltage at the poles. High-voltage accelerators are intended to accelerate light (electrons) as well as heavy particles (protons, ions). Their merits are as follows: continuous operation, high energy stability of the accelerated particles and small energy spread in the beam ($\Delta E \leq 0.01\%$). High-voltage accelerators with energies up to 10 — 20 MeV are still in current use for preliminary acceleration in large accelerators.

As the attainment of high potential difference (for example, 10^6 V) presents technical difficulties, it is expedient to cause the accelerated particles to rotate in the magnetic field, progressively accelerating them by low voltage pulses applied to the accelerating electrodes at certain instants of times and at a frequency equal to the particle rotation frequency. In this way acceleration is imparted only to the particles entrapped into the accelerating gap at proper times. Because of this, only parts of the beam are accelerated (particle clusters) rather than the whole beam. This principle was used for the creation of the first cyclic accelerator (cyclotron) constructed by E. Lawrence in 1931. In cyclotron the charged particles are accelerated from zero to maximum energy under the effect of an alternating electric field with the constant period T_a. Curving of the orbits is provided by a constant magnetic field directed perpendicular to the orbit plane. The rotation period of a particle is determined by the expression:

$$T^f = \frac{2\pi mc}{eB\sqrt{1-v^2/c^2}} = \frac{2\pi E}{ecB}, \tag{8.1}$$

where B is a magnetic field induction, v is a particle motion velocity on an orbit, and E is a total particle energy. It should be noted that Eq. (8.1) is written in CGS. In this section, for the sake of obviousness, it is convenient to use an ordinary system of units. For nonrelativistic velocities $E \approx mc^2$ the period is constant. Provided in this case T^f is a multiple of T_a, a prolonged resonance may be observed between the particle rotation in the magnetic field and variations in the accelerating voltage. Accelerated particles are moving along the spiral orbits with ever growing radius

$$R = \frac{mcv}{eB}. \tag{8.2}$$

And the acceleration takes place until the particle motion is in resonance with the accelerating field. When relativistic velocities are attained by the accelerated particles, the total particle energy begins to grow causing the resonance disturbance, and hence acceleration of the particles is terminated.

Independently of one another, V. I. Vecsler (USSR, 1944) and E. McMillan (USA, 1945) have proposed the phase stability principle which allows to ensure the resonance condition at any relativistic velocity. By Eq. (8.1), the relationship between the total particle energy and accelerating field frequency ω_a should be as follows:

$$E = \frac{ecBq}{\omega_a}, \tag{8.3}$$

where $q = 1, 2, 3....$ According to the stable-phase mechanism, the particle energy automatically takes the value close to the resonance one, with a relatively slow time variation of the accelerating electric field and magnetic field induction. The finding of the stable phase principle has resulted in the advent of the new type accelerators. As follows from (8.3), an increase in the equilibrium (resonance-associated) energy of a particle requires a decrease of the accelerating field frequency (synchrocyclotron) or increased induction of the twisting magnetic field (synchrotron), or else variation of both the indicated frequency and induction (synchrophasotron) or, finally, an increase in the acceleration multiplicity, i.e. in the value q (microtron).

The stable phase ensures stability of the particle motion in azimuthal direction (in the direction of a particle orbit). The transverse (orbit-perpendicular) motion stability, or focusing, is realized by an adequate selection of radial changes in the magnetic field. To achieve stability in both transverse directions, it is necessary to attain a minor radial decrease of the magnetic field.

Synchrotrons are used for the formation of a beam of relativistic electrons. Synchrophasotrons serve as accelerators of heavy particles (protons, ions). Accelerators of both types feature fixed radius of the equilibrium orbit, making possible the ring form of a magnetic system. The principal limitations of ring accelerators are as follows: (1) accelerated particles in the twisting magnetic field sustain the energy loss due to magneto-bremsstrahlung (synchrotron radiation) whose power is given by the formula:

$$P = \frac{C_\gamma c E^4}{R^2}, \tag{8.4}$$

where

$$C_\gamma = \frac{2e^2}{3(mc^2)^4},$$

(2) bulky magnetic system. To increase the energy of an accelerated particle considering these limitations, it is required to enlarge the ring radius of the accelerator. Indeed, with growing R losses by radiation are decreased together with the magnetic field value necessary for retention of the particle on the orbit. So, proton synchrophasotron with a maximum energy of 500 GeV, constructed in the Fermi Laboratory (FERMILAB) in 1972, was 2 km in diameter.

Proton and ion linear accelerators are based on the same principle as cyclic: in the process of its resonance motion a particle falls within the accelerating voltage phase in every gap. Nevertheless, the particle motion proceeds along a straight line, and the gaps along this line are arranged at certain intervals in order that the particle transit time from gap to gap be equal to the period of the accelerating electric field T_a or be a multiple of this period. Besides, phase stability is essential for matching of the transit time between the gaps with T_a and for focusing in the transverse direction.

Linear electron accelerators are considerably differing from the proton ones. Considering that the velocity of relativistic electrons is practically constant

$$v = \frac{pc^2}{E} = \frac{pc^2}{\sqrt{p^2c^2 + m_e^2c^4}} \approx c,$$

synchronism is ensured because the accelerating electromagnetic wave propagates at a velocity of light thus excluding the necessity for the phase-stability mechanism. As

$$\frac{dp_\perp}{dt} = 0, \qquad \text{then} \qquad \frac{mv_\perp}{\sqrt{1 - v^2/c^2}} = \text{const},$$

and the transverse motion velocities $v_\perp$ are also rapidly falling with an increase in v. Consequently, there is no need in focusing too. The transverse Coulomb repulsion of electrons in the beam is insignificant due to a nearly absolute compensation, owing to magnetic attraction of the currents. The energy losses by synchrotron radiation in a linear accelerator are also insignificant. To transit to high energies, however, calls for increasing of an accelerator length. For example, the linear electron accelerator at the Stanford Linear Accelerator Center (SLAC) constructed in 1966 and having a maximum energy of 25 GeV was 3 km in length.

For the most part, the primary electron source in accelerators is represented by the so-called electron gun including a thermionic cathode and electron-optical system. A source of protons and weakly-ionized heavy ions is plasma, from where they are pulled by external electric field. Positrons, antiprotons and greatly-charged ions are generated due to interactions between the primary electron, proton or ion beam and the matter. Vacuum within the volume, where the particle motion takes place, in all accelerators is of the order of 10^{-5} — 10^{-7} mm Hg to lower particle scattering from the residual gas.

The concept of colliding-beam accelerators put forward by D. Kernst in 1956 has led to a revolution in the technology. Its realization enables one to attain a critical energy increase for the colliding particles and hence to go to investigation of the matter structure at still closer distances.

With a stationary target the kinetic energy of an incoming particle (shell) is only partly transferred to the reaction energy; some part of the kinetic energy is spent for the target recoil energy. In case the target is more massive than a shell the recoil energy is low, otherwise the collision efficiency decreases drastically. Because of a relativistic increase in mass, the energy loss by recoil is growing with the particle velocity approaching the velocity of light.

A character of interaction is determined by the particle energy in the CMS rather than in the LS as the major part of the energy in the LS is transformed into the kinetic energy of the reaction products. In the present-day accelerators the colliding beams are not strictly opposite, intersecting at a small angle. During processing the results of such experiments all kinematic characteristics are transformed to the CMS for subsequent analysis. On collision of two particles with arbitrary momenta $\mathbf{p}_a$ and $\mathbf{p}_b$ the transition to the CMS is realized by Lorentz transformations at the appropriate velocity

$$\mathbf{v} = \frac{(\mathbf{p}_a + \mathbf{p}_b)c^2}{E_a + E_b}. \tag{8.5}$$

Actually, directing the axis x towards the sum of momenta $\mathbf{p}_a + \mathbf{p}_b = \mathbf{p}$ and using Lorentz transformations for the four-momentum, in a new coordinate system (marked with asterisk) we get:

$$p_x^* = \frac{p_x - (E_a + E_b)v/c^2}{\sqrt{1 - v^2/c^2}}, \qquad p_y^* = p_y, \qquad p_z^* = p_z. \tag{8.6}$$

Setting the vector $\mathbf{p}^*$ to zero, we can find the CMS velocity relative to the frame of reference, where

$$\mathbf{p} \neq 0, \qquad v = \frac{\mathbf{p}c^2}{E_a + E_b},$$

in accordance with Eq. (8.5). The CMS velocity with respect to the LS (b particle at rest) may be determined by the expression

$$v = \frac{\mathbf{p}_a c^2}{E_a + m_b c^2}. \tag{8.7}$$

With the use of (8.7) it is possible to relate the particle energies and momenta for both systems. For instance, in case of the particle a we have:

$$\mathbf{p}_a^* = \frac{\mathbf{p}_a - E_a v/c^2}{\sqrt{1 - v^2/c^2}} = \frac{m_b \mathbf{p}_a}{\sqrt{m_a^2 + m_b^2 + 2E_a m_b/c^2}}, \tag{8.8}$$

$$E_a^* = \frac{m_a^2 c^2 + m_b E_a}{\sqrt{m_a^2 + m_b^2 + 2E_a m_b/c^2}}. \tag{8.9}$$

The relation (8.9) enables one to be aware of the gain on going from ordinary accelerators to accelerators of the collider type.

For simplicity, we consider collision of identical particles. Passing to kinetic energy in formula (8.9) we find:

$$T = 4T^* + \frac{2(T^*)^2}{m_b c^2}. \tag{8.10}$$

From (8.10) it follows that to produce the energy $T^* = 200$ GeV one can realize the variant of a stationary-target electron accelerator at the energy of $T \approx 1.6 \times 10^8$ GeV. As regards the energy, the merits of colliders are obvious. At the same time, one should never forget that cheese may be free-of-charge in a mousetrap only. The principal drawback of colliders is low frequency of the reactions. This peculiarity is easily comprehended when we compare shooting at a large stationary target and shooting at the bullets flying towards.

The efficiency of colliders may be improved using a higher particle density of the beam. This is attained by the use of storage rings, where the accelerated particles are stored throughout many acceleration cycles. Moreover, the focusing system may be constructed so that maximum beam compression can be provided at the point of collision, contributing to the enhanced beam density and higher probability of the reaction.

We introduce the value known as accelerator luminosity that defines the number of events in a unit time at the unit interaction cross-section (1 cm^2). For stationary-target accelerators the luminosity L is equal to

$$L = n_0 l N, \tag{8.11}$$

where n_0 is the particle density in the target, l is a target thickness along the beam, and N is an outgoing particle flux. In case of colliding beams the accelerator luminosity is given by the expression:

$$L = \frac{N_1 N_2 l f}{l_c S}. \tag{8.12}$$

Here N_1 and N_2 is the total particle number in the beams, l — cluster extent, l_c — collision length ($l_c > l$), f — rotation frequency of the particles in accelerator in terms of s^{-1}, S — cross-section area of a larger cluster measured in cm^2. Luminosity dimension is $\text{cm}^{-2} \cdot \text{s}^{-1}$. The luminosity multiplied by the process cross-section σ in cm^2 gives the relevant number of events per second

$$R = \sigma L. \tag{8.13}$$

The modern high-energy accelerators are of the collider type. Also, these accelerators may be operated in the stationary-target mode. All of them represent synchrotrons, with the exception of the linear electron-positron collider at SLC (Stanford). This collider allows to obtain the energy up to 100 GeV at the luminosity $L = 2.5 \times 10^{30}$ $\text{cm}^{-2}\text{s}^{-1}$ (here and hereinafter the data are given as of 1998). In linear accelerators there is a single region for interaction of colliding particles, and therefore its cross-section area S should be exceptionally small with a diameter of the order of $\sim 10^{-4}$ cm. In SLC the beam of accelerated electrons (positrons) is divided into hyperdense clusters (4×10^{10} particles) 0.1 cm in length, with a cross-section width of 1.5×10^{-4} cm and height of 0.5×10^{-4} cm. SLC is about 1.5 km in length.

The operation of the most powerful cyclic electron-positron collider LEP (Large Electron-Positron) located at Geneva was terminated in 2001. Its maximum energies were in excess of 200 GeV. To decrease synchrotron radiation, the perimeter of LEP was increased to 26.66 km. Its luminosity was amounting to $L = 5 \times 10^{31}$ $\text{cm}^{-2}\text{s}^{-1}$. In this collider the clusters of accelerated electrons and positrons were denser by the order of magnitude (6×10^{11} particles), whereas their space dimensions were much greater: extent of 1 cm, height — 8×10^{-4} cm and width — 2×10^{-2} cm). The rotation period was 22×10^{-6} s. During acceleration time of 550 s the particles covered a distance of the ring approximately 2.5×10^7 times, however, keeping their orbits to an accuracy of 2×10^{-3} cm.

The energy of LEP is a limit for circular electron-positron colliders. Further energy enhancement for e^-e^+-machines is possible only in case of linear facilities. Such colliders are already at the stage of conception; their putting into service is expected in 10-12 years. The typical energy for such a linear accelerator so-called linac will reach 500 — 1000 GeV. The energy of linear colliders cannot be increased without limit. This is associated with the material structure of the accelerator. In order that a particle to be accelerated to energies of the order of 1000 TeV or higher on the typical distances of order of 100 km, it is required to create an accelerating gradient of the electric field in the region of 10^8 V/cm. Unfortunately, such strong fields will break away electrons from atoms, altering the structure of any material. An effort to realize acceleration using a field of this strength will result in destruction of the accelerator. It is hoped that the deadlock may be resolved with the help of nanotechnologies. There is a reason to believe that advances in nanotechnology will enable the creation of microscopic accelerator cells with the requisite accelerating gradient. In this case, these cells will have such property that after their disruption they could be regenerated in a short time.

It is clear that for the effective acceleration the gain in energy of a particle per cycle should be higher than the total radiation loss $\mathcal{P}$ determined by formula (8.4). Since $\mathcal{P} \sim m^{-4}$, in case of protons the attainable acceleration energies will be much higher. Presently, the highest energy (2 TeV) is provided by the FERMILAB proton-antiproton collider. Its performance is as follows: luminosity — 2.1×10^{32} $\text{cm}^{-2}\text{s}^{-1}$, rotation period — 3.8×10^{-6} s, total acceleration time — 10 s, proton-antiproton cluster extent — 38 cm, radius of $p(\overline{p})$-beam — 34×10^{-4} cm (29×10^{-4} cm). The ring length of this accelerator is equal to 6.9

km.

A new proton LHC (Large Hadron Collider) is currently constructed on the basis of LEP to reach fantastic energies 14 TeV by today's standards. However, this is not the limit as in principle the circular proton machines are capable of providing the energies from 100 to 1000 TeV. Because of this, the creation of another proton supercollider is technically possible. At the present time this idea is put forward for discussion. A tentative name for this machine is VLHC (Very Large Hadron Collider). Its contemplated running into operation should not be expected earlier than in 20-30 years.

As regards LHC, its start-up is scheduled for 2007. The research program associated with the use of LHC involves an experimental search for Higgs bosons and superpartners of ordinary particles. Also, it is planned to realize a search for preons and heavy gauge bosons W' and Z', additional to the SM gauge bosons. Another problem is associated with a search for the formation of quark-gluon plasma using the potentialities of LHC. The detection of such processes is a real challenge for researchers as the cross-sections of the expected processes are extremely small. Moreover, some of the background events possess the intensities by milliards higher than that of an event under study. Reliability of the obtained results will be ensured by the simultaneous use of two different detectors operated by different research teams. Two all-purpose detectors ATLAS and CMS are constructed to facilitate the solution of the principal tasks of LHC. It should be noted that the detectors are constructed on the basis of dissimilar conceptions. Their magnetic systems, the general construction, detecting devices are considerably differing. It is obvious that coincidence of the results obtained at ATLAS and CMS should be indicative of their maximum reliability.

A new ALICE detector that is also created at the present time is intended for investigation of collisions with ultrarelativistic energies: nucleus-nucleus (Pb-Pb, Ca-Ca) as well as proton-proton and proton-nucleus. Since these collisions exhibit a hyperhigh energy density (5.5 TeV per each pair of colliding nucleons), the occurrence of quark deconfinement and the formation of quark-gluon plasma may be expected.

Electron and proton beams may be ejected from accelerators and directed to the external targets, both hydrogen and nuclear. The produced charged $\pi^{\pm}$-mesons, K-mesons or $\overline{p}$ may be focused into the secondary beams for their further use during the experiments. This process may be continued to produce muon and neutrino beams. Since the decay of π^- results in two weakly-interacting particles μ^- and $\overline{\nu}_\mu$, after its passage through the absorber a π-meson beam is turned to the muon beam contaminated with neutrino. At a sufficiently high density of the absorbing material the muons disappear, leaving a pure muon antineutrino beam.

A neutrino (antineutrino) source is represented by K^+- and π^+-mesons (K^-- and π^--mesons) resultant from the proton bombardment of a beryllium oxide target. In essence, the beam comprises muon neutrinos and antineutrinos, the electronic component being strongly suppressed. The energy of neutrinos within the beam is uniformly distributed from zero to $E_{max} = r_{K,\pi} p_\mu c$, where p_μ is a muon momentum and

$$r_{K,\pi} = \frac{m_{K,\pi}^2 - m_\mu^2}{m_{K,\pi}^2},$$

($r_K = 0.954$, $r_\pi = 0.427$). The Earth shield serves as a muon absorber. An absorber 1 km in length provides absorption of muons with the energy up to 200 GeV. Increasing maximum energy of muons necessitates further growth of the absorber length. As neutrino is electrically neutral, focusing of the neutrino beam is accomplished indirectly. Muon storage rings are used as neutrino fabrics, where electron neutrinos and muon antineutrinos (or electron antineutrinos and muon neutrinos) are produced approximately in equal proportions already within the beam. The production of focused high-energy neutrino beams is required mainly for so-called "long-baseline" oscillation experiments, in the process of which a neutrino beam created by the accelerator penetrates the Earth's thickness and is detected by an underground detector. The principal objective of such experiments is a study of neutrino oscillations (transitions $\nu_i \leftrightarrow \nu_{j\neq i}$, where $i, j = e, \mu, \tau$).

Compton scattering of high-energy electrons from laser photons makes it possible to produce γ-beams with the energy amounting to 80% of that for the primary electrons (this procedure was first realized at SLAC in 1963). This opens up possibilities for the creation of $\gamma\gamma$-colliders with the luminosity to $10\% L_e$, where L_e is the luminosity of a parent electron collider (e^+e^- or e^-e^-). Besides, a γ-source may be provided by the classical photon bremsstrahlung of e^- or e^+ beam. The next generation electron-positron colliders, for example, NLC (Next Linear Collider) with a maximum energy of 500 GeV, are so designed that they can be operated both in the $\gamma\gamma$ and $e^{\pm}\gamma$ modes. In this way combined colliders may be constructed in addition to the available electron and proton colliders. Presently, this class of accelerators is represented by the DESY synchrotron (Hamburg), version HERA (e^-p-collider), that is used to study collisions of electrons with the energy 30 GeV and protons — with the energy 820 GeV.

Probably, our reader is of the opinion that only stable particles may be the candidates for the function of collider-accelerated particles as the limited life-time of unstable particles will preclude their acceleration to very high energies. However, this is a mere delusion, so to speak, a tribute to the nonrelativistic style of thinking. Just recall the time deceleration phenomenon associated with the moving clock. In a laboratory system, an unstable particle covers to its decay the distance that is much greater than may be derived from nonrelativistic considerations by means of simple multiplication of its velocity on its lifetime. Precisely this principle is the basis for the construction of muon colliders (MC). The greatest difficulties during the construction of MC are associated with the fact that over the lifetime of muon, that is equal to τ_μ=2.2 ms in an intrinsic coordinate system, muon beams should be stored, cooled, accelerated, and brought into interaction with each other. In the LS the muon lifetime is increased due to the relativistic factor by the value

$$\kappa_0 = \frac{1}{\sqrt{1-v^2/c^2}} = \frac{m_\mu c^2}{m_\mu c^2\sqrt{1-v^2/c^2}} = \frac{E_\mu}{m_\mu c^2}.$$

Then the intensity of muon decay along the beam trajectory may be given in the form:

$$\frac{dN}{dl} = -\frac{N}{L_\mu \kappa_0}, \tag{8.14}$$

where $L_\mu = c\tau_\mu \approx 660$ m and it is assumed that $v \approx$. In the absence of the acceleration mode this leads to the ordinary exponential form of the beam loss:

$$N = N_0 \exp(-\frac{l}{L_\mu \kappa_0}). \tag{8.15}$$

In the acceleration section κ is not a constant

$$\kappa = \kappa_0 + \kappa' l = \kappa_0 + \frac{eV_g}{m_\mu c^2} l, \tag{8.16}$$

where eV_g is the acceleration gradient. Substituting expression (8.16) for κ_0 in (8.14), we can obtain

$$N(l) = N_0 \left(\frac{\kappa_0}{\kappa_0 + \kappa' l} \right)^{1/(L_\mu \kappa')},$$

resultant in the relation

$$\frac{N(l)}{N_0} = \left(\frac{E_i}{E_f} \right)^{1/(L_\mu \kappa')}, \tag{8.17}$$

where E_i (E_f) is the initial (final) muon energy.

From (8.17) it follows that the decay losses in the muon beam are minimal for

$$L_\mu \kappa' \gg 1. \tag{8.18}$$

Substitution of the numerical values into (8.18) demonstrates that the normal functioning of a muon collider requires the existence of the acceleration gradient $eV_g \gg 0.16$ MeV/m throughout the whole muon system.

It is well known that the creation of e^-e^+ colliders with multiTeV energy is restricted by two factors: (1) increasing loss by synchrotron radiation; 2) drastic increase in the material costs as two linear accelerators will be required to avoid considerable synchrotron radiation in the storage rings. The bremsstrahlung of muons is negligible. They may be accelerated and stored in the rings, whose radius is considerably less as opposed to hadron colliders with comparable energies. Unlike hadron colliders, where the background appears at the point of particles interaction and comes from the accelerator as well, the background of MC may be found in the detectors only. Also, MC exhibits high monochromaticity. The roof-mean-square deviation of R from the Gaussian energy distribution in the beam falls within the interval from 0.04% to 0.08%. Owing to cooling of the muon beam, R may be decreased down to 0.01%. Thus, the energy resolution of the beam in MC is much higher than that in e^-e^+ colliders. Another advantage of MC is its fast rearrangement for operation in the $\mu^-\mu^-$ or $\mu^+\mu^+$ mode. Since the construction of MC includes special storage rings to provide optimization of the luminosity for some energy, MC is an ideal instrument for investigation of resonances with an extremely small decay width (e.g., Higgs boson of the SM).

A quantity determining the energy spread in the beam $\sigma_{\sqrt{s}}$ is essentially important characteristic for the collider. In case of MC this quantity is given by the expression

$$\sigma_{\sqrt{s}} = (7\ \text{MeV}) \left(\frac{R}{0.01\%} \right) \left(\frac{\sqrt{s}}{100\ \text{GeV}} \right). \tag{8.19}$$

Note that the detection and examination of a particular particle in s-channel may be successful provided $\sigma_{\sqrt{s}}$ is of the same order as the total decay width of this particle.

At present two projects are investigated for the construction of MC. The first, First Muon Collider, allows for building MC having the CMS energy $\sqrt{s} = 0.5$ TeV and luminosity $L \sim 10^{33}$ cm^{-2}s^{-1}. The second, Next Muon Collider, is a collider of much higher power with the following characteristics: $\sqrt{s} \sim 4$ TeV, $L \sim 10^{35}$ cm^{-2}s^{-1}.

8.2. Particle Detectors

For the progress in experimental particle physics the development of accelerators is an essential but insufficient requirement. Another prerequisite is the development of experimental methods, creation of more and more sophisticated detectors, computerization and other modern technologies. In order to propose an experiment for testing of a particular hypothesis or explain experimental results a physicist must necessarily know the advanced elementary particle detection techniques. Below is a short summary of most extensively employed types of detectors.

The functioning of a detector is based on ionization or excitation of the atoms within a detector material by the accelerated charged particles. A charged particle leaves a track of the ionized or excited atoms that enables one to judge about the form of a particle path. Uncharged particles (photons, neutrons, neutrinos, etc.) do not ionize the material, exhibiting themselves by the secondary charged particles, produced as a result of interactions between the uncharged particles and detector material. For instance, photons are detected by the produced electron-positron pairs and Compton recoil electrons; neutrinos — by the generation of μ^+-mesons upon their collision with protons and electrons, by the formation of recoil electrons or due to the inversed β-decay within the detector material. Pass of the particles through the material is accompanied by the formation of free electrons, ions, positrons subsequently annihilating with electrons to photons, and also by different reactions, thermal phenomena, and the like. Because of this, the particles may be detected by the output electric pulses from the detector, by photoemulsion blackening or else by changes in the structure of solid material in the detector. Electric signals of the particles are generally so weak that an additional system is required for the signal amplification. Most important characteristics of a detector are as follows: efficiency, i. e. detection probability for a particle finding its way into the active volume of a detector; detector memory time, i. e. time period when the changes within the detector volume due to transit of the detected particle are retained; dead time of a detector, i. e. time of returning to the initial sensitivity after the regular action; energy, spatial and temporal resolution, i. e. the energy, position and time determination accuracy for a particle detected. According to the information obtained about the particles, all the detectors may be conventionally subdivided into two classes. The first class comprises detectors signaling about the particles by short electrical pulses. This class, in turn, may be further subdivided into two subclasses:) detectors of single particles, and b) detectors of electromagnetic and nuclear cascades. The second class is represented by track registration detectors providing direct observation of particle tracks.

1 . Single Particle Detectors

Ionization Chamber (IC). IC is the simplest form of such a detector. It comprises two electrodes, and the interelectrode space is filled with gaseous, liquid or solid material. Under the effect of moving charged particles, the material within the chamber is ionized to produce free electrons and ions creating a current pulse, due to their motion between the electrodes, in the external circuit of the chamber. IC may be used to detect both the particle fluxes (measurement of the average output current) and single particles (measurement of output pulses). Actually, IC may be used for the detection of all particle types, although by the appropriate selection of the detector material and electric field one is enabled to adjust the chamber for the detection of the certain particle type.

Proportional Multiwire Chambers (PMC). PMC represent a modern type of proportional counters, gas-discharge detectors, where the electric signal amplitude at the output is proportional to the energy spent by the particle on gas ionization. PMC consists of numerous parallel, small-diameter ($\sim 2\times10^{-3}$ cm) anodic wires fixed between two flat cathodes, solid or wire-type but with wires of greater diameter. Each of the anodic wires is functioning as an independent detector. Under the effect of an electric field, the primary electrons formed by the particle entering in the detector are moving to the anode to get into a high-strength field, where they are greatly accelerated causing the secondary gas ionization, i. e. electron avalanches are observed. The spatial resolution of PMC is minor: $\geq 7\times10^{-2}$ cm. However, short dead time $\geq 3\times10^{-8}$ s make them most widely used position detectors.

Drift Chambers (DC). DC are used as position detectors. These gas-discharge devices involve wire electrodes, where the particle positions are determined by the drift time of electron in a homogeneous and constant electric field, from the place of their origination to anodic wires. The field around the anode is inhomogeneous, resulting in electron acceleration and hence in the formation of electron avalanches. The spatial resolution of DC amounts to 10^{-6} cm. As the dead time (electron drift time) of DC is long ($\sim 10^{-6}$ s), these chambers are inoperable in high-load conditions.

High-precision Position Detectors (HPD). HPD are used to reconstruct the particle positions and paths at the vertex of the event under study or in its neighborhood. A typical task for HPD is a search for the "second vertex" resultant from the decay of the short-lived (with a life-time of 10^{-12} — 10^{-13} s) particle that was produced at the first vertex. Most popular are microstrip semiconductor detectors structured as follows: strips of a conducting material are deposited as electrodes on one of the surfaces of a silicon monocrystal, whereas the other surface is metallized. The voltage applied to these electrodes makes up a few Volts. An ionizing particle in transit through the crystal is forming the electron-hole pairs migrating towards the electrodes to create there the current pulses. The spatial resolution d of microstrip detectors is determined by the strip width and interstrip gaps, reaching 10^{-7} cm (higher resolution is demonstrated only by nuclear photoemulsions $d \sim 10^{-8}$ cm). The temporal resolution of these detectors is of the order of 10^{-8} s.

Scintillation Counter (S). SC comprises a scintillating material (special liquids, plastics, crystals, noble gases), where a charged particle effects both ionization and excitation of the atoms and molecules forming this scintillator. Recovering to ground state, these atoms and molecules are emitting photons incident on the cathode of photomultiplier (PM) to knock off photoelectrons from the cathode. Owing to these electrons, a pulse with amplitude proportional to the energy transferred by the particle to the scintillator is formed at the anode of PM. The accuracy of the measured particle energy provided by SC is within 10%. Since under the effect of charged particles the majority of scintillators reveal the characteristic luminescence time of about 2×10^{-8} s, the transit time of a particle through the counter may be determined with high accuracy. The detection efficiency for the charged particles is close to 100%. Neutrons may be detected (by recoil protons) with the use of hydrogen-containing scintillators, and photons — with the use of higher-density scintillators like iodide sodium. Detection of neutrinos requires the use of combined scintillators. So, in their experiment (1953) F. Reines and K. Cohen have used the hydrogen-containing scintillator with an addition of a cadmium salt for the detection of electron antineutrino in the reaction

$$p+\overline{\nu}_e \to n+e^+. \tag{8.20}$$

The first scintillation flare was caused by positron-electron annihilation, while the second flare occurring in $(5-10)\times 10^{-6}$ s was due to the cadmium atom returning into the ground state after absorption of a neutron.

Cherenkov Counters (CC). The operation principle of CC is based on the detection of Cherenkov radiation. This radiation is generated when charged particles are moving in a transparent medium at the speed v exceeding the speed of light in the medium

$$v \geq \frac{c}{n}, \tag{8.21}$$

where n is the refractive index of this medium. Light is emitted only in forward direction along the motion path of a particle, forming a cone with an axis in the direction of v, and the cone angle, emission angle, is determined by the relation

$$\cos\theta = \frac{c}{vn}. \tag{8.22}$$

This phenomenon is similar to acoustic cone of the airplane flying at a supersonic speed. A light flare following the particle motion in the medium is detected by PM. The counter is used for the detection of relativistic particles as well as estimation of their charge, speed, motion direction (to within 10^{-4}). Measuring the particle momentum by deflection in a magnetic field, one is enabled to measure its speed with the help of CC and hence to determine a mass of this particle. The relation (8.21) forms the basis for the operation of threshold or integrating CC, capable to detect all particles having the speeds above threshold $v > v_t = c/n$. And the counter provides radiation measurements over the whole range of angles from 0 to $\theta_{max} = \arccos(/vn)$.

The operation of angular or differential CC is based on relation (8.22). These counters detect particles at the speeds from v_0 to $v_0+\Delta v$. Emission of the particles parallel to the optical axis of the counter is collected only in a narrow range of angles, from θ_0 to $\theta_0+\Delta\theta$.

CC comprise a radiation-generating medium, collecting optics that directs this radiation to the cathode of PM, PM and recording system. A new type of gas CC, RICH (Ring Image Cherenkov detector), has been proposed recently. This detector provides detection and imaging of the particles by their Cherenkov radiation. The energy region, where CC offer mass separation for the particles, has an upper limit as the difference in the speeds of particles distinguished by their masses is decreased with growing energy. For instance, the speeds separation of π- and K-mesons by threshold gas CC is possible up to the energies amounting to several dozens of GeV, whereas by differential gas CC with a compensation of radiation dispersion — up to several hundred GeV.

Transition Radiation Detectors (TRD). The operation of these detectors is based on the formation of an electric field by the moving charged particle. The electromagnetic field of this particle changes its configuration as the particle leaves a medium with the dielectric permeability ε_1 for that with ε_2. This process is accompanied by the emission of transient radiation. The distinguishing feature of this radiation is the fact that its properties are determined by the relativistic factor

$$\gamma = \frac{1}{\sqrt{1 - v^2/c^2}} = \frac{E}{mc^2}.$$

In this case the radiation intensity is proportional to the particle energy, and radiation is concentrated within a cone of angle $\theta = \gamma^{-1}$. In layered structures, where a particle crosses the interface repeatedly, the intensity of transient radiation may be resonantly amplified providing a means for the detection and identification of ultrarelativistic particles with $\gamma > 10^3$. The Lorentz factor $\gamma \sim m^{-1}$ and intensity of transient radiation at constant energy E will be greater for particles with lower mass. This allows for mass separation over the energy range hardly accessible for gas CC.

TRD consists of a layered medium, usually comprising a multitude of light-weight foil plates (Li, Al) perpendicular to the particle direction, and a radiation detector. In some TRD the radiator may be represented by ordinary porous materials like foamed plastics. The radiation detection is most commonly done by proportional wire chambers filled with heavy gases.

1 b. Electromagnetic and Hadron Cascade Detectors

The operation principle of these detectors, also referred to as Total-Absorption Detectors (TAD), is based on total absorption of the cascades created by the detected particles within the detector material. TAD enable detection of the integrated Cherenkov radiation for all the particles forming the electron-photon shower (total-absorption CC) or integrated energy spent by all the particles for ionization (calorimeters). In electromagnetic cascades this energy is practically equal to the energy of the primary electron or photon. In hadron cascades ionization requires the major part of the energy possessed by the primary particle but some part of its energy (up to 20 — 30%) is spent for nuclear disintegration, then carried away by neutrinos formed as a result of particle decays, and is not detected by calorimeters.

Total-Absorption Cherenkov Counters. These detectors provide detection of photons and electrons with estimation of their energy. The blocks of lead glass serve as radiators. Their size must be sufficient for absorption of the main part of the shower produced by the primary particle. Cherenkov radiation is detected by PM.

Calorimeters (C). C are intended to measure the energy of particles, both charged and neutral, beginning from 10^2 GeV and higher. Interacting with the nuclei within the material of C, a high-energy particle produces a cascade of the high-energy secondary particles, in turn, interacting with the material to generate new particles. Because of this, electron-nuclear shower occurring in the active volume of the device rapidly moves in a direction of the primary particle, and its energy is spent for ionization of the material. Provided a layer of the material in C is sufficiently large, all shower particles are left in the material, and the number of created ions is proportional to the energy of the primary particle. Then these ions are collected at the calorimeter electrodes, and their total charge is measured to determine the primary particle energy with an accuracy of 10—15%. The use of C makes it possible to locate the shower origination and determine its spatial development. The simplest C are constructed as "sandwiches" consisting of alternated layers of heavy material and ionization detectors. In electromagnetic C such a sandwich comprises thin layers of lead and scintillator. Its total thickness may reach a few dozens of centimeters. Hadron cascades are slowly developing in the majority of materials, penetrating deeper than the electromagnetic ones. Because of this, hadron C possess much greater thickness, up to several meters, with thicker layers of the material (commonly iron) and scintillator, that may be replaced by other ionization measuring detectors.

2. Track Detectors

These detectors enable observation of particle tracks. Track detectors, being subjected to magnetic fields, make it possible to determine a sign of the electric charge for the particles and to measure their momenta by the path curvature to a high degree of accuracy. The first track detector was constructed by Ch. Wilson in 1912 and received the name Wilson chamber. Its operation is based on condensation of supersaturated vapor and the formation of visible little drops of liquid at the ions originating along the path of a fast charged particle. The device comprises a closed vessel, having windows intended for the track observation, filled with gas and saturated vapors of some liquid substance, e. g., methyl alcohol. On rapid adiabatic discharging this gas is cooled, whereas the vapor becomes supersaturated. After photographing of the tracks, the chamber regains its initial state due to fast gas compression, causing evaporation of the droplets at the ions with the formation of saturated vapor, and the former ions influenced by the electric field are collected from the tracks at the electrodes.

Bubble Chamber (BC). BC is one of the main types of track devices in high-energy physics. It comprises a large container several meters in diameter filled with a transparent superheated liquid. Its boiling is delayed owing to the high pressure that is 5 — 20 times higher than the atmospheric pressure. Abrupt decrease of the pressure results in superheating of the liquid, and in case an ionizing particle is passing through the chamber an additional heating leads to drastic boiling of the liquid in a narrow channel along the particle path. Its trajectory is marked by a chain of vapor bubbles. These bubbles are allowed to grow for a period of 10 ms, afterwards they are photographed by stereoscopic cameras. Subsequently, the starting pressure is applied to the liquid causing collapse of the bubbles, and the chamber is ready for operation again. In BC most common is liquid hydrogen. Deuterium, propane and such heavy liquids as xenon and Freon are more rarely used. The latter are of a special interest, especially for detection of neutrino interactions. BC is usually placed in a strong magnetic field, and by the track curvature one can measure the particle momenta to a high accuracy. The spatial resolution of BC comes to 10^{-2} cm.

The main advantage of BC is the possibility to use its working liquid both as a target for the incoming particles and as a detector for the reactions proceeding upon the particle collisions with electrons and nuclei of the liquid. This advantage is especially marked in studies of complex processes involving large numbers of particles. The principal disadvantage of BC is its "uncontrollability": it is impossible to realize its response on a signal of fast detectors which previously select the required events. The response of BC with a period of ~ 1 s is synchronized with points in time of fast beam ejection from the accelerator.

Spark Chambers (SC). SC are the controlled gas discharge detectors including a series of parallel metallic plates placed into the container filled with inert gas. These plates, alternately, are connected to the high-voltage source or have an earth connection. Provided an ionizing particle crosses the working volume of SC, by command of the monitor counters a short high-voltage pulse (10 — 20 kV/cm) is applied. Originating at the points of particle pass, spark discharges parallel to the electric field are photographed or located by the magnetostriction method.

Streamer Chambers (StC). StC are the counter-controlled gas discharge detectors, where discharges are formed exclusively along the tracks. These chambers contain two flat parallel electrodes positioned at a distances measuring several dozens of centimeters. To the electrodes a very short ($\leq 10^{-8}$ s) high-voltage (10 — 50 kV/cm) pulse is applied. In these conditions the discharges, originating at ionizing particle passage, are terminated and take the form of short ($\sim 10^{-1}$ cm) luminous channels (streamers) aligned with the field. Their photographs are made to obtain the track images. As contrasted to SC, StC are isotropic, i. e. they are capable of reproducing the tracks of every spatial orientation and allow for the particle ionization measurements.

As a rule, all the available particle detectors are combined detector systems (CDS) featuring a series of detectors integrated in one detecting unit. CDS represent the major element of modern accelerator. Their size measures dozens of meters, mass amounts to $\sim 10^4$ t, and the number of information channels may be as great as 10^6. The personnel required for their operation runs into hundreds of people, whereas the construction expenditures comprise a significant part of the total cost for the whole accelerating complex.

The majority of CDS are similar in structure, though the choice, amount, dimensions and arrangement of elements are dependent on the specific task at hand. Most typical elements are as follows: target, vertex detector surrounding the target that indicates the reaction products and determines their escape direction; position detectors localizing trajectories of primary and secondary particles; spectrometric detectors measuring the momenta of secondary particles or their energy; identifiers of secondary particles. Large-scale CDS are given proper names as ATLAS, ALICE, DELPHI, etc. Now the particle fluxes passing through CDS are as great as 10^8 s^{-1}. Unfortunately, the difficulties in processing of the measurement results in case of numerous information channels and high detection rate generally prevent a real-time analysis. Considering this situation, the information is recorded and processed on completing the experiment.

8.3. Neutrino Telescopes

Among the fundamental particles, neutrino holds a special place since it plays an important role in large-scale events of the Universe. The finding that neutrino has a mass makes this particle a worthy candidate for the role of a particle constituting the hot dark matter and hence enables one to evaluate the average matter density in the Universe, age of the Universe and its further fate. The problems associated with detection of cosmic neutrinos are the subject matter of neutrino astrophysics. Neutrino astrophysics may be considered as a compound part of the elementary particle physics. And this is related not only with the fact that both physics divisions are concerned with the Universe structure. The other important aspect is that solution of the problems concerning the generation and detection of neutrinos depends upon the character and intensity of interaction between the elementary particles. Because of this, it is obvious that neutrino telescopes (NT), being basic instruments of neutrino astrophysics, are useful for studies of particle physics too.

Depending on the detection technique, all the available NT may be subdivided into two classes: NT operating in the continuous counting mode and NT operating in the discrete counting mode. The first class includes NT using the radiochemical methods. The second class involves NT intended for a real-time detection of the particles, the production of which is initiated by the interaction between neutrinos and the counter material.

The operation of radiochemical NT is based on investigating the process of inverse β decay due to the incoming neutrino

$$\nu_e + X \to e^- + Y, \tag{8.23}$$

where X are nuclei of the elements determining the initial composition of NT detector. As a rule, nuclei of Y originating in the detector are radioactive, their half life $T_{1/2}$ determining the duration of an active measurement stage $t_a \sim (2\text{—}3)T_{1/2}$. The formation rate of daughter nuclei is given by the expression

$$R = N \int \Phi(E)\sigma(E)dE, \tag{8.24}$$

where $\Phi(E)$ is the neutrino flux incident on the detector, N are numbers of detector atoms, σ is the process cross-section (8.23). At the incident neutrino flux 10^{10} cm$^{-2}\cdot$s^{-1} (approximately amounting to the flux of solar neutrinos incident on the Earth) and σ of the order of 10^{-45} cm^2 the provision of a single useful event a day necessitates about 10^{30} atoms. Consequently, a mass of this detector should be in the region of several kilotons. Chemical analysis of the detector material takes place in time t_a, and the nuclei number Y is indicative of the capture rate for neutrinos. The advantage of radiochemical NT is the possibility for varying the reaction energy threshold with changes in the detector material. This makes them indispensable in studies of low-energy neutrino fluxes. At the same time, radiochemical NT features inability to measure such neutrino characteristics as hitting time for the detector, energy and trajectory direction. The latter is rather discouraging as we have no chances to distinguish between, for example, solar neutrino and neutrino produced by the terrestrial source.

As an example of a radiochemical NT, we consider Homestake facility that was the first to study the fluxes of solar neutrinos (1967 — 2001). Neutrinos were detected with the use of the chlorine-argon method, i. e. the operation of this NT was based on the chemical reaction

$$^{37}Cl + \nu_e \to {}^{37}Ar + e^-. \tag{8.25}$$

Isolation of the useful event was realized using decay of

$$^{37}Ar \rightarrow {}^{37}Cl + e^- + \bar{\nu}_e, \tag{8.26}$$

whose half life is 35 days. This facility, representing a vessel with a capacity of 390 l filled with 610 t of perchloroethylene (C_2Cl_4), was located in the gold-bearing mine (Homestake, South Dakota, USA) at a depth of 1 480 m. As argon atoms are produced in the form of a volatile compound, they were isolated approximately once a month. The obtained argon was subjected to multistage processing. At the final stage, special small-size proportional chambers were filled with this argon. Then the chambers were shielded with lowest-activity lead to provide thc obscrvation of decays (8.26).

The operation of the second-type NT may be based, for example, on the detection of elastic scattering

$$\nu_l + e^- \rightarrow \nu_l + e^-, \tag{8.27}$$

where $l = e, \mu, \tau$. For a spectrum of recoil electrons NT gives the following expression

$$\frac{d\sigma}{dT} = \sigma_0 \left[g_1^2 + g_2^2 \left(1 - \frac{T}{E_\nu} \right)^2 - g_1 g_2 \frac{m_e T}{E_\nu^2} \right]. \tag{8.28}$$

Here T is the kinetic energy of recoil electrons, $\sigma_0 = 8.8 \times 10^{-45}$ cm^2, where "+" sign is associated with ν_e-scattering, while "-" sign — with ν_μ- and ν_τ-scattering. Because of this, in the second case the scattering cross-section is approximately one-sixth of that in the first case making it possible to distinguish between the neutrino kinds. The cross section of ν_e-scattering integrated with respect to the energy is simple in form

$$\sigma(\nu_e e) = 9 \times 10^{-44} \frac{E_\nu}{10\text{MeV}} \text{ cm}^2. \tag{8.29}$$

The angular distribution of recoil electrons is characterized by a sharp maximum in forward direction (relative to the incoming neutrino) $\Delta\theta \sim \sqrt{m_\nu / E_\nu}$ in width. This enables one to determine the direction of a neutrino source by the trajectories of scattered electrons. Another advantage of such facility is the possibility for the detection of individual acts when neutrino is hitting the detector, permitting measurements of time and energy for the neutrinos (real-time operation). A limitation of NT based on recoil electrons resides in the presence of a considerable background as the detected electrons may be produced during the processes of elastic scattering involving any neutral particles. However, in the case of high-energy neutrinos the background elimination is greatly facilitated.

Also, the operation of the second-type NT may be based on the detection of particles produced in reactions with high-energy neutrino, for example,

$$\nu_l(\bar{\nu}_l) + N \rightarrow l(\bar{l}) + N', \qquad \nu_l(\bar{\nu}_l) + N \rightarrow l(\bar{l}) + N' + X, \tag{8.30}$$

$$\nu_l(\bar{\nu}_l) + {}^{16}O \rightarrow l(\bar{l})\,{}^{16}O + \pi^+(\pi^-), \tag{8.31}$$

where $N = p, n$ and X denotes a hadron collection. Since the main free path of the secondary high-energy electrons within the detector material is very short, they are indistinguishable from hadrons, both being responsible for the nuclear-electromagnetic shower. Compared to electrons, the mean free path of high-energy muons is very long as their energy losses for bremsstrahlung, formation of e^-e^+-pairs and nuclear interactions are small. Of a special importance is the fact that muons are moving practically in the same direction as the neutrinos producing them. The average angle between ν_μ- and μ-trajectories expressed in degrees is determined by the expression

$$< \theta > \approx 2.6 \times \sqrt{\frac{100}{E_\mu(\text{GeV})}}.$$

In the active volume of NT, bremsstrahlung photons, e^-e^+-pairs and hadrons originate along the muon trajectory to initiate the nuclear-electromagnetic showers. With $E_\mu \approx 100$ TeV particular minor showers are overlapping, and the whole muon trajectory is glowing due to Cherenkov radiation of these showers. Based on the direction and intensity of Cherenkov radiation, one can determine the trajectory and energy for muon.

Hadrons generated in the reactions described by (8.30) and (8.31) also initiate nuclear-electromagnetic showers, the direction and energy of which is determined by Cherenkov radiation. The detection of showers may be performed using the acoustic method. In this case the detected signal is represented by the pressure pulse in the active volume of NT conditioned by drastic heating of a narrow channel within the shower due to ionization energy losses of the electrons. For instance, in water an acoustic signal is propagating in the form of a thin disk, with a thickness of about the shower length $s \sim 5$ m and characteristic radius $R \sim 1$ km. By this method the detecting element is represented by hydrophones detecting signals perpendicular the shower axis.

In NT based on detection of the secondary muons the effective volume of the detector is considerably greater than the physical volume owing to the detection of muons generated within a thick layer of the material surrounding the detector. When neutrinos are detected with the use of a detection mechanism on the basis of hadron showers, however, this is not the case. Short lengths of hadron showers enable their detection by Cherenkov radiation only within the physical volume of NT.

As a working material (detector) for NT of the second type one can use water or arctic ice. Arctic ice represents a sterile medium with lower concentration of radioactive elements than in sea or lake water. The use of arctic ice as a detector contributes considerably to the sensitivity of NT. For instance, an NT positioned at a depth of about 1 km makes it possible to separate the background muon (atmospheric) flux that is 100 times greater compared to the limiting flux for the deep-sea DUMAND (Deep Underwater Muon and Neutrino Detector) which was positioned in sea water at a depth of 4.5 km. NT AMANDA located at the South Pole is intended for studies of high-energy neutrino fluxes. Deep-water NT on muons, BAIKAL NT-200 (the Baikal Lake), is an example.

Among NT of the second type one may name the SuperKamiokande facility (Japan, Kamioka) constructed jointly with the USA specialists (1996). This NT is located in the mine with a shielding depth of 2 700 m in water equivalent; absorption of particle fluxes by the rock is equivalent to a water thickness of 2700 m. The principal element of this facility is a water Cherenkov detector in the shape of cylinder, 39 m in diameter and 41 m in height, that contains 50 000 t of water and provides ring imaging of the detected particles. The detector is optically subdivided into the internal (working) volume scanned by 11 200 PM, and also the outer (shielding) volume containing 2 200 PM and operating in the anticoincidence mode. This NT can investigate the fluxes of both solar and atmospheric neutrinos.

The main source of solar neutrinos is a series of thermonuclear fusion reactions at the central part of the Sun, resultant in hydrogen-to-helium transformation without catalysts, hydrogen cycle. This chain may be represented as a multistage process

$$4p \to {}^4He + 2\nu_e + 2e^+ + 26.73\,\text{MeV} - E_\nu, \tag{8.32}$$

where E_ν is the energy carried away by electron neutrino, its average value being ~ 0.6 MeV ($E_{max} < 18.8$ MeV). In SuperKamiokande the detection of solar neutrino is realized using the neutrino elastic scattering reaction from electrons with the energy threshold 5.5 MeV.

Cosmic rays interacting with the atomic nuclei initiate in the atmosphere surrounding the Earth the production of pions, kaons and muons, the decay channels of which involve electron and muon neutrinos as well as antineutrinos

$$\left.\begin{aligned} \pi^\pm &\to \mu^\pm + \nu_\mu(\overline{\nu}_\mu) \to e^\pm + \nu_e(\overline{\nu}_e) + \nu_\mu(\overline{\nu}_\mu), \\ K^\pm &\to \mu^\pm + \nu_\mu(\overline{\nu}_\mu) \to e^\pm + \nu_e(\overline{\nu}_e) + \nu_\mu(\overline{\nu}_\mu). \end{aligned}\right\} \tag{8.33}$$

The neutrino flux is formed in the region of 10 - 20 km altitudes above sea level, its energy varying from 100 MeV to 1 000 GeV. Since the dominant interaction type for neutrinos with such high energies is interaction with the target nuclei, in case of Superkamiokande the detection of atmospheric neutrinos is performed using reactions (8.30) and (8.31).

Besides, the second-type NT, Sudbury Neutrino Observatory (SNO), came into use in Canada in May 1999. At this facility the detector is represented by 1000 t of heavy water (D_2O) enabling investigation of solar neutrino with the help of the following processes

$$\nu_e + d \to p + p + e^-, \tag{8.34}$$

$$\nu_l + e^- \to \nu_l + e^-, \tag{8.35}$$

$$\nu_l + d \to \nu_l + n + p, \tag{8.36}$$

Reaction (8.34) is sensitive to ν_e-neutrino, whereas reactions (8.35) and (8.36) are sensitive to neutrinos of all three kinds. For reactions (8.34) and (8.35) the energy threshold equals 5 MeV, and that for reaction (8.36) is 2.225 MeV.

The multipurpose NT named KamLAND (Japan, Kamioka) using 1000 t of an ultra-pure liquid scintillator as a detector came in operation in Spring 2001. Although solar neutrino may be detected by KamLAND, its main function is to observe the oscillations in the total neutrino flux coming from ten reactors localized in the region at 80 — 350 km from the detector.

The flux of solar neutrinos originating in reaction

$${}^7Be + e^- \to {}^7Li + \nu_e,$$

(so-called beryllium neutrinos) is especially sensitive to the neutrino characteristics. Real-time measurements of this monoenergetic ($E_\nu = 0.86$ MeV) flux are the principal objectives of NT named BOREXINO (Gran Sasso, Italy) that is based on recoil electrons with the threshold 250 keV and came to operation early in 2002. Recoil electrons caused by $\nu_l e$-scattering (cross sections for ν_μ and ν_τ are smaller than for ν_e) produce light flare in the bulk of the liquid scintillator, that is detected by PM. Nylon sphere contains 300 t of ultra-pure pseudocumene, and 100 t of pseudocumene contained in the central region comprise the effective (sensitive) volume. The nylon sphere is, in turn, surrounded by pseudocumene filling the corrosion-proof steel sphere 13.7 m in diameter that contains optical elements surrounding the nylon sphere. The whole construction is submerged into the reservoir with purified water having a mass of 2500 t.

The experimental threshold is set as 0.25 MeV because the energy spectrum of recoil electrons is continuous up to 0.66 MeV. At these low energies the control of natural radioactivity caused by radioactive isotopes being present everywhere is the greatest problem. By the present time extensive research has been conducted with the aim to select materials and realize their purification to extremely high levels of radioactive purity. Simultaneously, the measuring techniques for ultralow radioactivity levels have been developed. The attained results are quite impressive: $10^{-16} - 10^{-17}$ (gram of contaminant per gram of material) for ^{232}Th and ^{238}U.

Chapter 9

Macroworld

9.1. Models of Universe Evolution

Actually, the principal purpose of science is a search for unification. The latest discoveries in Physics enable one to describe all the nature phenomena within a single descriptive scheme making it possible to establish the links between the macrocosmos, with galaxies and galactic clusters scattered as dust particles, and microcosmos of elementary particles. Two poles of the Universe! The giant Universe, on the one hand, and nearly ephemeral "construction blocks" of the matter, invisible despite the use of any available microscope, on the other hand. The Early Universe (when its size was about milliard times smaller than the atom size) was found to have the properties of a microparticle, while it is not improbable that now some microobjects (for example, microscopic black holes) include quite a number of worlds in their totality.

According to the modern observations, a radius of the Universe is $\sim 10^{28}$ cm. Our Galaxy, the Milky Way, is no greater than a tiny dust particle of the infinity that is still beyond human understanding. Indeed, the Milky Way is a plane disk, formed by the stars $\sim 7.5 \times 10^{22}$ cm in diameter and $\sim 5.6 \times 10^{21}$ cm thick, incomparable with the Universe size. The Milky Way has a spherical astral halo with a diameter of about 10^{23} cm. This disk rotating at the speed that amounts to 250 km/s has giant spiral arms. The Universe involves even larger formations: clusters and superclusters of galaxies. To illustrate, such a constellation as Veronica's Hair includes more than 3×10^4 galaxies. Astrophysical data indicate that all directions of the Universe are equivalent, with galaxies, galaxy clusters and superclusters uniformly distributed over the Universe space at scales exceeding $R_0 = 10^{26}$ cm, on the average. In this way, at the scales $R > R_0$ the Universe is uniform and isotropic.

Most important postulate of cosmology is the principle that the basic laws of Nature, those of Physics in particular, established and tested in laboratory conditions on the Earth are valid for the whole Universe and hence all the phenomena observed in the Universe may be explained proceeding from these fundamental laws. The cosmological knowledge has been changing with extended spatial and temporal scales for the part of the Universe apprehended by the mankind. The Ptolemaic geocentric system (IIth century A.D.) may be considered as the first cosmological model substantiated mathematically. This system prevailing for a period of about 1.5 thousand years was changed by the Copernican heliocentric one (XVIth century A.D.). Owing to the advent and improvement of telescopes, further explorations have resulted in our notion of the Universe as a totality of stellar objects. And at the beginning of the XXth century the Universe was considered to be a galactic world (Megagalaxy). It is obvious that each of the proposed "world systems" was actually a model for the greatest system of bodies sufficiently well studied at that time. So, the Ptolemaic model gives an adequate representation of the structure including the Earth and Moon, whereas that of Copernicus is a model for the Solar system.

Modern cosmology stems from the general relativity (GR). The first model of the Universe based on this theory, relativistic cosmological model, was advanced by A. Einstein in 1917 on the basis of the gravitational field equations (see, **1.4**)

$$R_{\mu\nu}(\eta) - \frac{1}{2}R(\eta)\eta_{\mu\nu}(x) = \frac{8\pi G_N}{c^4}T_{\mu\nu}(\eta). \tag{9.1}$$

Proceeding from the considerations conventional for the classical science, Einstein suggested that the Universe, as a totality, should be eternal and invariable. However, Eqs. (9.1) were inadequate to describe the stationary Universe. Because of this, Einstein has introduced the Λ-term, now known as the cosmological constant, in Eqs.(9.1)

$$R_{\mu\nu}(\eta) - \frac{1}{2}R(\eta)\eta_{\mu\nu}(x) = \frac{8\pi G_N}{c^4}T_{\mu\nu}(\eta) - \Lambda\eta_{\mu\nu}(x), \tag{9.2}$$

where $\Lambda > 0$. As a result, the last term in (9.2) describes the repulsive gravitational forces complementary to the attractive gravitational forces of the normal matter ($T_{\mu\nu}$ tensor). Nominally, the cosmological term is equivalent to the additional term of the energy-momentum tensor. Recalling the analogy between Poisson equation for the gravitational potential in the Newtonian theory and Einstein equation, the emergence of a similar term in the Newtonian gravitation theory were equivalent to the introduction of an additional force acting on the body from an object having a negative mass M_0

$$\mathbf{F}_0 = G_N\frac{mM_0}{r^3}\mathbf{r}.$$

As regards Eq. (9.2), in case the cosmological term is conditioned by the particular substance V, for the energy density $\rho_V c^2$ of this substance we obtain:

$$\rho_V c^2 = \frac{c^4}{8\pi G_N}\Lambda. \tag{9.3}$$

Thus, just from the beginning, the function of the cosmological constant was to create or, what is more accurate, to describe antigravitation. Einstein assumed that in this way it was possible to balance gravitation of the Universe matter and ensure an immovability of matter distribution, i.e. stationarity of the Universe. Such a model gives no answer, how and where had originated the Universe. This theory only passed this over in silence. Nevertheless, no longer than 15 years later the astrophysical observations made the scientists to give up a model of the stationary Universe.

At the beginning of the twenties of the last century A. Friedmann demonstrated that by the appropriate selection of a metric the GR equations have nonstationary solutions with the cosmological term present as well. Friedmann models have formed the basis for further development of cosmology. The previous theories have mainly described the observable structure of the Universe, whereas those of Friedmann were evolutionary, relating the current state of the Universe to its prior history. Beginning from the fourties of the XXth century, ever growing attention in cosmology has been centered at the physics of the processes proceeding at different stages of cosmological expansion. By a theory of hot Universe put forward by G. Gamow in 1946 — 1948, at the very beginning of expansion the matter was characterized by enormous temperature. Modern cosmology features active investigations into the problem of the initial cosmological expansion associated with tremendous matter density and particle energies. And the guiding ideas rely on the established laws in the behavior of elementary particles at very high energies. In Friedmann models based on the homogeneous and isotropic Universe the matter is considered as a continuous medium, uniformly filling the space and having specific values of the density ρ and pressure P at every instant of time. To analyze the motion of such a medium, the co-moving frame of reference is usually used, similar to the Lagrangian coordinates in the classical hydrodynamics. In this system, the matter is motionless, deformation of the matter being reflected by that of the reference system, and hence the problem is reduced to the description of the reference system deformation. The three-dimensional space of the co-moving frame of reference is referred to as a comoving space. For a homogeneous and isotropic space the square of the four-dimensional interval ds may be represented in the form:

$$ds^2 = \eta_{\mu\nu} dx^\mu dx^\nu = c^2 dt^2 - a^2(t) \frac{dx^2 + dy^2 + dz^2}{1 + k(x^2 + y^2 + z^2)/4}, \tag{9.4}$$

where x, y, z are dimensionless space coordinates, $a(t)$ is a radius of a space curvature and $k = -1, 0, 1$. It should be noted that for the selection of metric we have already assumed that the Universe is nonstationary. The space curvature is positive at $k = 1$ and negative at $k = -1$. Provided $k = 0$, the space is Eucledian (flat), and $a(t)$ has the meaning of a scale factor. The variation of $a(t)$ in time describes expansion or compression of the co-moving reference system and hence the matter. The metric in (9.4) is known as Friedmann — Robertson — Walker metric that forms a basis for the modern cosmology. It is convenient to rewrite the expression of (9.4) in spherical coordinates:

$$ds^2 = c^2 dt^2 - a^2(t) \left[\frac{dr^2}{1 + kr^2} + r^2 d\theta^2 + r^2 \sin^2\theta d\phi^2 \right], \tag{9.5}$$

that is, the non-zero components of the metric tensor $\eta_{\mu\nu}$ ($\mu, \nu = t, r, \theta, \phi$) have the form

$$\eta_{tt} = 1, \qquad \eta_{rr} = -\frac{a^2(t)}{1 + kr^2}, \qquad \eta_{\theta\theta} = -a^2(t) r^2, \qquad \eta_{\phi\phi} = -a^2(t) r^2 \sin^2\theta. \tag{9.6}$$

Using

$$\eta_{\mu\nu}\eta^{\nu\sigma} = \delta^\sigma_\mu,$$

we obtain

$$\eta^{tt} = 1, \qquad \eta^{rr} = -\frac{1 + kr^2}{a^2(t)}, \qquad \eta^{\theta\theta} = -a^{-2}(t) r^{-2},$$

$$\eta^{\phi\phi} = -a^{-2}(t) r^{-2} \sin^{-2}\theta. \tag{9.7}$$

To solve the problem about the deformation of a reference system, it remains only to find the unknown function $a(t)$. Dynamics of the homogeneous and isotropic Universe may be described similarly to a model for ideal liquid with the density $\rho(t)$ and pressure $P(t)$, averaged over all the galaxies, their clusters and superclusters. Then, a hydrodynamic energy-momentum tensor for the matter is given by:

$$T_{\mu\nu} = P\eta_{\mu\nu} + (P + \rho c^2)U_\mu U_\nu, \tag{9.8}$$

where $U^t = 1$, $U^i = 0$. As seen from the calculations, for the metric (9.6) the following components of the Christoffel symbols (see definition (1.7)) are nonzero:

$$\Gamma^t_{ij} = \frac{a\dot{a}}{a^2}\eta_{ij}, \qquad \Gamma^i_{tj} = \frac{\dot{a}}{a^3}\delta^i_j, \qquad \Gamma^i_{jk} = \frac{1}{2}\eta^{il}\left(\frac{\partial\eta_{lj}}{\partial x^k} + \frac{\partial\eta_{lk}}{\partial x^j} - \frac{\partial\eta_{jk}}{\partial x^l}\right). \tag{9.9}$$

Taking into account Eq. (9.9) we obtain the following expression for the components of the Ricci tensor (see Eq. (1.16))

$$R_{tt} = \frac{3\ddot{a}}{a}, \qquad R_{ti} = 0, \qquad R_{ij} = \frac{1}{a^2}\left(a\ddot{a} + 2\dot{a}^2 + 2k\right)\eta_{ij}. \tag{9.10}$$

The time components of the Einstein's equation give

$$\frac{\ddot{a}}{a} = -\frac{4\pi G_N}{3}\left(\rho + \frac{3P}{c^2}\right) + \frac{\Lambda c^2}{3} \tag{9.11}$$

while the purely space components of that lead to the relation

$$\frac{\ddot{a}}{a} + \frac{2\dot{a}^2}{a^2} + \frac{2k}{a^2} = 4\pi G_N\left(\rho - \frac{P}{c^2}\right) + \Lambda c^2. \tag{9.12}$$

Omitting $\ddot{a}$ from Eqs. (9.11), (9.12) we arrive at the first order differential equation for $a(t)$

$$\frac{\dot{a}^2}{a^2} + \frac{k}{a^2} = \frac{8\pi G_N}{3}\rho + \frac{\Lambda c^2}{3}, \tag{9.13}$$

Eq. (9.11) describes changes in the expansion speed of the Universe under the effect of gravitation. From this equation it follows that gravitation is due not only to the matter density but also to its pressure in the combination $\rho c^2 + 3P$, referred to as the effective gravitating energy of the matter ρ^G_{mat}. It is obvious that in this case the cosmological term will result in antigravitation since its effective gravitating energy is negative. Note that the stationarity of the Universe in Einstein model is provided by the requirement that the total effective gravitating energy is zero

$$\rho^G_{tot} = \rho^G_{mat} + \rho^G_V = 0. \tag{9.14}$$

To find the function $a(t)$ and determine a cosmological model by this means, it is necessary to know for some t the values of density $\rho(t_0) = \rho_0$ as well as the cosmological constant $\Lambda(t_0) = \Lambda_0$. Usually, instead of ρ_0 one uses the quantity $\Omega = \rho_0/\rho_c$, where $\rho_c = 3H^2/(8\pi G_N)$ is a critic matter density in the Universe and H is Hubbl constant which value is defined by experiments.

Let us consider for the Universe the contributions into the matter density made by various components of the cosmological medium. Whatever the scale, the mass calculations for the Universe will reveal the deficiency of mass. Dynamically, a behavior of the galaxies themselves as well as (super)clusters is so as if they contain much more matter than it is really available in their apparent components, known as luminous matter or baryon matter (not forgetting the presence of electrons). The present-day value of the cosmic density (average over the whole observable world) for baryon material is determined by:

$$\Omega_B = \frac{\rho_B}{\rho_c} = 0.02 \pm 0.01 \tag{9.15}$$

Apart from the baryon matter, in the Universe there is a hidden mass (lately referred to as a dark matter). Two special types of dark matter exist: hot dark matter and cold dark matter. The cold dark matter (CDM) is composed of nonrelativistic objects and its present density is given by:

$$\Omega_D = \frac{\rho_D}{\rho_c} = 0.3 \pm 0.1 \tag{9.16}$$

The CDM is forming a vast invisible corona, or halo, around the stellar disk of the Milky Way. Similar dark halos seem to be present in all sufficiently massive isolated galaxies. The CDM is also contained in galactic clusters and superclusters. As with our Galaxy, it makes up about 90% and sometimes more of the total mass for all these systems. There is no emission or absorption of electromagnetic waves by the CDM that manifests itself exclusively through the created gravitation. Owing to its gravitational effect, the CDM was discovered in the thirties of the last century by F. Zwicky, who has studied the kinematics and dynamics of galactic superclustering in the Veronica' Hair constellation. Observation of the rotation curves (rotation speed v_c of the galactic matter as a function of the distance r to the center of this galaxy) makes it possible to determine the mass distribution of the galaxy over the radius with the use of a simple relation:

$$\frac{v_c^2}{r} = \frac{G_N M(r)}{r^2},$$

where $M(r)$ is a mass located inside an orbit with a radius r. Unfortunately, a nature of dark matter has not been conclusively established up to the present. A wide variety of the possibilities is considered: from weakly interacting massive elementary particles to massive (exceeding a mass of the Sun) black holes, etc. In this way masses of the candidates differ by full 60 orders of magnitude representing a real measure of the existing ambiguity in this problem.

The third component of the cosmological medium is a hot dark matter (HDM). The HDM comprises ultrarelativistic particles with masses equal to zero or of the order of $\sim$ eV. The density of this medium may be determined by the expression:

$$\Omega_R = \frac{\rho_R}{\rho_c} = 0.8 \times 10^{-5} \alpha \tag{9.17}$$

where the constant factor $1 < \alpha < 10 - 30$ includes the contribution of neutrinos, gravitons, other possible ultrarelativistic particles, and also fields of cosmological origin, that is additional to the adequately well measured contribution of relict photons. As seen, there is a considerable ambiguity in the estimate of this contribution.

Provided $\Lambda = 0$, from Eq. (9.13) it follows that a sign of k is determined by the sign of:

$$\rho - 3H^2/(8\pi G_N) = \rho - \rho_c.$$

In case $\rho < \rho_c$ we have $k > 0$, with $a(t)$ increasing infinitely to denote unbounded expansion of the reference frame and the matter. In this case the gravitational forces are too weak to slow down or stop the Universe expansion. In the process the density is varying from $\rho = \infty$ for $t = 0$ to $\rho \to 0$ for $t \to \infty$. Provided $\rho > \rho_c$, we have $k > 0$, i.e. the gravitational forces are sufficiently large and in some time the Universe expansion should be changed by compression. The density ρ is first falling from an infinitely large value (at $t = 0$) to a minimum; then it is growing again to infinity. The case with $k = 0$ is intermediate; as it takes place, unbounded expansion is proceeding. The sing of the difference $\rho - \rho_c$ is invariable in the process of the model evolution, while ρ and ρ_c are changing in time. For $k = 0$ the space volume is infinite at any instant of time. For $k > 0$ the space is also infinite in volume. The models, where the spaces are infinite, are called open. In case $k < 0$ the space is not bounded but has a finite volume $V = 2\pi^2 a^3(t)$. Such models are termed as closed.

Let us consider the principal features of a Theory of Hot Universe (THU). By this theory the whole observable stellar world was created at some initial time $t = 0$ from the initial singular state, with $\rho = \infty$ and $a = 0$, owing to the Big Bang. All the symmetries and all the laws determining further dynamics of the Universe have been programmed in this starting singularity in much the same way as DNA molecules predetermine the future of people. The explosion occurring at $t = 0$ results in a fire ball, with an infinitely high temperature and energy density, that begins to expand and cool down initiating the generation of all the constituent material for the present stars, planets and all the living matter. At the time of this explosion the system symmetry was so that all four interactions were unified, that is, the system was described by the symmetry group G_{UFT} corresponding to the "Unified Field Theory". It is a pity that nowadays we have no true information about such a theory. Because of this, we are forced to leave out of consideration a time interval equal to the Planck time 10^{-43} s [1]. The initial symmetry of a system has already passed the stage of its breaking, i.e. the gravitational interaction has separated from the interaction of the Grand Unification. By that time, the temperature was tremendous $\sim 10^{32}$ K. None of the components of the normal matter (molecules, atoms, atomic nuclei and even nucleons) could survive at such a high temperature. Instead, the matter scattered after the explosion was composed of various elementary particles. Apart from the well-known particles as quarks, leptons, carriers of electroweak and strong interactions, very heavy gauge bosons were also available, through which quarks could be transformed into leptons, and vice versa [2]. At that time the matter was representing a particular cocktail of quarks, leptons and bosons with an extremely high density. It is likely that the number of particles and antiparticles of each kind was identical. All these particles were created and annihilated continuously. The number of the created particles for each kind was exactly equal to that of the annihilated ones, i.e. all the particles were in a thermodynamic equilibrium. Once the temperature has fallen down to $kT < m_i c^2$, i-particles were out of the state of thermodynamic equilibrium,

[1] By that time the Universe size was equal to the Plank length L_P and, as we know, the gravitational effects may be safety neglected at the distances greater than L_P.

[2] When as an example one chooses the $SU(5)$ theory then X- and Y-bosons play the role of such bosons.

and the process of their burnup was initiated. This stage of the Universe evolution is called the radiation-dominance phase.

With a temperature of the expanding Universe falling down below 10^{28} K [1], spontaneous symmetry violation G_{GUT} took place

$$G_{GUT} \to SU(3)_c \times SU(2)_{EW} \times U(1)_Y,$$

(as an example of the electroweak interaction the WSG model is used). And heavy gauge X- and Y-bosons were out of the state of thermal equilibrium. In other words, as the energy was inadequate for their creation, the decay processes became dominant for these particles. At this stage we are forced to make some hypothetical assumptions.

A significant predominance of the matter over antimatter (the antimatter fraction comes to $< 10^{-4}$) is observed in the galactic cluster under study. A measure for such asymmetry of the Universe (baryon asymmetry of the Universe) is the value:

$$\delta = \frac{n_B - n_{\overline{B}}}{n_\gamma},$$

where $n_B, n_{\overline{B}}$ and n_γ are concentrations of baryons, antibaryons and relic photons respectively. According to the present-day measurements, δ has the value of the order of $\sim 6 \times 10^{-10}$. The quantity δ is the basic characteristic of the Universe, an explanation for its origin being one of the key problems of cosmology. Two approaches to the solution of this problem are possible. By the first approach it is supposed that the Universe was globally asymmetric from the very beginning, and the value of δ is given as the initial condition. The second approach seems to be more appropriate, being based on the assumption that at some stage the initial symmetry of the Universe has passed the violation phase. Such a violation should be caused by interactions breaking both charge (C) and space (P) symmetry (CP — noninvariant interactions). In the $SU(5)$ theory similar interactions are caused by the X- and Y-bosons. This leads to the situation when due to the decay of X-, Y-bosons the formation of quarks (ΔN_q) is somewhat greater than the formation of antiquarks. It should be noted that other sources for the occurrence of baryon asymmetry (baryogenesis) are also possible. For instance, the certain models of the GUT predict baryogenesis due to the formation of leptons (leptogenesis) under decays of superheavy neutral particles. At energies of $E \sim 300$ GeV ($t \sim 10^{-13}$ s) the symmetry breaking occurs down to the present-day level

$$SU(3)_c \times SU(2)_{EW} \times U(1)_Y \to SU(3)_c \times U(1)_{em},$$

that is, all the interactions became to be divided on four classes. For $t \approx 10^{-6}$ s ($T \sim 10^{13}$ K) the annihilation of quarks and antiquarks takes place. It is obvious that in the process the fraction of surviving quarks is equal to ΔN_q as before. Further phase transition occurs within approximately 10^{-5} s after the explosion, or at energies from 100 to 300 MeV characterized by Λ_{QCD}. It is associated with breaking of a chiral symmetry of strong interactions [2] and with quark confinement. So, at this stage free quarks, forming previously a part of the quark-gluon plasma, unify (forever?) to form hadrons (of course, the protons and the

[1] The law of changing the temperature for the early epoch of the Universe expansion (within bounds of few hundred years after the Big Bang) is written in the form $T = 10^{10}/\sqrt{t}$.

[2] If one neglects the quarks masses, the QCD Lagrangian (6.30) will be invariant with respect to the rotations

neutrons are the most interesting for future fate of the Universe). A few number of quarks ΔN_q had ensured the baryons abundance, which led to the formation of a minor admixture of the normal matter in the sea of light particles, a starting material for the formation of all future celestial bodies.

Let us consider the fate of leptons according to this scenario. With cooling and decreasing the reaction rates, there is a moment when the reactions involving particular particles cease to proceed, making these particles free, i.e. the Universe becomes transparent for them. In this manner, neutrinos get free (first ν_τ then ν_μ and ν_e) during a period of 10^{-2} — 10^2 s, i.e. the background cosmic neutrino radiation is initiated. At the same time, τ-leptons and muons are disappearing, whereas the electron-positron pairs are practically extinct, being transformed to photons. It is important that after getting free the particles still persist in "cooling", with the reduction of their energy due to the Universe expansion. This is caused by the fact that a free flying particle passes from one volume of the matter into the other removed from the first. Because of this, its energy with respect to the second volume is greater that the energy relative to the first volume, and so on so forth. Subsequently, in the Universe one can find only neutrinos and antineutrinos of all kinds, photons and a small amount of the normal matter in the form of plasma (mixture of baryons and electrons).

For further evolution of the Universe of particular importance are those physical processes which proceed in the matter subsequently forming the galaxies, stars, planets. At $T \approx$ few $\times 10^{11}$ K baryons exist in the form of protons and neutrons. These particles are rapidly interconverted under the effect of the surrounding primary particles ($e^{\pm}$, ν_e, $\overline{\nu}_e$):

$$n + e^+ \leftrightarrow p + \overline{\nu}_e, \qquad n + \nu_e \leftrightarrow p + e^-, \tag{9.18}$$

and thermodynamic equilibrium between the numbers of neutrons and protons is reached. Neutron-to-proton ratio within the unit volume at equilibrium is determined by the following expression:

$$\frac{N_n}{N_p} = \exp\left[-\frac{\Delta mc^2}{kT}\right],$$

where $\Delta m = m_n - m_p$. For t of the order of a few seconds the reactions (9.18) are practically terminated, and the ratio of the neutrons number and the total number of nucleons $N_n + N_p$ within the unit volume "is frozen" at the value:

$$\frac{N_n}{N_p + N_n} \approx 0.15.$$

With further decrease in T, in several minutes after the onset of expansion intensive nuclear fusion reactions of neutrons and protons result in the formation of ^4He. There is

in the quarks flavor space. In this case, thanks to the vector character of the interaction between the quarks and the gluons one may independently rotate the left-hand and the right-hand components of the quarks fields q_L, q_R. The transformations of such a kind are featured by eight independent parameters α_{aL} (see Eq. (6.21)) for left-hand particles and those α_{aR} for right-hand particles

$$q'_{L(R)} = (1 + i\alpha_{aL(R)}\lambda_a/2)q_{L(R)}. \tag{I}$$

If $\alpha_{aL} = \alpha_{aR}$, then the transformation (I) conserve the parity. From the mathematical point of view the invariance with respect to the transformation (I) means the chiral $SU(3)_L \times SU(3)_R$ symmetry (equal status of the left and the right) strong interaction. However, since the masses of the quarks are not equal to each other, the chiral symmetry has been violated in Nature.

no fusion of heavier elements as the nucleus ^4He fails to attach neutrons and other particles available. As a result, nearly all neutrons form a part of the nuclei of ^4He to give a relative content about 25% by mass of the whole matter. The remaining protons by mass account for about 75%. The content of other elements is negligible. Subsequently, the matter with such a composition is involved in forming celestial bodies, specifically stars of the first generation.

After the lapse of the first 5 minutes, all nuclear reactions in the Universe are terminated, the matter proceeds in expansion and cooling. But only after about 1 million years following the Big Bang comes the time for another critical stage in the evolution of the Universe. A temperature of plasma goes down to $T \approx 3000$ K, unification of electrons and protons takes place, and plasma is converted to a mixture of neutral atoms of hydrogen and helium. Prior to this situation, photon in its path should have encountered enormous numbers of free electrons capable of the effective photon scattering or absorption (just scattering with electrons is the dominant process for photons). Sudden disappearance of free electrons leads to the transparency of the Universe for photons.

Approximately at the same period the Universe passes from the phase of the radiation dominance to that of the matter dominance. This process is accompanied by the enhanced density fluctuations and hence the formation of large-scale structures. Due to the effect of gravitational compression, first-generation stars are created from the produced hydrogen and helium. Notice, that these stars also contain negligibly small admixture of deuterium and lithium. As the stars undergo condensation, the potential gravitational energy is released, with a temperature at the star center growing until the initiation of the thermonuclear reaction (burnup of hydrogen to form helium). The advent of a new energy source causes retardation of the compression process as the radiation exerts pressure on the outer layers of the star. Finally, the release rate of thermonuclear energy is increased so that the radiation pressure within any volume of the stellar material is in equilibrium with the effect of gravitational forces. With exhausted hydrogen at the center of a star it is compressed, leading to the temperature growth and burnup of helium. Since the process of helium transformation to hydrogen proceeds with a great release of energy, the stellar luminosity is increased. The energy release results in greater radiation pressure on the outer layers of a star leading to their expansion. Because of the expansion, gas is cooled making the light of a star more red. This expansion and reddening persists so long as the stellar diameter is increased by a factor of 200-300. In case of less massive stars such a star is known as a red giant, and otherwise — red supergiant. Future progress of the stellar evolution is mainly determined by its mass M.

Nuclear combustion of stars with $0.8M_{\odot} < M < 8M_{\odot}$ is terminated after the formation of carbon-oxygen core with a mass of $\sim 1M_{\odot}$. Once the whole shell surrounding this core is released, the star core is transformed to a "dead" star or so-called white dwarf. Massive stars ($M > 10M_{\odot}$) undergo their evolution path of combustion up to the formation of core of the most stable element ^{56}Fe. Release of the nuclear energy in such a core is impossible, increase in pressure is not compensating an increase of the gravitational forces with growing density, and slow quasi-static compression is changed by a sudden collapse — a supernova explosion takes place. Fast compression to a density that is close to the matter density within the atomic nuclei initiates release of a huge amount of energy, the major part of which is carried away by neutrinos. Following the explosion and shell release, the remainder is formed as a neutron star representing the second type of dead stars.

The stars with intermediate masses ($M \approx 8M_{\odot}$) are characterized by the formation of a degenerate carbon-oxygen core, whose mass is so enormous that it could not exist as a white dwarf any longer, being continuously compressed so long as the temperature and density growth results in explosive combustion of carbon and complete breaking of the whole star. Also, this breaking is observed as a supernova explosion leaving no remnant.

For stars with the greatest mass ($M > (40 - 50)M_{\odot}$)the collapse may proceed beyond the neutron star stage, developing further to form a relativistic object known as a black hole[1]. Such a collapse should be accompanied by neutrino radiation with extinction of the star that was extant before the collapse.

Explosions of supernovas were followed by synthesis of heavy elements, ejected subsequently into the interstellar space together with the elements synthesized in the process of prior evolution. All these factors have created the conditions for the formation of planets rings of dust and gasses around the stars similar to our Sun. Unification of these regions that followed, as well as their displacement under the effect of gravitational forces, has resulted in the formation of galaxies, galactic clusters and superclusters.

Now we turn our attention to two very important experiments providing support for the principal statements of THU.

The catalogue "Nebulae and Stellar Clusters" published by Ch. Mercier in 1781 includes 103 objects, the classification numbers of which are still used in modern practice. Even in the XVIIIth century it was clear that these distant objects are different. Some of them were obvious star clusters. But the others, about one third of the objects, were representing white nebulae with regular elliptical form, Andromeda Nebula being the most apparent (M31). Owing to the improvement of telescopes, thousands of the like nebulae have been revealed. By the end of the XIXth century it has been found that some of them including M31 have arms. At the same time, even with the use of the best telescopes available it was impossible to classify elliptical and spiral nebulae into the constituting stars. And their nature has been unknown until the advent of a 100-inch telescope at the Mount-Wilson laboratory. Using this telescope, 1923 E. Hubble succeeded in separation of the particular stars in the Andromeda Nebula. He has found that spiral arms of this nebula contain several bright variable stars, characterized by the same type of the luminosity alternation as was known for certain star classes of our Galaxy and referred to as cepheids (pulsating supergiants). The brightness of cepheids is changed regularly with a period of 1 — 100 days, the luminosity variation period being directly proportional to the absolute value of luminosity. The typical representative of this class is the Delta Cephei star. Its brightness is varying approximately by a factor of two with a period of 6 days. In this way cepheids in distant galaxies enable one to measure their distance R on the assumption that their apparent brightness is inversely proportional to R^2. When observing cepheids in the Andromeda Nebula, Hubble has found that the distance to this nebula is equal to 8.5×10^{21} cm ($\sim 1.9 \times 10^{22}$ cm by modern data), i.e. by the order of magnitude greater than the distance to most remote objects of our Galaxy. Thus, in 1923 it became obvious that the Andromeda Nebula and thousands of similar nebulae, are galaxies resembling ours and occupying the Universe in all directions up to enormous distances.

[1]Black holes are space-time ranges with so strong gravitational field that even light could not leave them. They were predicted by J. Mitchell in 1783.

By W. Slipher in 1910 — 1920 it was found that spectral lines of many nebulae are slightly shifted to the red or blue. These shifts were immediately interpreted as conditioned by Doppler effect, from whence it follows that the nebulae are moving in the direction of the Earth or in the opposite direction. To illustrate, it has been established that the Andromeda Nebula is approaching the Earth at a speed of about 100 km/s, while a more distant galactic cluster in the Virgo constellation is moving from the Earth at a speed of about 1000 km/s. Subsequent observations have demonstrated that, except of some nearby galaxies, all others are flown away our Galaxy. It looks as if the Universe were experiencing an explosion after which each galaxy is flying apart from any other galaxy. As a result of his astronomical observations, by 1931 Hubble has established the proportionality between the motion speed of a galaxy V and its distance R (Hubble red-shift law):

$$V = HR.$$

The Hubble constant H should be better called the Hubble parameter. It is constant only in the sense that the proportionality between the motion speed and distance is identical for all the galaxies at the present moment, i.e. H is independent of all the directions and distances. Nevertheless, H is variable in time with evolution of the Universe. At the matter-dominated stage it is decreased as $1/t$ but, as we see later, at the vacuum-dominated stage it turns a constant that is independent of time. So, the Hubble parameter is growing as time goes backwards, being infinite at the initial cosmological singularity. In this case the initial singularity is characterized by two infinities: infinite density and infinite Hubble parameter.

If the galaxies are flown away each other, it is probable that sometime they were positioned closer. More accurately, at constant speed the time required for any galactic pair to reach the present-day distance between them should be equal to the present day distance divided by their relative speed.

Provided the speed is proportional to the present-day distance between the galaxies, time should be the same for any pair of the galaxies, and, consequently, in the past they all should have been positioned closely at the same time. Using the present-day value of the Hubble parameter

$$H = (74 \pm 4)\frac{\text{km}}{\text{s} \cdot \text{Mpc}}$$

(1 parsec=3.0867×10^{18} cm), we obtain age of the Universe $(13.7 \pm 0.2)10^9$ years.

A conclusive demonstration of the fact that galaxies are flown away precisely so as indicated by their red shifts may be any other evidence in support of the established Universe age. Actually there is a great number of such evidences. Let us consider a few examples.

Meteorites age. The half-life period $T_{1/2}$ of most abundant uranium isotope ^{238}U is 4.5×10^9 years. It is found together with a more rare isotope ^{235}U ($T_{1/2} = 0.7 \times 10^9$ years), whose abundance comes to 0.7 of that for ^{238}U. A series of radioactive transformations in case of ^{238}U is terminating in the isotope ^{206}Pb, whereas in case of ^{235}U — in ^{207}Pb. Assuming that these uranium isotopes were formed at the same time and in the same amounts, we can use the above data to define the time during which these isotopes were decaying. Measurements of the ratio $m_{^{235}U}/m_{^{238}U}$ in meteorites with due regard for $T_{1/2}^{^{235}U}/T_{1/2}^{^{238}U}$ enable one to determine the age of meteorites as $(4.5 — 5) \times 10^9$ years.

Age of the Earth and the Moon. The age of the Earth is understood as a period that had elapsed since the time when the rock and the entire Earth were in the molten state. As suggested by Rutherford, the scientists are studying uranium and thorium ores, and also other kinds of rock containing these elements. When the rock is molten, lead formed due to the radioactive decay of uranium and thorium may be isolated from the praelements. But as soon as the rock solidifies, all the components of this rock become frozen, lead being found together with parent thorium and uranium decaying to this lead all the time. The values determined for the ratios $m_{^{207}Pb}/m_{^{235}U}$, $m_{^{206}Pb}/m_{^{238}U}$ and $m_{^{208}Pb}/m_{^{232}Th}$ in case of the eldest mineral monazite yield an age of 2.7×10^9 years.

The age of oceans may be determined by the method put forward by astronomer E. Halley. The method is based on estimation of the salinity of ocean water at the present time (3%) and the rate at which salt is carried out into the ocean by the rivers. According to this method, the age of oceans is estimated as 3×10^9 years.

As is known, the Moon was once integral with the Earth. Its velocity of recession from the Earth makes up 125 mm/year, and it is caused by the friction effects on ocean tides on the Earth under the influence of the Moon. As this takes place, the length of a lunar month is progressively increased. Considering these factors (as suggested by D. Darwin), the age of the Moon comes to 4×10^9 years.

Age of the Milky Way. Let us take stars of the Milky Way as the molecules of gas. Assume that at some initial instant of time their kinetic energies were different. Then in the course of time the energy distribution of the stars should finally reach its equilibrium due to the gravitational interaction, and a star in our Galaxy has the velocity inversely proportional to its mass. An astronomer F. Gondolach has examined the velocity profile of the stars close to the Sun and found that such an energy equidistribution is realized by 98%. On the basis of this fact he came to the conclusion that the age of the Milky Way ranges from 2 to 5 milliard years.

These results are associated with the age estimates for stars and star clusters. Since there is no direct relation between the above-mentioned phenomena used for the age estimation of the objects in the Universe, and the red shifts of distant galaxies, such coincidence is a convincing evidence for reliability of the age estimates (close to ideal values) derived from the Hubble parameter.

Let us proceed to the second experiment providing support for a theory of the Big Bang. Its history has very nearly the same plot as the fascinating history of the electrons diffraction discovery[1]. The same beginning — routine industrial experiment, the same scenery — premises of the Bell Telephone company, the same long period of doubt in the interpretation of the obtained (possibly by chance?) results, and the same happy ending — awarded Nobel prize in Physics.

[1]In 1922 D. Dawisson, an employee of the Bell Telephone, was studying electron scattering with the energy $E = 54$ eV from nickel crystals. The preliminary results were in a complete agreement with the predictions of the classical physics. During the experiments a nickel target was subjected to the high-temperature annealing to remove the forming oxide. As it was found later, this process has resulted in the formation of a series of diffraction gratings (Bragg planes) within the target bulk, with a period on the same order as the de Broglie electron wavelength ($\lambda = h/p = h/\sqrt{2mE} = 1.65 \times 10^{-8}$ cm). The experiments with that target became to give the results which were anomalous from the viewpoint of a corpuscular character of electron. Only in 1926, during his visits to Oxford and Gettingen, Davisson found the information about the works of De Broglie and inferred that the anomalies observed were associated with electron diffraction. On his arrival to America, Davisson together with A. Germer has provided support for de Broglie's hypothesis performing a series of the experiments, awarded with Nobel prize in 1937 by the Sweden Academy of Sciences.

In 1964 the Bell Telephone laboratory became possessor of an extraordinary antenna designed for the communication through the Echo satellite. A 20-feet horn reflector with an exceptionally low noise level made this antenna a promising astronomical instrument. Astronomers A. Penzias and W. Wilson have decided to use this antenna for the intensity measurement of radio waves emitted by our Galaxy at high galactic latitudes, i.e. beyond the Milky Way plane. They were surprised to find (in summer of 1964) that at a wavelength of 7.35 cm (4080 MHz) they receive an appreciable microwave noise. Moreover, the measurements have demonstrated that this "statistical background" was invariable both in time and with direction. Such a radiation could not be produced by our Galaxy, since otherwise the galaxy 31 should have the same strong radiation at this wavelength and such a microwave noise should have to be detected already. Besides, the fact that the observed radiation was invariable with the direction pointed to a much larger volume of origination: origin of the radiation was in the giant Universe rather than in the Milky Way. Penzias and Wilson have found that an equivalent temperature of the detected radiation was approximately equal to 3.5 K, or to be precise, its interval was from 2.5 to 4.5 K.

In March 1965 E. Peebles, physicist-theoretic from the University of Princeton, conducted studies of radiation that should have to be present in the early Universe. He has demonstrated if there were no intense radiation within a few minutes of the Universe existence, nuclear reactions would be extremely fast to make the majority of the available hydrogen "boiling" to form heavier elements, and that is at variance with the experiment. A fast process of nuclei forming could be prevented only on condition that the Universe were filled with radiation having an extremely high temperature to crash nuclei into parts at the same rate as they were formed. This suggests that the present-day Universe must be filled with radiation but at a considerably lower equivalent temperature compared to the temperature of the first minutes of its existence. By the estimates of Peebles, the present-day temperature of this radiation should be at least 10 K. Later this value was lowered owing to the calculations of Peebles, Dicke, Roll and Wilkinson. Exchanging their results, both theorists and experimenters have decided to publish two letters in "Astrophysical Journal" at the same time: to present the evidence found by the experimenters and to provide a cosmological interpretation for this evidence by theorists. However, Penzias and Wilson were still doubtful, and the title of their work was more than modest: "Measurement of an Excessive Antenna Temperature at a Frequency of 4080 MHz". They simply declared, " the measured value of the effective zenithal temperature of noise ... was by 3.5 K higher than expected". No reference to cosmology was made, except of the phrase that "possible explanation of the observed excessive noise temperature may be found in this issue in the accompanying letter by Peebles, Dicke, Roll and Wilkinson".

Nevertheless, a character of the radiation found by Penzias and Wilson and even the fact of its occurrence was open to question as long as (several month later) the group of R. Dicke made independent measurements at a wavelength of 3 cm. These measurements provided support for the results of Penzias and Wilson in full. As in the case of 7.35 cm wavelength, the radiation intensity was the same in all directions, invariable in time and in line with the predicted intensity of radiation for a blackbody with a temperature of about 3 Ł.

The microwave radiation discovered by Penzias and Wilson was represented by the photons, remaining from the times of the early Universe when electrons, protons and neutrons were combined to form helium and hydrogen atoms [1]. With sudden disappearance of free electrons the thermal contact between radiation and the matter was being disturbed, and, as a result, this radiation became to expand freely. By that moment, the radiation field energy at different wavelengths was conditioned by thermal equilibrium, and hence it may be determined using the Planck formula for a blackbody with a temperature equal to that of the matter $\sim 3 \times 10^3$ Ł. At this stage there were no generation or annihilation of single photons, the average distance between them was increasing with the Universe size. In this case the wavelengths of all individual photons were also growing in proportion to the Universe size. Because of this, the distance between photons was left equal to the average wavelength, following the pattern for radiation of a blackbody. Quantification of these arguments enables demonstration of the fact that Planck's formula of a blackbody still holds for the description of radiation filling the Universe in the process of its expansion, despite a lack of the thermal equilibrium between this body and the matter. The only expansion effect is an increase in the average wavelength of photons, proportional to the Universe size. A temperature of the blackbody equilibrium radiation is inversely proportional to the average wavelength and hence is decreasing in the process of the Universe expansion, inversely to its size.

By more recent and accurate measurements, a temperature for the relict radiation is determined as $T = 2.736 \pm 0.003$ K. This means, in turn, that each cubic centimeter of the Universe contains about $n_\gamma \approx 400$ of relict photons. As it turned out, detection of the relict background is the most important cosmological discovery since the red shift has been revealed. In 1978 Penzias and Wilson were awarded the Nobel prize for their "discovery of a background microwave radiation from outer space".

Despite obvious advantages of the Theory of Hot Universe, a number of problems still remain to be unsolved. Within this theory, most unified theories of elementary particles result in the cosmological inferences inconsistent with the observations. By the standard model, for instance, at the earliest stages of hot Universe there should be the production of numerous ultraheavy particles possessing a magnetic charge, so-called magnetic monopoles. By the present moment, the matter density due to these particles should have been by 15-orders of magnitude higher than the observable density of the matter in the Universe. This theory fails to provide an adequate answer for the following questions: what was before the Big Bang?; why Riemann geometry describing the space properties of our Universe with an enormous accuracy is so close to the Euclidean geometry of a flat world?; why the observable part of the Universe is, on the average, homogeneous?; what is the origin for the initial inhomogeneities required for the creation of galaxies in this homogeneous world?; why different portions of the Universe formed independently are so alike at the present time?; and, finally, what is the reason for simultaneous expansion of all the parts constituting the infinite flat or open Universe?. On condition that the Universe is closed, it is not clear what provided resorts for its survival for $\sim 10^{10}$ years, in spite of the fact that a typical lifetime of the closed hot Universe should not be considerably greater than Planck time t_P.

The majority of these problems may be solved within the scope of so-called Inflation Models of the Universe (IMU). The general feature of a variety of these models is the stage of exponential (or quasi-exponential) expansion of the Universe which was in a vacuum-like

[1] By the reason this radiation is often called the relict radiation.

state with high energy density. This stage is termed as the inflation phase. After this phase, the vacuum-like state disintegrates and the created particles interact with each other leading to the thermodynamic equilibrium, while the following evolution proceeds according to the THU.

A cosmic vacuum in this case is the same vacuum as in microphysics, where it represents the lowest energy state of quantum fields. This is the same vacuum, where the interactions of elementary particles take place and whose manifestations may be observed in direct experiments. According to quantum mechanics, the lowest energy of quantum oscillator is nonzero and equals $\hbar\omega/2$. These "zero oscillations" result in a nonzero energy at the lowest energy state of quantum fields. But a quantum field theory fails to provide real calculations for the total energy density associated with zero oscillations. Considering an assembly of quantum oscillators as a model of physical fields and taking a sum for the energy of zero oscillations over all the frequencies possible up to the infinity, we obtain an infinite energy of vacuum as a result. To eliminate these infinities, one can set an upper limit to the frequency range at some value, i.e. the energy cutoff is used. It is possible to assume that the cutoff frequency conforms to the Planck energy M_P, so that $\hbar\omega \sim M_P c^2$. Such a choice of the cutoff frequency is attested by the fact that for energies in excess of the Planck's the standard notions of physics, the concept of frequency including, lose their meaning.

According to the simplest variants of IMU, at the initial vacuum-like state there is a space filled with a rather homogeneous slowly varying scalar Higgs field φ, already encountered by us when studying a model for the electroweak interactions by Weinberg, Salam, Glashow. Expansion of the Universe decelerates the process of varying the field φ. In consequence, the energy density $V(\varphi) = m\varphi^2/2$ is remaining nearly constant for a long period of time: compared to the density of the normal matter, it is hardly decreasing with the Universe expansion. In the end this results in the exponential growth of the Universe regions filled with a large field $\varphi \geq M_P$

$$a(t) \sim \exp\left[t\sqrt{\frac{8\pi V(\varphi)}{3M_P^2}}\right].$$

In typical models the inflation phase is not long $\sim 10^{-35}$ s. But during this period the Universe has a chance to increase its size by 10^{10^5} — $10^{10^{10}}$ (exact numbers depend on the choice of a specific model for elementary particles and on the mechanism responsible for the Universe inflation).

As soon as φ becomes sufficiently low ($\varphi \leq M_P$), the expansion speed and the associated decelerating force affecting φ decrease. Rapid oscillations of the field begin close to a minimum of its potential energy $V(\varphi)$. In the process the field φ generates pairs of elementary particles, donating its energy them and hence heating the Universe. Subsequent to the inflation phase, the space geometry within the inflation region of the Universe becomes practically indistinguishable from the Euclidean geometry of flat world, in analogy to the geometric properties of the balloon surface more and more resembling those of a plane as the balloon is blown. Due to inflation of the Universe, most of the monopoles and other inhomogeneities are beyond its presently observable part of the size of $a_0 \sim 10^{28}$ cm. Because of this, the problems associated with the observable Universe homogeneity and small numbers of monopoles are solved at a time. As the entire observable Universe has been formed owing to the inflation of a single region negligibly small in size, no surprise that the features of different spaced apart regions of the observable world are, on average, the same.

The majority of extensions of the SM have several Higgs fields rather than only a single one (φ_i, where $i = 1, 2, ...N$). Provided such models with an extended Higgs sector present a true theory of electroweak interactions, an inflation theory predicts the following scenario. The field fluctuations φ_i generated during the inflation process will result in the creation of exponentially large regions, occupied by various fields φ_i corresponding to all possible energy minima $V(\varphi_i)$. Quantum fluctuations in the regions with extremely great field values φ_i may be responsible for the formation of inflation regions with other types of the space-time compactification. As a result, the Universe is subdivided into N exponentially large regions, where the space-time dimensions, compactification type and properties of elementary particles may be different (domain structure of the Universe). According to the inflation theory, these regions are separated apart at a distance greater than the size of the observable Universe part by many orders of magnitude. But if the principal statement of the inflation theory concerning the creation of the Universe from vacuum is true, the problem is, what is the fate of this vacuum. It must be present in the contemporary Universe as well. And its density should be appreciably lower than the initial one.

A series of experiments conducted by two big research Collaborations of astronomers [1] in 1998 — 1999 provided support for the vacuum occurrence in the Universe. As it turned out, vacuum (cosmic vacuum is often referred to as a dark energy) predominates in the Universe, with the energy density making it superior over all the "ordinary" forms of cosmic matter taken together

$$\Omega_V = \frac{\rho_V}{\rho_c} = 0.73 \pm 0.01. \tag{9.19}$$

This means that 73% of the total energy of the world is due to vacuum, 23% — due to cold dark matter, about 4% — due to baryon matter, and less than 0.03% — due to radiation. The discovery was based on a study of distant supernovae outbursts. Owing to their exceptional brightness, supernovae may be observed at enormous, really cosmological distances. The data used are associated with the supernovae of the certain type, conventionally considered as "standard candles". Their self-radiant exitance lies within fairly narrow limits, making it possible to trace the relationship between the visually registered brightness of the sources and their distance. Now these supernovae occupy in cosmology the place previously taken up by less bright Hubble's cepheids. The observation presents difficulties, as supernovae are few in number. On the average, a typical galaxy exhibits approximately one supernova outburst some 100 years, the outburst itself being very short: a few months or even weeks. Two quantities are directly measured during observations of supernovae, namely the energy coming from supernova to the Earth in a unit time per unit square J (visual luminosity), and the red shift ς. The red shift caused by the distance to the observable galaxy is given by the formula:

$$\varsigma = \frac{\lambda - \lambda_0}{\lambda_0} = \frac{V}{c}\sqrt{1 - \frac{V^2}{c^2}},$$

[1] A. G. Riess was a leader of one group while the other group is guided by S. Perlmutter.

where λ is a wavelength of a light detected, λ_0 is a wavelength of a light emitted by a source, and V is a source speed. According to Hubble law, short distances are associated with low ς, and long distances — with great ς. Fig. 45 shows in the coordinate plane (J, ς) two theoretical curves describing accelerating and decelerating expansions of the Universe.

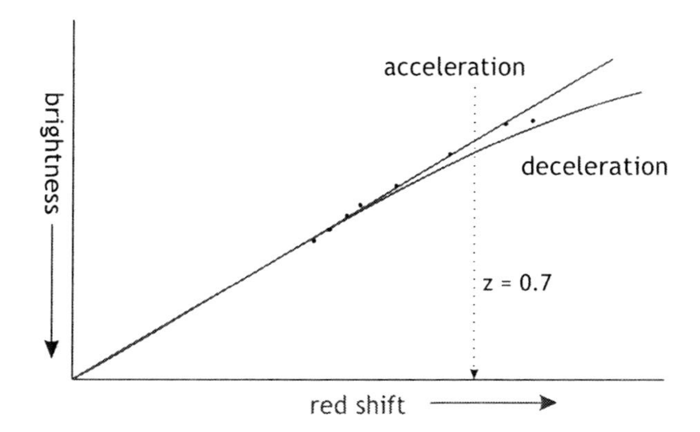

Figure 45. Two theoretical curves describing accelerating and decelerating expansions of the Universe.

At short distances these curves are coincident, whereas at long distances the curve representing an accelerating expansion goes above the curve for decelerating expansion. To determine a character of the Universe expansion, one should observe supernovae to the distances, where the theoretical curves shown in Fig. 45 are moving apart. The observations indicate that experimental points are located at the upper curve, i.e. cosmological expansion proceeds with acceleration. On the other hand, acceleration may be exclusively due to antigravitation whose origin is the cosmological term Λ in Einstein equations. Immediately the question arises: "What cosmological substance is described by the term Λ?"

Already at the outset of the relativistic quantum theory G. Gamow has suggested that the Dirac vacuum should manifest itself through gravitation. It is presently universally acknowledged that cosmic vacuum is described by the cosmological constant. A density of vacuum is related to the value of this constant by the following relation:

$$\rho_V = \frac{\Lambda}{8\pi G_N}.$$

From this point of view, we can relate evolution of the Universe to its genesis: expansion of the matter stems from antigravitation of cosmological vacuum, the matter per se appearing as a result of quantum fluctuations of the same vacuum. As expected, vacuum exhibits rather unusual properties. A state equation for vacuum, i.e. the relationship between the pressure and density, is derived in a quantum field theory and has the form:

$$P_V = -\rho_V c^2. \tag{9.20}$$

Eq. (9.20) is relativistically invariant, that is, is valid in any frame of reference. Also, from Eq. (9.20) it follows that the energy density of vacuum is invariable with expansion of the Universe. Indeed, under expansion of the Universe the energy density should have to be decreased as:

$$d(\rho c^2) = -\rho c^2 dV,$$

where dV is an increase of a volume element. But this decrease is compensated for by a negative work done in this case by the expanding volume

$$PdV = -\rho c^2 dV.$$

Since the effective gravitating energy of vacuum

$$\rho_V^G = \rho_V c^2 + 3P_V = -2\rho_V c^2$$

is negative for a positive density, this vacuum is responsible for antigravitation. Thus, in any reference system the cosmological vacuum has a density invariable in time and space. In principle, in its properties a vacuum differs from all other forms of cosmic energy, whose density is inhomogeneous in space, decreases in time with cosmological expansion, and may be different in various reference systems. In any arbitrary reference system a vacuum seems absolutely identical, every system being co-moving. In other words, two frames of reference may be moving with respect to each other at any speed, but a vacuum will be co-moving with each of them.

The vacuum has another unique property: influencing all the natural bodies through antigravitation, the vacuum is immune to their gravitational effects. Consequently, it is not governed by the third law by Newton. In terms of dynamic observables, the vacuum has a negative active gravitational mass, while its passive gravitational and inertial masses are zero. At the same time, all the above is true for weak fields only. In intense fields one can observe a number of effects such as vacuum polarization, *particle+antiparticle* pair production, etc.

Let us consider Fig. 45 again. We observe a star as it has been in the process of light emission. Approximately at the epoch of $t_V = (6 - 8)10^9$ years acceleration $\ddot{a}$ has changed its sign. Because of this, in case we are observing a supernova at distances on the order of $t_V c$, the corresponding point should be located at the lower curve. At the same time, to travel such a distance, the speed of the galaxy should be very high and hence the red shift should be great too. As seen from the calculations, this takes place at $\varsigma > 0.7$. Thus the theory predicts that for great ς there are observation points associated with the lower curve too. Actually, at the top of the graph shown in Fig. 45 there is such a point that has evidently "descended" from the upper curve.

For the present-day value of the Hubble constant, the obtained densities of the components of a cosmological medium are consistent with open and flat as well as closed cosmological models. A flat model is associated with:

$$\Omega = \Omega_V + \Omega_D + \Omega_B + \Omega_R = 1,$$

In open model this sum of relative densities is below unity and in closed — over unity.

It should be noted that some scientists hold another viewpoint of the acceleration source for the Universe expansion. They supposed that the cosmological acceleration is produced by a quintessence, so far unknown and totally hypothetical, rather than vacuum. This source is understood as a special form of cosmic energy described by the state equation $P = q\rho c^2$, where q is a constant parameter with the values falling within the interval $-1 < q < -1/3$. Since the effective gravitating density is negative for this energy type, quintessence is creating antigravitation too. The idea of quintessence is deeply rooted in ancient times. According to Greek philosophers, quintessence represents the fifth element that is complementary to the earth, water, air and fire and forms the basis for celestial bodies.

Now we consider a modern model of the Universe in the light of new discoveries. Let us begin with the dynamics of cosmological expansion. Eq. (9.13) is rewritten in the following form:

$$\frac{1}{2}\dot{a}^2 = C_V^{-2}a^2 + C_D a^{-1} + C_B a^{-1} + \frac{1}{2}C_R^2 a^{-2} - \frac{1}{2}k, \tag{9.21}$$

where constants C (Friedmann's integrals) are given by the common relation

$$C = \left[\left(\frac{1+3w}{2}\right)^2 \frac{8\pi G_N \rho a^{3(1+w)}}{3}\right]^{1/(1+3w)}, \tag{9.22}$$

$w = P/(\rho c^2)$ and for vacuum $w = -1$, for cold dark matter and baryon matter $w = 0$, for radiation $w = 1/3$. Knowing the densities for some a, one can find the constants C. In this way these integrals are used to set the initial conditions for the Friedmann theory. As seen from (9.22), all integrals C have the dimensions of length. Their numerical values are close in the order of magnitude and amount to 10^{26} — 10^{28} cm. It is obvious that the left-hand side of Eq. (9.21) contains the kinetic energy attributed to the unit mass. Therefore, a sum of the first four terms taken with an opposite sign is nothing else but the potential energy. Taking into account that the first integral in equations of motion represents energy, the value $-1/(2k)$ in (9.21) should be identified with the total mechanical energy of a particle. The total energy may be positive, negative or equal to zero, the associated motion types usually being referred to as hyperbolic, elliptic and parabolic, respectively. The sign of the space curvature in (9.21) k is opposite to that of the total energy . So, we have a one-to-one relation between the curvature of the three-dimensional space and dynamic type of cosmological expansion.

From Eq. (9.21) it follows that a dynamic role of vacuum is different under evolution of the Universe. At the early expansion stages of the Universe, the effect of vacuum is insignificant as for small $a(t)$ ($a \to 0$) the vacuum term on the right-hand side is less than four others ($\rho_V a^2 \to 0$). By virtue of the fact that gravitation of the normal matter (understood as nonvacuum components of a cosmic medium) leads to a negative acceleration, $\ddot{a} < 0$, cosmological expansion in this case will be realized with deceleration. The role of vacuum becomes significant for big times. As follows from (9.21), sooner or later there comes an instant for the dynamic vacuum domination, that is,

$$C_V^{-2}a^2 > C_D a^{-1} + C_B a^{-1} + \frac{1}{2}C_R^2 a^{-2}.$$

Now (formally at $a \to \infty$) we can neglect gravitation of the nonvacuum components, and acceleration $\ddot{a}$ turns out to be positive. A solution for (9.21) is easily found and takes the form:

$$a(t) = C_V f(t), \tag{9.23}$$

where

$$f(t) = \sinh\left(\frac{t}{C_V}\right), \qquad f(t) = \exp\left(\frac{t}{C_V}\right), \qquad f(t) = \cosh\left(\frac{t}{C_V}\right),$$

for $k = -1, 0, +1$, respectively. In this manner for all three variants of Friedmann model a solution of Eq. (9.23) describes the cosmological expansion accelerating in time. In the long times limit the expansion varies exponentially for all the three variants. A change from deceleration to acceleration, and transition to the vacuum domination in the dynamics of cosmological expansion is associated with zero total gravitating energy $\rho^G_{tot} = 0$. However, in Friedmann model, as distinct from the Einstein static model, this is possible only for a single time $t = t_V$ when $\ddot{a} = 0$. Fig. 46 shows the density of the cosmological components as a function of time.

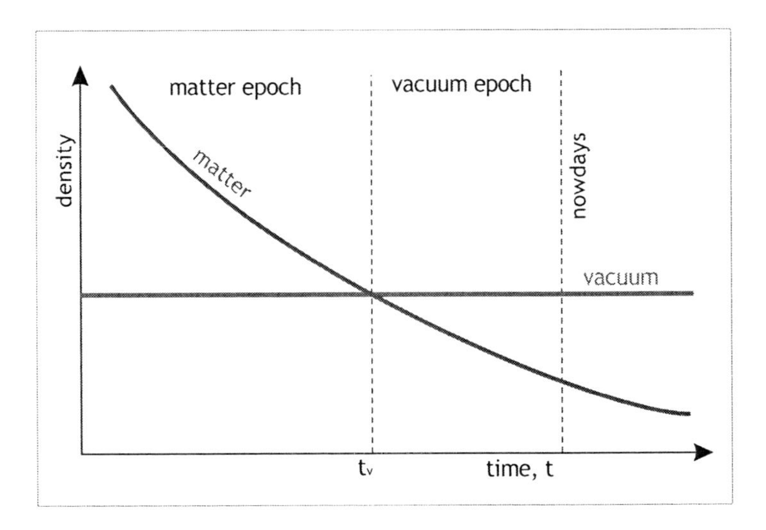

Figure 46. The density of the cosmological components as a function of time.

Integrating Eq. (9.21), we derive a relation that holds true for any instant of time:

$$\int \left[C_V^{-2}a^2 + 2C_D a^{-1} + 2C_B a^{-1} + C_R^2 a^{-2} - k\right]^{-1/2} da = t, \tag{9.24}$$

where we have accepted $a(0) = 0$ and taken into account that the sign plus corresponds to the cosmic medium expansion. Based on the derived solution, it is possible to find, for instance, time t_V as well as an age of the Universe t_0.

It is of interest that in both limiting cases, $a \to 0$ and $a \to \infty$, the dynamics of cosmological expansion is independent of a sign of the total energy or sign of the space curvature k as it follows from Eq. (9.24). For all these variants the expansion begins in the parabolic mode; then, during a finite period of time, possible difference between the expansion dynamics and this mode may be exhibited, and finally the dynamics is characterized by the parabolic mode again, maintaining this type of motion for infinitely long time.

Assuming for $a(t)$ the exponential time dependence that is associated with the dynamic vacuum domination, we can change the Friedmann solution by the solution of de Sitter.

Substituting $a(t) = C_V \exp[t/C_V]$ into Eq. (9.5) and changing coordinates t, θ by τ, z, we come to one of the possible expressions for the de Sitter interval:

$$ds^2 = (1 - z\sqrt{K})^2 d\tau^2 - z^2 d\Omega^2 - (1 - z\sqrt{K})^{-2} dz^2, \qquad (9.25)$$

where a four-dimensional world curvature K is defined by the Friedmann's integral for vacuum $K = C_V^{-2}$. It should be emphasized that for any form of writing de Sitter's interval the differential geometry of the four-dimensional world is identical to the geometry in case of (9.25): this is a four-dimensional space-time geometry of the constant and positive curvature K.

The Friedmann's integral for vacuum C_V (let us call it by Friedmann constant) plays in cosmology the same role as Planck constant in the microworld. Indeed, the observable size of the Universe (Friedmann length)[1] is simply equal to $C_V \sim 10^{28}$ cm. An age t_0 (Friedmann time) and a mass M_U (Friedmann mass) of the Universe are determined by the relations:

$$t_0 = C_V/c \sim 10^9 \text{ years}, \qquad M_U = \rho_V C_V^3 \sim 10^{55} \text{ g}.$$

In the cosmological solution for (9.23) the Hubble constant is $\dot{a}/a = H \sim C_V^{-1}$ practically for any k shortly after the transition to vacuum domination. At the stage with complete predominance of vacuum H is independent of time, being determined by the value of the Friedmann constant only. Nonzero curvature of the real four-dimensional world Ł representing nearly the major constant of the Nature is also associated with the Friedmann constant

$$Ł = C_V^{-2} \sim 10^{-56} \text{ cm}^{-2}.$$

Thus, in the expanding Universe a change from domination of the normal matter to the vacuum-dominated stage points to gradual vanishing of the dynamics in the four-dimensional space-time. Actually, the stronger acceleration of the galaxies, the lower their distribution density and hence the weaker their influence (through gravitation) on the properties of space-time. And the influence of vacuum exerted by its antigravitation becomes more and more prevalent. The space-time structure of the world freezes, ceases to vary in time, and remains frozen forever. All the processes, events, transformations of the matter have actually no effect on metrics of the four-dimensional world now, and this effect will be still greater attenuated in the future. With increasing acceleration of the cosmological expansion under the effect of antigravitating vacuum, our four-dimensional world is nearing absolute statics, invariability and rest. These are the most important dynamic and geometric effects of vacuum in cosmology. Obviously, such new conditions make the conventional problem, what is the type of a real cosmological model (open, closed or flat), or what equals k, less acute and principal than it has been previously. It is clear that selection of a certain model out of the three available variants for the three-dimensional geometry is not critical when solving the problem, what awaits the world: could its expansion be perpetual or this expansion would be changed by compression. The solution for (9.19) demonstrates that cosmological expansion is infinitely long for all the above-mentioned models. The models are distinguished by different approaches to separation of the three-dimensional spaces

[1] As a light passes finite distances during finite time, there is a fundamental limit for observations range. It means one may not observe of that lies far beyond the distance a light could travel during the Universe existence time t_0. Notice, that a visible range of modern telescopes has just the same order ct_0.

from the same unified four-dimensional space-time with the constant and positive curvature K.

A theory of Friedmann, where the dynamics is given by Eq. (9.24) and geometry is determined by the interval of (9.4), together with the observable cosmic densities and the Hubble constant represent a present-day standard cosmological model (SCM). By this model, antigravitation may be caused both by vacuum and quintessence. However, Occam's razor[1] is not in favor of such a new degree of freedom as quintessence. Moreover, inclusion of vacuum into a model for the Universe inflation has been proved reasonable. Giving preference to a hypothesis of cosmic vacuum, one can state that evolution of the Universe was initiated at dynamic domination of vacuum. It is evident that the initial and the present-day densities of vacuum are different values, the first value being much higher than the second.

May one state that the modern cosmological model, the standard cosmological model (SCM), gives the answers to all questions? Unfortunately, experimental facts do not all find explanations within the SCM. For example, the confused ambiguity concerning the initial singularity is being conserved till now. There are no theoretical foundations allowing to calculate the present day vacuum density. Moreover, the relict radiation, the detection of which has played the important role in the Big Bang model formation, seriously puzzles theorist-cosmologists. Recently the data on the background microwave radiation fluctuations was obtained with the help of automatic cosmic station MAP (Microwave Anisotropy Probe). Collected by MAP information allows to build the most detail map of small temperature fluctuations in microwave radiation distribution within the Universe. At present the microwave radiation temperature comprises nearly 2.73 K differing by only million part of degree in different sites of heavenly sphere. It turned out that on the heavenly sphere the "cold" and "warm" regions found by the cosmic telescope are located not by accidental manner, as it would wait, but by ordered one (see Fig. 47). So, the relict radiation map has the symmetry axis which penetrates all the observed Universe. The SCM fails to account for the existence of this phenomenon and, as a result, the axis derived the semimystical name "Axis of Harm".

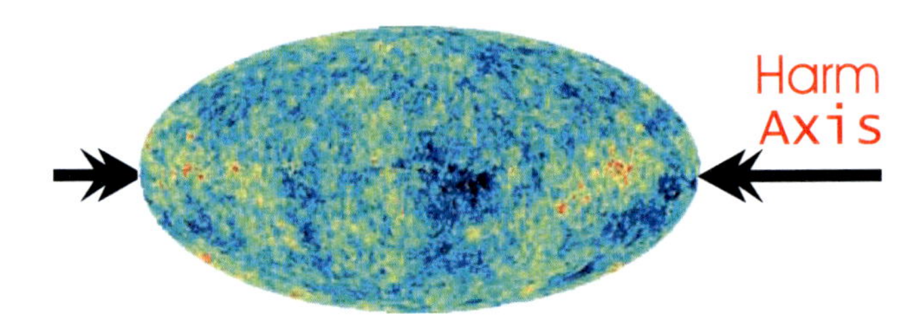

Figure 47. The Harm Axis.

In the nearest future the increase of the observations precision both in γ-astronomy and in neutrino astronomy will lead to the enhancement of number of reliable experiments. The time will show what kind of reconstruction is needed for the SCM to explain these new experiments. However, one should not forget that a great number of existing experimental

[1]According to the statement formulated by English philosopher W. Occam, "the essences should not be multiplied without necessity".

facts have found their solutions within this model. One should also remember that the SCM will be considered as the completed theory only after the consecutive gravitation theory will have been built. Nowadays it may only state that we have established the SCM basic outlines and behind them we could make out a pre-image of the true model of the Universe evolution.

9.2. Neutrino Astronomy

The data obtained during investigation of solar neutrino under SNO in 2002 directly supported the transitions ν_e in ν_μ and ν_τ. This result presents a conclusive evidence of the validity of a neutrino oscillation hypothesis used for explanation of the solar neutrino deficiency, measured by Homestake, SAGE, GALLEX, GNO Super-Kamiokande neutrino telescopes (NT). And the problem of solar neutrino stimulating physical studies of neutrino for a period of 35 years was successfully solved. The same year the results of SNO were supported by the experiments with a strictly controlled beam of reactor antineutrinos. All this has made it possible to conclude that neutrino has a mass, and mixing is a feature of the lepton sector. A year of 2002 was unanimously declared an "annus mirabilis" in physics of solar neutrinos. This triumph was the first great success of such a recently developed branch of science as neutrino astronomy (NA).

The history of NA is dated from 1967 when the first Homestake NT was put into operation. NA presents studies of cosmic objects on the basis of incoming neutrino fluxes. Without NA, electromagnetic waves (visible light, IR and UV radiation, short wavelength radio waves) were the only accessible type of radiation falling down on the Earth from the outer space. Note that electromagnetic waves are emitted only from the surface layer of celestial bodies. In the process of motion from their source to the terrestrial observer, electromagnetic waves interact with cosmic rays, particles of the Earth atmosphere and so on, loosing the major part of information about their praobjects. Following the steps of the conventional gamma-astronomy that covers a very wide range of the wavelengths $\lambda = (1 \div 10^{-12})$ m, NA shows tendency to expansion of its potential to cover the whole energy range of neutrinos occurring in the Universe.

By their generation sources and energy range these neutrinos may be subdivided into cosmological (relict), stellar and high-energy cosmic neutrinos.

A temperature of gas comprising relict neutrinos is presently equal to ~ 1.9 Ł, and the average energy amounts to 5×10^{-4} eV. With the use of modern experimental techniques detection of relict neutrinos is very difficult due to extremely small cross-sections of their interaction with the matter.

Two classes are distinguished for the sources of stellar neutrinos. The first class relates to the quiescent (static) stars, similar to our Sun, producing neutrinos in the process of nuclear fusion reactions which provide the observable luminosity. The energy of these neutrinos ranges from a few fractions to several dozens of MeV. The second class is represented by the collapsed stars, i.e. those collapsing to a neutron star or to a black hole. At the final stage of stellar cores evolution their density is increased up to 10^{10} — 10^{15} g/cm^3, and their temperature — up to 10^{10} — 10^{12} Ł. The principal mechanism responsible for the energy loss in these conditions is emission of neutrinos through the reactions:

$$e^+ + e^- \to \nu_l + \bar{\nu}_l, \qquad e^- + p \to n + \nu_e, \qquad e^+ + n \to p + \bar{\nu}_e.$$

The energy carried away by neutrinos (all types of neutrinos are radiated) may amount to tens of the star mass percentage. The duration of neutrino radiation comes to 10 — 20 s, and the average energy is 10 — 12 MeV. The outbursts of the supernova SN 1972E, SN 1987A, SN 1993J may be taken as an example of such radiation. Obviously, at the modern stage of cosmology a registration of supernova outbursts is the principal goal.

Cosmic neutrinos are defined as those produced by cosmic rays. The energies especially convenient to search for the local sources of cosmic neutrinos are tens of GeV and above. The lower limit of this range is determined by the requirement of a smallness of an angle between the momentum directions of an incident neutrino and outgoing particle (e.g., muon) during the reaction used for neutrino registration. This requirement is essential in determining the direction for the source. As the energy is reduced, the angle in question increases and the background of atmospheric neutrinos within a solid angle in the source direction grows too. The energy of cosmic neutrinos may be fantastic as compared to their accelerating counterparts. One of the sources of ultrahigh-energy neutrinos is represented by active galactic nuclei (AGN). As a typical luminosity of AGN is within the range from 10^{44} to 10^{47} Erg/s, it may be assumed that the evolution of AGN is determined by gravitation, i.e. supermassive black hole ($M \geq 10^6 M_\odot$) accretion of the matter. In the neighborhood of AGN the protons accelerated to superhigh energies are interacting with the matter or with radiation to produce in the process π-mesons, whose radioactive decay products include photons, neutrinos, and antineutrinos. Maximum neutrino energy of AGN is of the order of 10^{10} GeV. Another source of superhigh-energy neutrinos and antineutrinos are also the decay products of π-mesons, but now produced in the inelastic collision reactions of protons and photons forming the microwave cosmic-ray background. The energy of these neutrinos may be as high as 10^{12} GeV.

The effect of magnetic fields on the neutrino is minor. The cross-section of neutrino scattering from the interstellar matter is also small. To illustrate, for νN-interaction the characteristic cross-section in case of high-energy neutrinos ($E_\nu \sim 1 \div 10^3$ TeV) measures $\sim 10^{-35} \div 10^{-33}$ cm^2 (in case of low-energy neutrinos it is still smaller). If one assumes that the matter along the whole neutrino path has the density which equal to the galactic density (≈ 1 nucleon/cm^3), then the mean free path of neutrino amounts to $\sim 10^{33} \div 10^{35}$ cm, being well in excess of the Universe radius.

Neutrino radiation is the only radiation type that comes to the terrestrial observer from the extraterrestrial source carrying almost invariable information about the praparent object. Thus, NA is characterized by a number of unique features making it superior to gamma-astronomy.

First, with the use of high- and superhigh-energy ν-astrophysics there is a possibility to widen the horizon of the observable Universe and to deliver information about extremely distant cosmological epochs. γ-astronomy is inefficient in the high-energy region because of very small free path of the associated γ-quanta due to their scattering from relict radiation in the intergalactic space.

In the Universe one can find the objects radiating extremely low γ-fluxes, whereas their neutrino fluxes are very great. Such objects are called the hidden sources. Among these objects are young supernova shells, active galactic nuclei, black holes, etc. Consequently, the second merit of NA is its effectiveness in the detection of hidden sources. Also, NA is used in search for bright phases of galaxies and antimatter in the Universe.

Third, analysis of the high-energy neutrino spectra from cosmic sources, in principle, enables registration of the relict neutrino background of the Universe. Actually the calculations demonstrate that in case when high-energy neutrinos are scattered from the background neutrinos

$$\nu_l + \bar{\nu}_l \to Z^* \to l^- + l^+,$$

at the energy of $(E_\nu)_r = m_Z^2/(2m_\nu)$ one can observe the resonance associated with Z-boson. Then for the source-radiated neutrinos with the energy $E_\nu = (E_\nu)_r$, reducing the neutrino flux will be within the limits from 15 to 50%.

Fig. 48 presents the first neutrino image of the Sun (Sun neutrinography). However, the Sun in neutrinography is greater in size than in an ordinary photography. This stems from the fact that the direction of neutrinos arrival in modern NT is determined less accurately than the photons direction. At the same time, NA is still making the first steps, and its maturity may be attained only upon definite establishment of a structure of the neutrino sector and production of high-resolution NT.

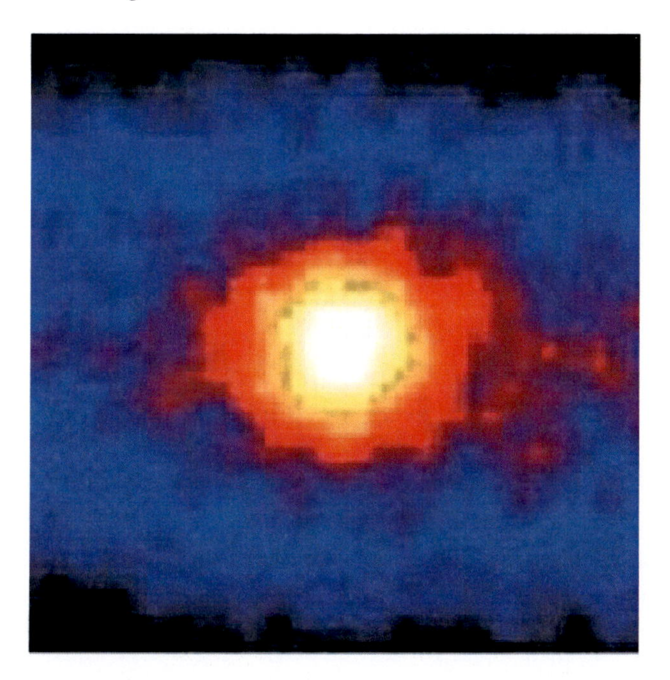

Figure 48. The Sun neutrinography.

No doubt that this should occur in the nearest decades. In the future the neutrinographies of supernovae located at enormous distances will present much more accurate data concerning the structure and evolution of the Universe.

Also, neutrino fluxes may be used to acquire information about the structure of stars and planets. For instance, a study of solar neutrinos enables one to obtain more accurate data about the Sun compared to those of helioseismology. The other example may be a neutrino geophysical tomography (geotomography).

All methods of geophysical tomography are based on acquiring the information about the physical properties of elements occurring in the Earth's thickness with the use of the summarized effects measured at its surface. Presently, the data on the Earth structure are acquired using seismic and gravitational tomography. Seismic tomography is associated with registration of time spent by seismic waves on covering the distance from the internal regions of the Earth to the detectors positioned on the surface at different distances from the source. Seismic waves are produced by earthquakes the deepest seismic foci of which are allocated at a level of about 700 km. An analysis of the measuring results makes it possible to determine the values of seismic speeds for bulk (transverse and longitudinal) waves, or so-called speed profiles. Nevertheless, the features of the outer Earth shells may be successfully established by seismic tomography, while three inner shells including the outer core, transition core region, and inner core are inaccessible as irradiance of these regions by seismic waves is extremely weak.

It is assumed that within the outer core the convection processes governing the magnetic field of the Earth are proceeding, and that it is rotating faster than the solid Earth by $1-3^0$ a year. Also, the inner core oscillations near the Earth center are expected. We can only guess about the state of the matter in the inner core. But considering that seismic waves are transmitted through the core, the aggregate state of the inner core is a solid. By now this core remains inadequately studied due to the shielding effect of the liquid interlayer (outer core) inaccessible for seismic waves, i.e. for seismic tomography there is no physical effect in this region. One of the latest hypotheses suggests that at the center of the Earth one can found a mixture of uranium and plutonium maintaining the continuous nuclear reaction. This core representing a "giant natural nuclear reactor" is almost 8 km in diameter. Due to the activity of this nuclear core, a high-power magnetic field formed around the Earth protects our planet against hazardous cosmic rays, capable to wipe out all forms of living biological objects in a few seconds. This natural reactor provides energy for continental drift and manifests itself as volcanic eruption.

Thus, applicability of seismic tomography is limited; its measurement accuracy is relatively low and, what is more, it fails to control the initial conditions. The potentialities of gravitational tomography are limited even more, as it is based on measurements of the terrestrial gravitational field by changes in the free fall acceleration values.

Neutrino tomography can outperform the seismic and gravitational tomography for accuracy by some orders of magnitude, enabling determination of the Earth structure at a radically new degree of quality. The development of neutrino tomography will be realized in two directions. The former is based on use of high-energy collider neutrinos. The scattering cross section of neutrinos from nucleons σ_ν turns out to be proportional to the neutrino energy E_ν, namely $\sigma_\nu \simeq 10^{-35} E_\nu$ cm^2. So, the part of neutrinos withdrawn from the initial beam through the interaction with nucleons of the matter nuclei is proportional to the nucleon number N_m at the beam path per unit area. On the other hand, N_m is determined by:

$$N_m = N_A m(L) = N_A < \rho > L,$$

where N_A is Avogadro number, $m(L)$ is a summarized matter mass in 1 cm^2, $< \rho >$ is a average matter density along a path L. Then, one can obtain the mass $m(L)$ by measuring the absorption degree of neutrinos in the path L. Detail information about $m(L)$ enables one to establish the in-depth variation of the matter density without any additional assumptions.

This is a great advantage of neutrino tomography compared to the seismic one, where the matter density may be recovered only on the introduction of additional assumptions concerning the values of the elastic modulus for the in-depth regions of the Earth. Since the section σ_ν is the same for nucleons of the matter, both at the surface and in-depth of the Earth, its value may be found by the laboratory measurements and hence neutrino tomography is free from the above-mentioned ambiguities.

Let us estimate the neutrino energies required for geophysical tomography. Attenuation of neutrino flux J through absorption is exponential in character:

$$J(L) = J(0)\exp\left[-\frac{L}{L_\nu}\right],$$

where $L_\nu = (N_A\rho\sigma_\nu)^{-1}$ is an absorption length in which the flux is lowered by a factor equaling to $e = 2.718281828....$ The total mass in the path of a neutrino beam passing along the Earth diameter comes to about 1.2×10^{10} g/cm^2. This value is associated with $\rho = 10\ \mathrm{g/cm^3}$ giving:

$$L_\nu = 1.7 \times 10^5\ \mathrm{km}/E_\nu.$$

Comparison between the obtained absorption length and the Earth diameter demonstrates that these values correlate at the neutrino energies of the order of TeV. For instance, at $E_\nu = 10$ TeV the Earth will absorb about a half of the initial neutrino beam. Registration of the neutrino beams transmitted through the Earth thickness with energies over the range from a few fractions to tens of TeV enables one to obtain a detail neutrinography of the Earth. This type of neutrino geotomography makes it possible to have information about the in-depth distribution of nucleons. Detector at the far side of the Earth, serving as a photographic film in X-ray radiography, will be used to record the withdrawal process of the initial neutrinos from the beam.

The second type of neutrinography is based on the Mikheyev — Smirnov — Volfenstein effect[1]. In this case there is no need in such colossal neutrino energies. The detector provides registration of the events connected with transitions of neutrinos from one flavor to the another ($\nu_l \rightarrow \nu_{l'}$). The sources may be both natural neutrinos — the neutrinos coming from the solar and stellar nuclear reactions, and artificial neutrinos — the reactor or collider neutrinos. Note that in this case the resonance neutrino conversions are influenced exclusively by the interaction with the matter electrons N_e. As the probability of resonance transitions is dependent on the energy as well, by the appropriate selection of the neutrino

[1]Consider the electron neutrino flux motion in a condensed matter with a variable electron density $N_e(z)$. If one is constrained by the two flavor approximation (only the electron and muon neutrinos exist) then probability of the transition $\nu_e \rightarrow \nu_\mu$ is defined by the expression

$$\mathcal{P}_{\nu_e\rightarrow\nu_\mu} = \frac{\mathrm{const}}{(N_e(z) - N_R)^2 + \Gamma^2}, \qquad (I)$$

where

$$N_R = \frac{\Delta m^2 \cos 2\theta_0}{2\sqrt{2}EG_F}$$

$\Delta m^2 = m_1^2 - m_2^2$, θ_0 is a neutrino mixing angle in vacuum, $\Gamma = \delta N_e$ is a resonance width being equal to $N_R \tan 2\theta_0$. As it follows from Eq.(I) at $N_e(z) = N_R$ a sharp increasing of $\mathcal{P}_{\nu_e\rightarrow\nu_\mu}$ takes place. The effect of the resonant oscillations increasing in matter was predicted by L. Volfenstein, P. Mikheyev, Yu. Smirnov and gave their names.

energy one is enabled to provide fulfillment of the resonant conditions for the particular regions of the Earth, thus making the measurements more sensitive to certain values of $<\rho>$ and less sensitive to some others. The principal merit of this method consists in the possibility to measure not only attenuation of the initial neutrino beam but also the flavor composition of the final neutrino beam. It should be emphasized that, as the first method provides the in-depth nucleon composition and the second method gives the in-depth electronic profile, combination of these two methods makes it possible to obtain a detail map of the Earth structure by neutrino geotomography.

The time is coming when neutrino tomography will be used for studies of other planets. Note that a detection system used in neutrino tomography need not be stationary. In case of collider neutrino the initial beam may have different orientation angles. Directing the beam in such a way that its outgoing to the surface takes place at the water spaces, one is enabled to use floating objects for a system of neutrino detectors. Neutrino detectors may be also mounted at artificial satellites.

Epilogue

Admiring a night sky, it is hard to imagine that magnificent stars and our planet were created as a result of the explosion of a fire ball compressed to the size of the Planck length. Similar to the midnight chimes of the Big Ben signaling a new day, this explosion was an announcement of the Universe creation. From the start, this all-embracing explosion has filled the available space that was closed on itself like the sphere surface. On expiration of ten milliard years, the aftereffects of this explosion are still material for the Universe: such enormous stellar clusters as galaxies are moving father apart at a speed close to the speed of light. The Universe has its starting time, and there is neither start nor ending for its space. The Universe will be expanding for a infinitely long time and it will be laid ahead an extinction in the boundless cold.

However, well before this dismal end the mankind will be threatened by other problems of cosmological scale. Nothing could be eternal, and our Sun is not an exception to this rule. Evolution of the Sun is governed by changes in its chemical composition due to thermonuclear reactions. According to the calculations, the present hydrogen content within the core amounts to 35% by mass, while in the beginning of evolution, judging by the surface layers where no thermonuclear reactions are proceeding, the hydrogen content was about 73%. In the process of its evolution, the solar core is compressed and its shell is expanded. As predicted by a theory of stellar evolution, at the stage when the Sun will be aged 9×10^9 years, hydrogen within its core will be exhausted leading to helium burning. And at the stage with a duration of $\approx 5 \times 10^8$ years a radius of the Sun will be considerably increased, and its effective surface temperature will be decreased making the Sun a red giant. As a red giant, due to the increased release of energy, the Sun will burn out our Earth first, subsequently absorbing its remainder as a result of huge expansion. Before it happens our descendants will be forced to leave the Earth and search for shelter at other planets of the Milky Way. However, there they will also face an other problem. The nearest galaxy, the Andromeda Nebula, is nearing the Earth at a speed of ~ 100 km/s. In five-six milliard years both galaxies must collide. It is evident that nobody is interested to take part in such an event, i.e. the mankind is expelled to find for living some other galaxy.

This frightening scenario that seems to be beyond human apprehension has been obtained on the basis of the existing standard model of elementary particles physics and irrefutable data of astronomical and astrophysical observations. Under the milestones of mathematics and experiment all other models of the Universe evolution (sometimes even more favorable for us) have turned to ashes. We must sadly accept that our wonderful planet is a tiny isle for temporary shelter in the boundlessly hostile ocean of the Universe. To survive, the civilized world must seek another shelter.

A more or less happy ending of this sad story is only possible with the advances in the basic sciences, most of all Physics. Since the times of Copernicus it has been understood that there is no force in the world which could stop scientific progress. Despite the enough wide spectrum cares even the inventive Farthers of the Inquisition were not able to do it. The developing of the fundamental research could not be stopped by numerous scientific officials, demanding an instantaneous practical yield for all types of research activities and wishing in no way to understand that the progress is the following nonseparable chain:

fundamental science⟶applied science ⟶production.

But the role of basic research is easily comprehended. Really, take any device or mechanism and trace its production history backwards in time. And you always make sure that its development was initiated owing to a certain law of the fundamental science. Unfortunately, the time interval between this law and the technological discovery may often last decades and even more. To illustrate, an electromagnetic field theory put forward by Maxwell in 1860 — 1865 has been embodied in technological discoveries not before the end of the nineties of the XIXth century. And a positron predicted by Dirac in 1928, whilst it was discovered in cosmic rays in 1932, has found no application as an energy source for our power stations up to the present.

Actually, by now the resources of the classical physics as a source of new technologies have been practically exhausted. New technology trends are based on the discoveries within the scope of the already-built standard model for strong and electroweak interactions. Controlled thermonuclear fusion, neutrino tomography, nanotechnologies, quantum computers, prospects of using collider neutrino for the disposal of nuclear ammunition may provide excellent examples. It is hoped that in the future a source for the development of new technologies may be found in the Grand Unified Theory with the Unified Field Theory to follow.

Appendix

Natural System Units

Typical velocities of elementary particles are close to the light velocity, moment of momentum represent multiples from $\hbar/2$, while energies, even if they reach the order of 10^{-5} J, are ranked among a category of superhigh ones. Consequently, when we are trying to imagine the elementary particles world visually one of the trouble in our consciousness is caused by the fact that constants whose values can not be laid in macroworld standards present in the microworld physics formulae. On the other hand, it is easy to appreciate their values compared with each other. So, we should break off, as the natural step, the connection with macroworld. It is achieved by the transition to the system units where the fundamental physical constants of microworld are used as the basic units. Such systems are called Natural System of Units (NSU). The first to suggest one of NSU kind was M. Plank. He chose $\hbar$, c, G and k (the Boltzmann constant) as the basic units. Under the NSU construction the fundamental constants, taking as the basic units, are formally assumed to be equal 1. So, the Plank NSU is defined by the relation

$$\hbar = c = G = k = 1.$$

Since the quantum field theory is symbiosis of the special theory of relativity and the nonrelativistic quantum mechanics then in the quantum field theory NSU should be defined by the relation

$$\hbar = c = 1.$$

In this system the dimensionality of any dynamical observable A is connected with the mass dimensionality, i.e.

$$[A] = [m^n] \qquad (n \in N).$$

For example, for velocity, action and moment of momentum n is equal to 0. From the Heisenberg uncertainty relations

$$\Delta p_i \cdot \Delta x_i \geq \frac{\hbar}{2} \qquad (A.1)$$

$$\Delta E \cdot \Delta t \geq \frac{\hbar}{2}, \qquad (A.2)$$

it follows that for coordinate and time n equals -1. In this system the electric charge is a dimensionless quantity as its linkage with the fine structure constant α is given by

$$\frac{e^2}{4\pi\hbar c} = \alpha.$$

Using the definition of Lorentz force

$$\mathbf{F} = e\mathcal{E} + \frac{e}{c}[\mathbf{v}\times\mathbf{H}],$$

one may be convinced that the strengths of the electric $\mathcal{E}$ and the magnetic $\mathcal{H}$ fields have the dimensionality of m^2. In NSU the value "eV" (1 eV=$1.6021892\cdot 10^{-19}$ J) and derivatives from it (keV, MeV, GeV, TeV etc.) are used as a mass unit.

To pass to some ordinary system of units we must have formulae connecting the basic units of both systems. Let us choose "GeV" as a basic unit in NSU. Then for CGS system the definitions of "g", "cm", and "s" may be found from the relations

$$1\text{GeV} = 1.6021892\cdot 10^{-10}\text{J} = 1.7826759\cdot 10^{-24}\text{g}\cdot c^2, \qquad (A.3)$$

$$\hbar c = 1.9732858\cdot 10^{-14}\text{GeV}\cdot\text{cm}, \qquad (A.4)$$

$$\hbar = 6.582173\cdot 10^{-25}\text{GeV}\cdot\text{s}. \qquad (A.5)$$

Using Eqs. (A.3) — (A.5) one could express any derivative CGS unit through "GeV". For example, in CGS the force dimensionality is given by

$$1\text{dyn} = \frac{\text{g}\cdot\text{cm}}{\text{s}^2} = \frac{\cdot 10^{-12}\text{Gev}^2}{\hbar c},$$

that allows us to state 1 dyne is equal to GeV^2 in CGS.

To comparison some quantities it is necessary that they have identical dimensionalities. One of the advantages of NSU implies that some quantities with incoincident dimensionalities in ordinary system of units have one and the same dimensionality in NSU. As an example, it is instructive to consider the coupling constants G_F and G defining the intensity of weak and gravitational interactions respectively. The calculations give

$$G_F \sim 1.4\cdot 10^{-49}\text{Erg}\cdot\text{cm}^3 = 1.2\cdot 10^{-5}\hbar^3 c^3\text{GeV}^{-2} \overset{\text{NSU}}{\longrightarrow} 1.2\cdot 10^{-5}\text{GeV}^{-2},$$

$$G \sim 6.7\cdot 10^{-8}\text{cm}^3\cdot\text{g}^{-1}\cdot\text{s}^{-2} = 6.7\cdot 10^{-39}\hbar c^5\text{GeV}^{-2} \overset{\text{NSU}}{\longrightarrow} 6.7\cdot 10^{-39}\text{GeV}^{-2}.$$

The obtained result corresponds to the fact — in the region of energies reached to date, the intensity of weak interaction is as great as 10^{33} times than that of gravitational interaction.

After the formula for observable quantity A is obtained we should make the transition from NSU to some ordinary system units, to say, to CGS units. For this purpose we must rebuild the right dimensionalities of all the physical quantities entering into A (aside from energy, of course) with the help of the following replacements:

$$[m]_E \to [m]c^2, \qquad [l]_E \to [l](\hbar c)^{-1}, \qquad [t]_E \to [t](\hbar)^{-1}, \qquad (A.6)$$

where the subscript E shows that quantity dimensionality is taken in NSU.

References

[1] Aitchison, I.J.R. and Hey, A.J.G. (1982). *Gauge theories in particle physics*, Hilger, Bristol

[2] Bahcall, J. (1989). *Neutrino Astrophysics*, Cambridge University Press, Cambridge

[3] Bjorken, J.D. and Drell, S.D. (1965). *Relativistic Quantum field*, McGraw-Hill Inc., New York.

[4] Bogoliubov, N.N. and Shirkov, D.V. (1959). *Introduction to the Theory of Quantized Fields*, Interscience Publishers Inc., New York

[5] Cahn, R.N. and Goldhaber, G. (1989). *The experimental Foundations of Particle Physics*, Cambridge University Press, Cambridge

[6] Creutz, M. (1983). *Quarks, Gluons and Lattices*, Cambridge University Press, Cambridge

[7] Davies, P. (1985). *SUPERFORCE. The Search for a Grand Unified Theory of Nature*, Simon and Shuster, Inc., New York

[8] Feynman, R. P. (1972). *Photon-hadron interaction*, Reading, Massachusets, Benjamin

[9] Gasiorowicz, S. (1966). *Elementary Particle Physics*, Wiley, New York

[10] Gotfried, K. and Weisskopf, V.F. (1984). *Consepts of Particle Physics*, Oxford University Press, New York

[11] Green, M. and Schwartz, J. and Witten, E. (1987). *Superstring Theory*, Cambridge University Press, Cambridge

[12] Greiner, W. and Muller, B. (2000) *Gauge Theory of Weak Interactions*, Springer-Verlag, Berlin

[13] Halzen, F. and Martin, A.D. (1984). *Quarks and Leptons*, John Wiley and Sons Inc., New York

[14] Itzykson, C. and Zuber, J.B. (1980). *Quantum Field Theory*, McGraw-Hill Book Company, New York

[15] Jauch, J.M. and Rohrlich, F. (1976). *The Theory of Photons and Electrons*, Springer-Verlag, Berlin

[16] Layzer, D. (1984). *Constructing the Universe*, Scientific American Library, An imprint of Scientific American Books, Inc.

[17] Mott, N. F. and Massay, H.S. W. (1965). *The Theory of Atomic Collisions*, Clarendon

[18] Pilkuhn, H.M. (1981). *Relativistic Particle Physics*, Springer-Verlag, New York, Heidelberg, Berlin

[19] Peskin, M. E. and Schroeder, D.V. (1997). *An Introduction to Quantum Field Theory*, Addison-Wesley Publishing Company

[20] Pokorski, S. (2000). *Gauge Field Theories*, Cambridge University Press, Cambridge

[21] Ross, G.G. (1984). *Grand Unified Theories*, Benjamin/Cummings, Menlo Park, California.

[22] Ryder, L.H. (1984). *Quantum Field Theory*, Cambridge University Press, Cambridge

[23] Ryder, L.H. (1975). *Elementary Particles and Symmetries*, Gordon and Breach Science Publishers, New York, London, Paris

[24] Schweber, S.S. (1961). *An Introduction to Relativistic Quantum Field Theory*, Row, Peterson and Co Evanston Inc., Elmsford, New York

[25] Wainberg, S. (2000). *The Quantum Theory of Fields*, Cambridge University Press, Cambridge

[26] Wess, J. and Bagger, J. (1983). *Supersymmetry and supergravity*, Princeton University Press, Princeton

[27] Yndurain, F.J. (1983). *Quantum Chromodynamics*, Springer-Verlag, New York, Berlin, Heidelberg, Tokyo

Index

D

G

H

I

J

L

M

N

O

P

Q

R

S

W

X

Y

Z